D1415852

ABOUT ISLAND PRESS

Island Press is the only nonprofit organization in the United States whose principal purpose is the publication of books on environmental issues and natural resource management. We provide solutions-oriented information to professionals, public officials, business and community leaders, and concerned citizens who are shaping responses to environmental problems.

In 1994, Island Press celebrated its tenth anniversary as the leading provider of timely and practical books that take a multidisciplinary approach to critical environmental concerns. Our growing list of titles reflects our commitment to bringing the best of an expanding body of literature to the environmental community throughout North America and the world.

Support for Island Press is provided by Apple Computer, Inc., The Bullitt Foundation, The Geraldine R. Dodge Foundation, The Energy Foundation, The Ford Foundation, The W. Alton Jones Foundation, The Lyndhurst Foundation, The John D. and Catherine T. MacArthur Foundation, The Andrew W. Mellon Foundation, The Joyce Mertz-Gilmore Foundation, The National Fish and Wildlife Foundation, The Pew Charitable Trusts, The Pew Global Stewardship Initiative, The Rockefeller Philanthropic Collaborative, Inc., and individual donors.

*Biodiversity and
Human Health*

Biodiversity and Human Health

Edited by Francesca Grifo and Joshua Rosenthal

FOREWORD BY THOMAS E. LOVEJOY

ISLAND PRESS
Washington, D.C. ■ Covelo, California

Library of Congress Cataloging-in-Publication Data

Biodiversity and human health/edited by Francesca Grifo and Joshua
　Rosenthal
　　　p.　　cm.
　　Includes bibliographical references and index.
　　ISBN 1-55963-500-2 (cloth). — ISBN 1-55963-501-0 (paper)
　　1. Environmental health—Congresses. 2. Biological diversity—
Congresses. 3. Pharmacognosy—Congresses.　I. Grifo, Francesca.
II. Rosenthal, Joshua.
　　RA565.A2B56　1997
　　610—dc20　　　　　　　　　　　　　　　　　　96-31800
　　　　　　　　　　　　　　　　　　　　　　　　　CIP

Printed on recycled, acid-free paper ✛

Manufactured in the United States of America

10 9 8 7 6 5 4 3 2 1

Contents

PART I

Causes and Consequences of Biodiversity Loss for Human Health   5

PART IV

An Agenda for the Future: Conserving
Biodiversity and Human Health   265

Foreword

The links between biodiversity and human health have largely been ignored until the precedent setting symposium which has resulted in this volume. In retrospect, the linkage is so compelling it is hard to believe how neglected it was previously. The relationship was, nonetheless, not entirely neglected. Medicines have continued to be drawn from nature, and epidemiology more often than not has required consideration of organisms beyond the pathogen and the less than ecstatic host (ourselves).

Indeed, when the Rockefeller Foundation, at its inception, set out to eliminate the great scourges of humanity (essentially succeeding in the case of yellow fever), it early on discovered that it needed the talents of field naturalists to unravel the epidemiology. The investigations soon revealed a tremendous number of arthropod borne viruses (some of which are now recognized as major human afflictions like eastern equine encephalitis and dengue) and led to the establishment of a stellar set of virus laboratories around the world.

One of these was at Villavicencio in the Colombian llanos, then characterized by Nancy Bell Bates, the wife of its first director, as "east of the Andes and west of nowhere." It fell to the second director of that laboratory Jorge Boshell (with whom I was fortunate to share an office at Belém in the late 1960s) to make the insightful natural history observation which unlocked the secret of sylvan (or "jungle") yellow fever. Urban yellow fever, with a simple human–*Aedes aegypti* cycle, was early understood and brought under control by extremely thorough inspection and control practices, the exemplary value of which should not be forgotten. In contrast, the sylvan cycle, known to be nomadic and to involve howler monkeys and *Haemagogus,* a blue forest canopy mosquito, remained puzzling: how did it ever reach from this nomadic canopy existence to people tens of meters below on the rainforest floor?

The answer was obvious once Jorge Boshell watched woodcutters bring down a tree and become surrounded by little blue mosquitoes. Obvious it may seem now, but it took an informed naturalist's mind to make sense of it. The secret of jungle yellow fever, as I like to think of it, can be taken as symbolic of human disease emanating from human and natural disturbance of the biological assemblages called ecosystems. Examples are legion: El Niño weather change leading (perhaps assisted by human waste) to outbreaks of *Vibrio* cholera in warm coastal waters; malaria outbreaks linked to deforestation and different forest harvest schemes in Rondonia; and Lyme disease linked to land use change in the eastern United States.

Lyme disease is an interesting example of a multispecies relationship. The host of the adult deer ticks, *Ixodes scapularis,* are white-tailed deer, essentially a forest edge species favored (to a point approaching pest status) by land use changes which also bring human population closer to the cycle. A complex cycle involving oak mast (massive fruiting) years with super-abundant acorn crops, gypsy moths, and white footed mice, *Peromyscus leucopus* has only recently been illuminated. Particularly intriguing is the role of biological diversity with implications for beneficial management. Josh Van Buskirk and Richard Ostfeld have demonstrated that larval and nymph stages of the deer tick are less likely to become infected with the etiological agent, *Borrelia burgdorferi,* when there is a high diversity of hosts. When the host diversity is low and dominated by the super competent host for juvenile ticks, the white footed mouse, models indicate much higher transmission rates. The probable practical consequence (now in process of testing) is that transmission rates can be reduced more effectively by increasing the diversity of hosts of larval ticks rather than reducing populations of white-tailed deer which appear to be far from limiting to adult tick populations.

Biological diversity also provides the epidemiological equivalent to the Cold War's DEW (distance early warning) line. It is a vast array of highly sensitive indicators, very likely to show effects of environmental change (physical, chemical, and biological) long before humans are likely to be affected—if we are willing to pay attention. Vanishing and declining amphibian populations are one set of current signals which may well represent more than one kind of change. The reproductive disorders in wildlife are another which represent an imperative to take endocrine disrupters and the need for research on them most seriously.

The contributions of biodiversity to human health have even more potential today when the sciences are able to make extraordinary contributions at the level of the molecule. One of the greatest elements of this potential consists of biotechnology and genetic engineering. Genes from the winter flounder make the russet potato more frost resistant. The molecules which keep ice from forming at $-2°C$ in the blood of Antarctic fishes promise improvement in organ transplant and nontoxic deicing compounds. Probably no more dramatic example exists than the polymerase chain reaction: an extraordinary magnifying reaction that can multiply tiny amounts of genetic material a billion times over in a very short time. This Nobel Prize winning reaction depends on a heat resistant enzyme from a bacterium isolated from a Yellowstone hot spring. Valuable in a wide array of research, PCR is the fundamental underpinning of the human genome project which will characterize our entire genetic composition and render benefits for human health beyond estimation.

Improvements to nutrition from new strains of crops and biodegradable pesticides are nontrivial contributors to human health. Nutrition problems lead to additional health problems; nutrition solutions lead to health improvement. Bioremediation to clean up toxic wastes and make industrial ecology more possible represents a major potential for improvements in human health. The extent to which industry becomes more biologically based likely means a cleaner environment, a cleaner workplace, and a healthier workforce.

The traditional way in which biodiversity has contributed (and continues to do so) to human health is through important new medicines. A tradition as old (and probably older than) aspirin, it continues to make major contributions as revealed in the chapter by Grifo and colleagues on the origin of prescription drugs. While biochemical prospecting has always seemed something of a long shot, "informed" prospecting, using astute natural history observation as a guide to more selective testing, together with ever improving efficiency and speed of prospecting techniques, renders this an increasingly exciting scientific frontier.

In the end, the most powerful contribution of biological diversity is as the fundamental library for the life sciences. Knowledge advances in a partly serendipitous fashion in which fundamental knowledge and concepts emerge from observations and research on seemingly irrelevant and esoteric species. As a rainforest biologist, my favorite is capoten, the (1.3 billion dollars in annual sales) drug for hypertension. It derives from research on the venom of a new world rainforest viper. Literally tens of millions of people live longer, healthier, and more productive lives as a consequence.

This current and future potential of biodiversity is being eroded at frightening rates which need no chronicling here. Literally all environmental problems end up impinging negatively on biodiversity which is of such great value. Were an external threat like a large meteor to threaten biodiversity as happened at the end of the age of the dinosaurs, I can't help but believe society would recognize the crisis and leap to address it. Somehow, since the problem proceeds incrementally and also because we are the largely unintentional driving force, we seem unable to comprehend that the threats to biodiversity represent a true global emergency.

The editors of and contributors to this volume have made a vital contribution by spotlighting the multiple links between biodiversity and human health. Take the time to read this compendium, and you will forever think differently about the human condition.

THOMAS E. LOVEJOY
COUNSELOR TO THE SECRETARY
FOR BIODIVERSITY AND ENVIRONMENTAL AFFAIRS
THE SMITHSONIAN INSTITUTION

Preface

The staggering rate of irreversible biodiversity loss we are currently experiencing around the globe holds serious consequences for medicine and public health. Plants, animals, and microbes have provided a major source of therapeutic agents for human disease, including chemical leads toward treatments for cancer, glaucoma, malaria, Parkinson's disease, inflammatory disorders, high blood pressure, and a range of bacterial, fungal, and viral infections. Yet, only a tiny fraction of the world's biological wealth has been studied for potential therapeutic benefit. Biodiversity loss also may pose new risks for emerging diseases arising from human exposures to new pathogens and the adaptation of microbes to environmental change. These are viewed as possible root causes of the AIDS pandemic, as well as the recent outbreak of Hantavirus in the Southwest, Lyme disease, hemorrhagic fevers, and other emerging infectious diseases.

We often think of the study of biodiversity as the province of ecologists and systematists, but in fact a great deal of biomedical research is directly related to increasing our understanding of biodiversity. I would like to highlight a few examples of such research supported by the National Institutes of Health.

To begin with, this book has emerged from a conference by the same title that was sponsored by the National Institutes of Health (NIH), the National Science Foundation (NSF), the Smithsonian Institution, and the National Association of Physicians for the Environment and held in the spring of 1995 at the Smithsonian Institution.

From classical times to the early part of this century, the search for medicinal properties in plants and animals has been a bedrock of pharmacology and medicine. For several decades medical science moved away from biodiversity-based therapies in an attempt to use our growing knowledge of mechanisms of disease to design drugs "rationally." More recently we have returned to the natural world, prospecting for answers to human health problems among the diversity of plants, animals, and microorganisms that have spent several billion years evolving chemical responses to the challenges they face. Today, advanced bioassays, automation, and computer-aided analyses have made prospecting for bioactive compounds easier, faster, and far more productive.

One exciting example of this modern "bioprospecting" approach is exemplified by a program described in the chapter by Joshua Rosenthal. This program represents a partnership between the NIH, the NSF, and the U.S. Agency for International Development, to support interdisciplinary research of American and developing country scientists to discover new drugs from natural products derived from the planet's species-rich habitats. We believe this program offers an important paradigm—drug discovery, economic interests, and conservation needs can be mutually supportive, rather than competing—by demonstrating the value of biological resources from which medical products are derived, and sharing those benefits equitably among all stakeholders.

Other natural product drug discovery efforts by the NIH include programs of the National Cancer Institute, the National Institute of Allergy and Infectious Diseases, the National Institute of Diabetes and Digestive and Kidney Diseases, the National Institute of Mental Health, and the National Heart, Lung and Blood Institute.

The National Cancer Institute (NCI) has been bioprospecting for decades. It is currently testing natural products from more than 30 countries, searching for potential treatments for cancer, AIDS, and other diseases. The NCI was instrumental in the discovery of taxol from the Pacific Yew tree. Through NCI's Natural Products research program another possible anticancer drug, halichondrin, has been found in a species of New Zealand sponge. The same program has also found three compounds with anti-HIV potential: a vine from Cameroon yields michellamine, a Malaysian tree yields costatolide, and conocurvone has been derived from the West Australian smokebush.

Similarly, the National Institute of Allergy and Infectious Diseases supports research and screening for anti-HIV activity from natural products from Central and South America, East and Southeast Asia, West Africa, Israel, Australia, Bolivia, Japan, and Taiwan. Their work on anti-infective and antiparasitic diseases has turned up a compound designated as "SP 303,11" initially derived from the latex of an Amazonian tree, and now in Phase II clinical trials in humans. This isolate shows significant activity against watery diarrhea and recurring genital herpes.

Researchers at the National Institute of Diabetes and Digestive and Kidney Diseases have studied skin secretions from poison dart frogs—finding many alkaloids useful in both drug development and neurobiology research. This research may be especially timely; the entire family of poison dart frogs is threatened by destruction of tropical rainforests in Central and South America.

NIH also supports some fascinating work in its intramural and extramural research programs on the genetics of endangered species: cheetahs, Florida panthers, and humpbacked whales. Some of this work was described by Stephen O'Brien of NCI at the Biodiversity and Human Health conference from which this book derives. The work has advanced basic research on genetics, provided important management information for endangered species, and may lead to new gene therapy techniques.

Basic genome analysis of pathogens, vectors, and hosts is also supported by the NIH. This work frequently depends upon and provides data for our understanding and use of genetic diversity, an important component of biodiversity. A significant activity of the National Library of Medicine's National Center for Biotechnology Information is to organize this information into automated databases for research purposes.

Finally, the NIH conducts and supports research on changes in the environment that could facilitate the emergence or re-emergence of disease pathogens, including the impact of stratospheric ozone depletion and deforestation. One of the key scientific challenges in the next few years will be to better define the epidemiologic and biologic principles that determine the emergence of microbial diseases, and to design model systems for predicting their spread.

There is a tendency among laypeople and scientists alike to assume that our increasing technological sophistication moves us further away from dependence on the natural world. Biomedical science, like the discipline of physics, gains much of its information from a process of reductionism. This is the manifesto of molecular biology, which has revolutionized biomedical science: to examine progressively more minute details of biological structure and function in order to ultimately make sense of the whole.

However, as we face planet-wide losses of biodiversity it is urgent that we recognize that natural ecosystems and the species that maintain them are part of our protection against diseases. Perhaps the greatest biological challenge ahead of us is to make sense of the whole: to understand the complex matrix of ecological relationships between all the planets inhabitants, and to more fully appreciate why the diversity of life is critical to the sustenance of life.

PHILIP E. SCHAMBRA, PH.D.
DIRECTOR
FOGARTY INTERNATIONAL CENTER
NATIONAL INSTITUTES OF HEALTH

Acknowledgments

We list ourselves alphabetically as editors to indicate our equal contributions to this effort. It is with gratitude that we wish to acknowledge the cooperation and assistance of many individuals and institutions.

Many individuals helped to make the conference from which this book evolved possible. They include John Grupenhoff, who provided the original idea for the conference and other assistance along the way; the Conference Steering Committee; and Amar Bhat, Paula Cohen, and Uri Ratner, who provided logistical support. The Fogarty International Center, National Cancer Institute, National Center for Research Resources, National Institute of Allergy and Infectious Diseases, National Institute of General Medical Sciences, National Institute of Diabetes and Digestive and Kidney Diseases, National Institute of Child Health and Human Development, National Institute of Dental Research, National Institute of Environmental Health Sciences, National Institute of Drug Abuse, National Institute of Mental Health of the National Institutes of Health, the National Science Foundation, Smithsonian Institution, National Association of Physicians for the Environment, and Pan American Health Organization all sponsored the conference.

Our own institutions, the Fogarty International Center (FIC) and the American Museum of Natural History (AMNH), have both provided generous logistical support of the preparation of this volume. At FIC, we especially thank Ken Bridbord for all his help. At AMNH, Margaret Law and

Valeda Slade have provided unstinting logistical, research, and editorial assistance.

The individual chapter authors provided creative and interesting manuscripts and patient cooperation with our time lines, queries, and revisions. The following people made very useful comments on one or more of the chapters of the book: Ken Bridbord, Joel Breman, Gordon Cragg, Rob Eiss, Alex Fairfield, Charles Gardner, Otto Gonzales, William Hahn, Maurice Iwu, Stuart Plattner, David Simpson, and Martha Weiss.

Island Press was an enthusiastic partner from the beginning and cheerfully tolerated our inexperience as editors. And finally, we wish to thank our families for amiably enduring our extended absences even when some members were too young to understand them.

This book is dedicated to the memory of Matt Suffness. Matt was an ardent supporter of the infinite possibilities of natural products and their value to improve human health and conserve biodiversity.

Introduction

FRANCESCA GRIFO AND JOSHUA ROSENTHAL

Of all the environmental challenges faced by humankind at the threshold of the 21st century, none has such profound implications for the future of life on Earth as the accelerating loss of biological diversity. As pressures on aquatic and terrestrial habitats around the world have increased, nature's endowment of genetic, species, and ecosystem diversity has diminished. Still greater losses are predicted for the decades to come, with corresponding losses of the actual and potential benefits that we derive from biodiversity.

Biological diversity is threatened by many forces, including unabated human population growth, unsustainable resource consumption, and economic policies that do not fully account for biodiversity's value, all of which form the basis for habitat loss and degradation, pollution, overexploitation, introduction of alien species, and global climate change. In responding to these threats, conservationists face many constraints, including incomplete knowledge of biodiversity, lack of integration among the relevant sciences, and poor communication between scientists and policymakers. These very difficulties are now bringing scientists, conservationists, educators, economists, policymakers, and others more closely together to forge effective solutions. If the biodiversity crisis is unprecedented in human history, so too are the collaborative efforts that people are undertaking in response.

The urgency of the biodiversity crisis is reflected in the high levels of species extinction that scientists are now detecting and quantifying. Two

1

sets of data point unequivocally to an ongoing extinction event of geologic proportions: (1) the large body of observations that document human-induced extinctions, and (2) extrapolations based on available knowledge of diversity patterns in the landscape and the rate at which habitats are being lost due to human activities. Although various methods are being used and developed to estimate extinction rates, all essentially point to the same conclusion: that, given current trends, a substantial percentage of the world's species are likely to go extinct within the next several decades. According to analyses carried out by E. O. Wilson, 20% of the world's species are liable to be lost within the next 30 years, and at least 50% beyond that.

At least five mass extinction episodes are recorded in the planet's geological record. Paleontologists differentiate between the high extinction rates that marked these mass extinctions and the "background extinction rate" recorded in other periods. According to conservative assessments, the contemporary, human-exacerbated extinction rate is at least 1,000 times greater than the normal background extinction rate, placing us in the midst of the sixth mass extinction in the Earth's history. This figure reflects human impacts not just in recent decades, but over the last several millennia. For thousands of years human beings have been significantly modifying Earth's ecosystems, favoring some species while driving others to extinction. The paleontological record, along with more recent historical evidence, documents these extinctions in stark detail.

Although the phenomenon of human-induced extinction is not new, the circumstances under which species now face extinction are. The threats are not confined to particular localities or regions, but are global in scope. Those threats are affecting a broader range of organisms. In the past, large and conspicuous species (especially birds and large mammals) were at greatest risk; now invertebrates, other small animals, marine and other aquatic organisms, and, significantly, plants are also threatened. In addition, effects of the main cause of the current crisis, namely humanity, escalate with our growing population.

In the last two decades comprehensive documentation of the ramifications of biodiversity loss for man has emerged. Yet, the discussions of the specific relationships between species loss and human health have remained anecdotal and are spread throughout the literature. The goal of this volume is to bring discussion of the diverse health consequences together and make them accessible to both the conservation and biomedical communities.

This volume is the product of a conference held in March of 1995. The

National Institutes of Health, the National Science Foundation, Smithsonian Institution, National Association of Physicians for the Environment, and Pan American Health Organization sponsored a gathering of scientists from the fields of epidemiology and ecology, immunology and systematics, agriculture and biotechnology, lawyers and ethnobotanists, chemists and physicians, policymakers and citizens to open a dialogue on the significance of biodiversity to human health. The conference was attended by 250 registrants who participated in over 30 presentations and discussions.

Among the key findings which emerged were that ecosystem disruption and subsequent loss of species have had and will continue to have profound implications for human health. Ecosystem disruption can mean changes in food supply and water quality, with known and unknown implications for nutrition and health. In addition, damage to ecosystems can cause changes in the equilibria between hosts and parasites and between predators and prey. Under these circumstances, parasites may switch from now rare organisms to humans as hosts, accidentally introduced disease vectors and pathogens may thrive, predators on both pathogens and vectors may be reduced in numbers or even go locally or globally extinct leading to the "emergence" of "new" diseases. In addition, global warming, with its many negative effects on biodiversity is linked to many direct and indirect negative health effects. Widespread occurrences of agricultural and industrial chemicals may be correlated with multiple reproductive disorders in wildlife and humans. While there are many uncertainties about the causal relationships between endocrine "disrupters" and the observed changes in wildlife and in humans, it seems clear that these compounds have the capacity to compromise the health and survival of a wide variety of species. Last, living in a world devoid of the beauty and tranquillity which accompanies diverse, intact ecosystems can have profound effects on our mental health.

Some rainforests and coral reefs may harbor tens of thousands of species, and most other habitats contain close to a thousand species per square kilometer. Hence it is not surprising that habitat loss leads to species loss and with it come a host of additional ramifications for human health. Losing species means we lose the raw materials for existing and new potential weapons in the fight to alleviate human suffering and death. This is true for both pharmaceutical products, and for traditional medicines, which are still an important source of medical care for 80% of the world's population. Second, it means we lose models through which we learn about human physiology, the prerequisite for our understanding of health and illness. Third, it means we lose organisms related to the pathogens and vectors of disease whose study allows us to make predictions about the epidemiology and genetics of the causal agents and carriers of disease. Fourth, it means we lose future raw materials for new processes and products of biotechnology, many of which are critical to understanding, preventing, and curing disease. Last, it means we lose indicators of the ability of ecosystems to support life of all kinds, including humans.

The following chapters provide concepts that explain the complex relationships between biodiversity loss and human health and examples that demonstrate the gravity of their loss. Indications are that no matter how good we become at chemistry, or biotechnology, we will never outdo mother nature. Entirely new classes of compounds routinely emerge from studies of biodiversity. We know that if ecosystems continue to be disrupted and species lost at current rates, we will pay a horrible price in increased human suffering from illness and injury that inevitably will touch all of our lives. Despite the urgency of our needs to utilize biodiversity both as a source of raw materials and a place to learn physiology, biochemistry, natural products chemistry, epidemiology, and much much more, it is disappearing at an ever increasing rate. Furthermore, our ability to continue to use and protect biodiversity is constrained by our own ignorance of the identity, relationships, and natural history of most of the world's biodiversity and by the division of our study of the life on this planet into oftentimes artificial disciplines. The division of biologists into those focused on biomedical applications and those focused on other aspects of biology begins early in one's undergraduate career and leads to an unnatural separation. It is the sincere hope of the editors that this volume will accomplish two things. First, we hope that it will offer examples of the kinds of cross-disciplinary research necessary to better understand and solve the biodiversity/human health crisis and inspire more of them. Second, we hope that it will educate and empower the public to both bring these issues to the attention of policymakers and consider the consequences of all their actions on our life-support system.

PART I

Causes and Consequences of Biodiversity Loss for Human Health

The chapters in this section describe some of the most important causes of biodiversity loss and provide multiple examples of the diverse relationships between the loss of biodiversity and human health and well-being. In Chapter 1, Eric Chivian eloquently describes the linkages between environmental degradation, species extinction, and our resulting loss of medicines, medical models, and ecological protection from infectious disease. Robert Engelman (Chapter 2) explores the ways in which human population growth affects the capacity of the planet to support wild species and the production of healthy food for humans. Chapters 3 and 4 by Epstein et al. and Dobson et al., respectively, are important attempts to synthesize a great deal of ecological and epidemiological information regarding the possible relationships between biodiversity and infectious disease. Finally, John Vandermeer (Chapter 5) describes effects of increasing industrialization of agriculture on planned and associated biodiversity in agroecosystems.

In reading this section, we encourage the reader to recall two things. The first is that these chapters cover but a few of the many forces driving biodiversity loss and some of the medical and epidemiological consequences. For every useful and endangered species discussed here there are countless others on which that organism depends for survival, numerous genetically diverse relatives on which its continued usefulness depends, and many other organisms we have yet to discover. The second point is that the examples discussed here provide only a starting point from which much research is needed to understand the chemical, ecological, and epidemiological relationships described. The authors have outlined numerous important relationships and we hope their efforts will stimulate future collaborative work in this critical area.

CHAPTER 1

Global Environmental Degradation and Biodiversity Loss: Implications for Human Health

ERIC CHIVIAN

"The Planet is not in jeopardy. We are in jeopardy. We haven't got the power to destroy the planet—or to save it. But we might have the power to save ourselves."

—Michael Crichton, *Jurassic Park* (1990)

Introduction

There is abundant evidence that we are beginning to alter, for the first time in history, the chemistry and physics and physiology of the Earth. A basic understanding of biological systems and their dependence on the environment should alert people to the potential dangers these alterations pose for human beings. Yet, most people, including most policymakers, do not comprehend the human implications of global environmental change. Underlying this lack of comprehension is the widespread belief that human beings are separate from the environments in which they live, that they can change the atmosphere and oceans—and damage marine, aquatic, and terrestrial ecosystems in the process—without these changes affecting them.

In focusing on the human health dimensions of biodiversity loss, this volume helps people understand that human beings are an integral part of the environment, and that to protect their health and lives, and those of their children, people must learn to protect the environment.

This chapter will provide an overview about how environmental degradation leads to biodiversity loss, and what the implications of this loss are for human beings. This is an enormously complex topic, one that is in its infancy in terms of scientific understanding. Much of what can be said about it is uncertain and speculative. But the subject is also perhaps the most important one of all—namely the ways that global ecosystems support human health and make human life possible, and there is enough evidence available in the scientific and medical literature to justify looking at possible future scenarios, especially as they may serve as warnings.

Medicine brings a new perspective to discussions about global environmental change. It has a long tradition of acting decisively to prevent life threatening situations from occurring even when all the evidence is not in. The low threshold for performing appendectomies is a case in point (Gross 1956)—if surgeons waited until they were absolutely certain that their patients had appendicitis, it would often be too late to prevent serious illness and death. This is the situation we face today with global environmental degradation. If we wait until we have definite proof of its occurrence and of its consequences, it may be too late to avoid a medical catastrophe.

This chapter will cover the contributions to biodiversity loss from:

- global climate change,
- stratospheric ozone depletion,
- toxic substances in the environment, and
- habitat destruction;

and the possible effects on human health from this loss, including:

- the loss of medicines,
- the loss of medical models, and
- the emergence and spread of infectious diseases.

Central to these discussions will be two themes:

1. That the study of species and biodiversity may be the best means we have for recognizing future danger signals to human health from global environmental degradation, as some species may be so uniquely sensitive to specific assaults on the environment that they may serve as our "canaries," or so-called "indicator species."

2. That we must focus much greater attention on biodiversity loss, which looms as a slowly evolving, potential medical emergency of unprecedented proportions, still largely unappreciated by policymakers and the public.

When *Homo sapiens* evolved, some 100,000 years ago, the number of species on Earth was the largest ever, but current rates of species extinction resulting from human activities, at least 1000 times those that would have occurred naturally, rivalling the great geologic extinctions of the past, may be reducing these numbers to the lowest levels since the end of the Age of Dinosaurs, 65 million years ago (Wilson 1993). Paul Ehrlich and E.O. Wilson (1991) and others have predicted that one-quarter of all species now alive may become extinct during the next 50 years if these rates of extinction continue.

There are 100 times more people on Earth than any land animal of comparable size that ever lived, and we are the most voracious and destructive species that ever existed. We consume or destroy or co-opt, for example, as much as 40% of all the solar energy trapped by land plants (Ehrlich and Wilson 1991). Not only is this behavior morally indefensible, endangering in the process countless other species, it is fundamentally and ultimately self-destructive.

Several aspects of global environmental degradation have an impact on species populations and biodiversity. This chapter will briefly review some of the evidence.

Global Climate Change

The Intergovernmental Panel on Climate Change (IPCC), the United Nations working group of 2500 of some of the world's most eminent atmospheric, physical, and biological scientists, has stated that projected worldwide CO_2 (the main greenhouse gas) emissions will result in an increase in global mean surface temperatures of about 1.8 to 6.3°F by the year 2100 (IPCC 1995). The magnitude of these changes may not seem very large, but it must be remembered that the difference between the temperatures at present and those at the peak of the last ice age, 18,000 years ago, are only 5–9°F (Stevens 9/20/94), and that temperatures of 7°F higher than those of today have not been present since the Eemian interglacial period 130,000 years ago (Stevens 11/1/94), and perhaps not since the Eocene epoch, tens of millions of years ago (Webb 1992).

Even small changes in temperature can lead to enormous changes in global weather patterns and habitats. For example, during the years from about 1000 to 1350 A.D., known as the "Medieval Warm Period," when global mean surface temperatures were only slightly more than 1°F warmer than they are now, there were vineyards in England, and Greenland supported hundreds of farms (Ponting 1991). By contrast, from 1430 to 1850, the period known as the "Little Ice Age," when global mean surface temperatures were only slightly more than 1°F cooler than they are now, the Thames was often frozen in winter, and there were icebergs off the coast of Norway (Leggett 1990).

Since the late 19th century, global mean surface temperatures have already warmed on average approximately 1°F (Stevens 1995), consistent with the increases in atmospheric CO_2 concentrations during the past century, from preindustrial revolution levels of 280 ppm to current levels of 350 ppm (Maskell et al. 1993). And 1995 was the warmest year since 1856, when mean global temperatures were first systematically recorded (Stevens 1996).

There is growing evidence that the seas are undergoing a similar warming, from measurements off the coast of California (Barry et al. 1995) and deep in the Atlantic Ocean (Parrilla et al. 1994), and that sea levels worldwide have risen about 10–20 cm over the last century (Warwick and Oerlemans 1990). The IPCC predicts that with greenhouse warming, sea levels will increase an additional 15–95 cm by the year 2100 (IPCC 1995).

These predicted changes in climate and sea levels will pose enormous and unprecedented threats to plant, microbial, and animal species, including human beings.

Many paleontologists believe that climate change, both global warming and cooling, was the dominant factor in the great extinctions of the past (Eldredge 1991), both directly, because of shifts in temperature outside the ranges to which species could adapt, and indirectly because of changes in habitats, for example, the formation of glaciers or changes in sea levels (Stanley 1987). Fossil records indicate that many species have been able to adapt to climate changes by shifting their ranges—during warming periods, for example, species colonized new habitats nearer the poles or at higher altitudes, while during cooling periods they retreated back toward the equator (Peters and Darling 1985). During several Pleistocene interglacial periods, when mean surface temperatures in North America were 3–5°F higher than they are at present, Cape Cod had forests like those found in present day North Carolina, manatees swam off the coast of New Jersey, and osage oranges grew near Toronto, several hundred kilometers north of their present ranges (Peters and Darling 1985). But many other species could not adapt and were lost, either because their rates of migration were too slow, or because geographical barriers like oceans, mountains, or unsuitable habitat conditions prevented their advance (Peters and Darling 1985).

By contrast to the past major changes in climate, however, when temperatures warmed or cooled over thousands of years, the changes predicted over the next century will be an order of magnitude or more faster, and it is not at all clear, even if there were no barriers to migration, whether species can migrate fast enough to avoid extinction. But there *are* barriers everywhere people live—cities, roads, agricultural lands, and other human constructions would further complicate species migration. It is calculated, for example, that for each 1°C rise in temperature, land plants would have to shift their ranges toward the poles by 100–150 km (Roberts 1989). The warming predicted by 2100, for example, would mean shifts of a few hundred kilometers. Some species, propagated by spores or dust seeds, might be able to achieve these rates (Perring 1965); most others would not (Peters and Darling 1985). Some spruce tree species, for example, even though they have light, wind-carried seeds, disperse them no farther than 200 m from the parent tree, corresponding to a potential maximum migration rate of only 20 km per century (Seddon 1971).

Animals, while more mobile, would be limited by the distributions of the plants they eat or otherwise depend on, by their ability to adapt behaviors to climate-altered habitats, or by changes in the populations of their predators or competitors, even if they could adapt physiologically (Peters and Darling 1985).

We may already be seeing evidence of species migrations and potential losses, paralleling the increases in recorded temperatures—in large reductions in red spruce trees in New Hampshire over the last 200 years (Hamburg and Cogbill 1988), in the upward climb of several vascular plant species in the Austrian alps over the past 70–90 years (Grabherr et al. 1994), in the shift to the north of Edith's checkerspot butterfly in the western United States (Parmesan 1996), and in the dramatic shift northward of a large number of marine invertebrate species off the coast of California over the past 60 years (Barry et al. 1995). In one particularly alarming study, warming seas off the coast of San Diego have been linked to an 80% reduction in zooplankton since 1951 and to similar declines in sea birds and fish, with the creation of a veritable "biological wasteland" (Roemmich and McGowan 1995).

Other aspects of global climate change that could have major impacts on species and biodiversity include: algal blooms (fertilized by the discharge of sewage and by agricultural run-off) (Epstein and Colwell 1993); rising seas that may threaten species in coastal wetlands, mangrove swamps, and coral reefs (Grigg and Epp 1989); the very worrisome possibility of major alterations of ocean currents from sea warming and changes in salinity, with potentially enormous changes in climate and in marine ecosystems (Broecker 1987); and finally the increase in CO_2 itself, which may threaten ecosystems by altering carbon and nitrogen cycles fundamental to the interactions between plants, the atmosphere, and the soil (Hilchey 1993) [for example, by slowing photosynthesis in some land plants (Korner and Arnone 1992)]. Furthermore, global warming may increase turnover in tropical forests, favoring rapidly growing, light-demanding plants (that take up less CO_2) over denser, slower-growing, shade-tolerant plants (Phillips and Gentry 1994), thereby accelerating global warming (Pimm and Sugden 1994).

If one considers the human species as a part of biodiversity, it is clear that greenhouse gas warming will also eventually reduce human numbers, as human beings are exquisitely sensitive to high temperatures (Rogot and Padgett 1976, Kilbourne 1992, Kalkstein 1993). During the two-week summer heat wave in the eastern United States in July 1993, for example, 84 people died in Philadelphia as a result of the increased temperatures (*Morbidity and Mortality Weekly Reports* 1993). And during the five days in mid-July 1995 when temperatures over 100°F swept over the central plains of the United States, there were 700 excess heat–related deaths in Chicago alone (*Morbidity and Mortality Weekly Reports* 1995, Semenza et al. 1996, Kellermann and Todd 1996).

The indirect effects on human health secondary to global climate change—from infectious diseases caused by a spread of disease vectors (Shope 1991) or by changes in habitats [resulting, for example, in outbreaks of cholera (Epstein 1995)]; from crop failure; from violent weather patterns; and from the unavailability of drinking water—are likely to extract an even heavier toll on human beings.

With global climate change, we are clearly conducting a gigantic experiment with life on this planet, knowing almost nothing about the potential consequences, endangering perhaps not only countless microbial, plant, and animal species, but ultimately perhaps ourselves as well. No human subjects committee at any hospital or medical research facility would ever approve this experiment to be performed.

Stratospheric Ozone Depletion

Stratospheric ozone depletion may also threaten species, both on land and in the sea. It was the formation of the ozone layer 450 million years ago that permitted marine life forms to colonize the land, as it protected them from the lethal effects of ultraviolet radiation.

In the mid 1970s Mario Molina and Sherwood Rowland (1974) predicted that the continued use of chlorofluorocarbons (CFCs) would lead to a decrease in stratospheric ozone, but it was not until 1985 that the first conclusive evidence of ozone depletion was reported (Farman et al. 1985), when ozone levels over the Antarctic were observed to have declined by 40% between 1975 and 1984 compared to 1960 baseline levels. Since then, record low levels have been recorded over the Antarctic in 1993 [down by more than 70% (Wilford 1993)], and over the Arctic in 1995, when readings 40% below normal values were recorded (Zurer 1995). Significant ozone depletion has also been observed by NASA and NOAA scientists over the heavily populated middle latitudes of the Northern Hemisphere, with early 1993 levels down by 10 to 20% (*New York Times* 4/23/93), and with corresponding increases in ground level ultraviolet-B radiation (UV-B) (Kerr and McElroy 1993).

The increased UV-B reaching the ground as a result of ozone thinning will damage the DNA and proteins of all living things (Leaf 1993), and could become a cause of species extinction and of a loss of biodiversity. Food crops (Worrest and Caldwell 1986), wild plants, and marine phytoplankton (Bridigare 1989) may all be affected, with possible major implications for terrestrial and marine food chains. Animals would also be vulnerable. Laboratory studies have shown, for example, that UV-B can lead to suppression of the immune response in mice, rats, and guinea pigs (Kripke 1990). Some amphibians seem particularly sensitive (see below), and there are anecdotal reports of sheep developing cataracts in Punta Arenas, Chile, at the tip of South America, beneath the ozone "hole" (Sims 1995).

Again, human beings are also at substantial risk, with increased rates expected of nonmelanocytic skin cancers (Suarez-Varela et al. 1992), malignant melanomas (Kricker et al. 1994), and cataracts (Taylor et al. 1988). The incidence of malignant melanomas has been increasing faster than that of any other cancer in the United States and worldwide (Rigel 1994), growing in Western populations by 20 to 50% every five years over the past two decades, particularly in young adults (Coleman et al. 1993). Every 11% decrease in stratospheric ozone is expected to increase melanoma mortality by 0.8 to 1.5% (Hoffman and Longstreth 1987).

Of great concern also is the potential for UV-B to weaken the systemic immune response in humans (Kripke 1990, Jeevan and Kripke 1993), potentially impairing peoples' ability to fight infections and cancers.

Toxic Pollution

Toxic substances—air pollutants such as acid rain, nitrogen dioxide, and ground level ozone; long-lived chlorinated hydrocarbons such as DDT and polychlorinated biphenyls (PCBs); heavy metals such as lead, mercury, cadmium, copper, arsenic, and molybdenum; and other compounds that alter immune, endocrine, or reproductive systems—may also act to threaten species and reduce biodiversity. The enormous quantities of such substances released into the environment [in the United States in 1988, 0.36 billion pounds of toxic chemicals, for example, were released into lakes, rivers, and streams, 2.43 billion pounds into the air, and 0.56 billion pounds into landfills (U.S. EPA 1990)] coupled with the almost complete absence of toxicity information available for the tens of thousands of chemicals in current use (National Research Council 1984) make this an area of enormous research importance.

There are several mechanisms by which toxic substances may be harmful to those organisms exposed.

Airborne Pollutants

The emission of air pollutants high into the atmosphere, particularly by fossil fuel burning power plants, results in the formation of acidic aerosols [mostly sulfuric acid, metallic acids, and ammonium sulfates (Christiani 1993)], and the eventual deposition of these as acid rain or snow. There is good evidence that this acidic precipitation may reduce biodiversity in terrestrial ecosystems, for example, in freshwater habitats (Hutchinson and Havas 1980, Altshuller and Linhurst 1983, Blaustein et al. 1994). The complexity of the effects of such acidification on ecosystems can be seen in two recently studied examples. In one study acid deposition, when combined with climate warming, was shown to increase the penetration of UV-B radiation in freshwater ecosystems, exposing aquatic organisms at greater depths to harmful radiation (Schindler et al. 1996). In another, the spreading of lime in southwest Sweden, which was done since the mid 1980s to counteract the effects of acidification from power plants and factories, has been thought to have caused the death of hundreds of moose in that region (Line 1996). The deaths seem to have been the result of increased concentrations of the element molybdenum, secondary to the liming, in plants eaten by the moose, leading to an imbalance between molybdenum and copper in the moose livers and a resultant fatal copper deficiency in the animals.

Other forms of air pollution, such as ground level ozone, produced mainly by the combustion of gasoline in motor vehicles, can also be toxic to various life forms, including food crops (Brown and Young 1990) and some forest trees (Prinz 1987, McLaughlin and Downing 1995). And the role of heavy metals (carried by polluted air) in the decline of forests has only recently begun to be elucidated (Gawel et al. 1996).

Endocrine-Mimicking Substances

In 1991, twenty-one biologists met in Racine, Wisconsin, to share observations about a wide variety of amphibians, reptiles, birds, and some marine and land mammals that demonstrated endocrine and reproductive effects, thought to be caused by exposure to synthetic chemicals in the environment (Colborn and Clement 1992). A sense of urgency was expressed at this and subsequent meetings, as it was believed by those attending that some chemical pollutants mimicked hormone activity in these animals, and by doing so, disrupted embryonic development and reproductive functioning. The result, they believed, was an observable decline in many North American species (Colborn and Clement 1992).

There was also great concern that these effects could occur in humans (Colborn et al. 1993, Colborn et al. 1996), and that the increased incidences of breast and prostatic cancers in the United States between 1969 and 1986 (Hoel et al. 1992, Davis et al. 1993, Davis and Bradlow 1995), the 400% increase in U.S. ectopic pregnancies between 1970 and 1987 (Nederlof et al. 1990), the marked rise in the incidence of cryptorchidism (undescended testes) (Chilvers et al. 1984) and hypospadias (a malformation of the penis where the urethra opens on the underside) (Matlai and Beral 1985) in the United Kingdom over the past 20 years, and the decline in sperm counts (Carlsen et al. 1992, Auger et al. 1995) and in normal sperm *since* [with lower percentages of motile and morphologically normal sperm *disputed* (Auger et al. 1995)] in fertile men in recent decades—all could have resulted from exposure to endocrine-disrupting chemicals. Recent sperm count studies in the United States do not show such declines, raising the possibility of regional differences in counts, perhaps secondary to differences in environmental exposures (Fisch et al. 1996, Paulsen et al. 1996).

While there are many uncertainties about the causal relationships between endocrine "disrupters" and the observed changes in wildlife and in humans, it seems clear that these compounds have the capacity to compromise the health and survival of a wide variety of species, particularly when these species are exposed to combinations of these compounds, whose synergistic effects may be 1000 times as potent as those from exposures to single compounds (Arnold et al. 1996).

Substances That Alter Immune Functions

A large number of environmental agents can also impair immune system functioning and potentially lead to a loss of species. These include metals like lead, arsenic, and methyl mercury; halogenated aromatic hydrocarbons like PCBs and dioxin; airborne pollutants like nitrogen dioxide, ozone, and sulfur dioxide; solvents like benzene and toluene; and pesticides like chlordane and malathion (National Research Council 1992, Descotes 1988). This is an area of study that has not received the attention it deserves. These immunosuppressant effects could act synergistically with global climate change and ozone depletion to result in significant species morbidity and mortality, affecting humans as well, from infectious diseases and cancers.

Indicator Species

It is a fundamental fact of biology that all living things share a basic chemistry and physiology, and that the closer two organisms are on the evolutionary tree, the more they share. This principle underlies why biomedical laboratory research with animals can be applied, in varying degrees, to human beings. It is also the reason that the study of other species' viability may offer the best window we have for the potential health consequences from environmental degradation for our own species. Some species may be uniquely sensitive to specific environmental changes and thereby be our "canaries" or "indicator species," warning us of potential future harm. In some cases, their vulnerability may be due, not to a greater sensitivity, but to greater exposures, for example, for species at the top of the food chain exposed to unusually high concentrations of toxins through bioaccumulation. Some possible examples deserve mention.

The precipitous drop in common seal (*Phoca vitulina*) populations in the western part (the Dutch area) of the Wadden Sea between 1950 and 1975 (Reijnders 1986), and the reproductive failure among beluga whales (*Delphinapterus leucas*) in the St. Lawrence River in Canada during the 1980s (Beland et al. 1993), offer perhaps models of unusual sensitivity to environmental chemicals that affect fertility. In the case of the seals, ingestion of PCB-contaminated fish, the PCBs most likely coming from the Rhine river, seemed to be the culprit, while for the belugas, there were high tissue concentrations of PCBs, DDT, Mirex (an insecticide that is carried by eels, a favorite beluga food, migrating from Lake Ontario), lead, and mercury (Martineau et al. 1987). In both cases, these marine mammals, at the top of the marine food chain, concentrated the pollutants involved.

Belugas may also serve as an "indicator species" for immune system dysfunction secondary to pollution, as a high proportion of those studied by scientists at the St. Lawrence National Institute of Ecotoxicology in Canada were diseased, and many died from a variety of tumors and infections (Martineau et al. 1988), presumably because of weakened immune responses (De Guise et al. 1995), resulting from the high levels of known immunotoxins (National Research Council 1992) in their tissues.

With other marine mammals, there is also evidence of possible immunodeficiency states, perhaps caused by their bioaccumulation of toxic chemicals, notably PCBs. Between 1987 and 1992, there have been massive die-offs, involving tens of thousands of seals (Lake Baikal freshwater, common or harbor, grey, monk, ringed, and harp), dolphins (bottlenose, white beaked, and striped), and porpoises (common) in Lake Baikal, the Ionian Sea, the Mediterranean, the North Sea, the Baltic Sea, the Barents Sea, the Gulf of Mexico, and the Atlantic Ocean off the Eastern coast of the United States (Osterhaus et al. 1988, Kennedy et al. 1988, Grachev et al. 1989, Harwood 1989, Domingo et al. 1990, Garrett 1994). These studies have identified four newly discovered viruses of the morbillivirus family (which includes the human measles virus and that causing canine distemper) as the infectious agents responsible. By 1993, the epidemics ended, and although it is still not clear what triggered them [except in the case of Lake Baikal (Grachev et al. 1989)], it has been widely believed that environmental change was the cause. Some researchers have speculated that organochlorine pollution, particularly PCBs, was involved, impairing the immune systems of those animals exposed and resulting in their vulnerability to morbillivirus infections. Many of the carcasses did, in fact, contain high concentrations of these toxins. It was also hypothesized that unusually warm winters led to relatively smaller fish populations and therefore less food for the seals, dolphins, and porpoises, and that the undernourished animals, as a result, mobilized organochlorine immunotoxins into their bloodstreams from stores in body fat. If this were so, then the marine mammal kills might be a case where global climate change and chemical pollution acted synergistically to promote the emergence of a global epidemic.

The hope that mobilliviruses such as those that seemed to decimate marine mammal populations would not infect humans (the only one that was known to do so had been measles) was shattered when 14 racehorses in

Australia died from another new morbillivirus that also infected and killed their trainer (Murray 1995). The underlying cause of this viral outbreak remains a mystery.

Manatees, endangered marine mammals which have lived in waters off the coast of Florida for 45 million years, have also been dying in great numbers. During the first two and one-half months of 1996, some 61 manatees were found dead off southwestern Florida, a number estimated to be almost 10% of the total population in that area (*New York Times* 3/19/96). All of the animals that died had pneumonia, but it is not yet clear what killed the manatees, and what infectious agent or agents caused the pneumonias. The record northward migration of a Florida male manatee to the East River during the summer of 1995, perhaps the result of unusually warm air and ocean temperatures recorded at that time (Sudetic 1995), suggests that increases in ocean temperatures may have had some relationship to the manatee mortality, perhaps secondarily from an increase in toxic red tides from the warming (*New York Times* 4/12/96).

The third example involves amphibians and stratospheric ozone depletion. In recent years, populations of some frogs, toads, and salamanders have declined markedly in many parts of the world, even in pristine habitats far from people (Blaustein and Wake 1990, Wake 1991). Some species may have already become extinct (Blaustein et al. 1994, Blaustein and Wake 1995). These declines have been viewed with alarm by scientists, not only because amphibians are crucial species in some ecosystems, serving as apex predators of invertebrates (including mosquitoes that may carry infectious diseases affecting humans), and as an important food for a wide variety of other species (Pierce 1985) including birds and mammals, but because they have been on the planet for over 100 million years, surviving other great extinctions (such as that of the dinosaurs), so that their current reduced numbers worldwide indicate that global ecosystems may be dangerously unhealthy.

Habitat destruction, particularly in tropical rainforests, is probably the predominant cause of amphibian loss (Blaustein and Wake 1995); other possible causes include acid rain, endocrine-disrupting chemicals (Colborn and Clement 1992), and the introduction of predators and competitors (Blaustein and Wake 1995).

The observation that many of the amphibian species known to be in decline live at high altitudes and lay their eggs in open, often shallow, water led to investigations of the role of UV-B radiation (Blaustein and Wake 1995) in their disappearance. And in 1994, the first conclusive evidence of

UV-B radiation as the likely cause of mortality in some amphibian species was published (Blaustein et al. 1994). The proposed mechanism, that the increased UV-B from stratospheric ozone depletion led to the amphibian eggs' succumbing to opportunistic fungal infections (Blaustein and Wake 1995), raises important questions about possible consequences for other species from thinning of the ozone layer.

The seeming disappearance of Basking sharks—the second largest fish, reaching lengths of 40 feet or more and living for as long as 100 years (*New York Times* 6/27/95)—from many coastal areas around the world may also be an indication of environmental degradation. Some of the loss may be explained by the high prices shark fins command for use in soup [up to $200 a pound (*National Geographic* 1995)] leading to their slaughter, but it is possible that environmental change is also part of the reason. As plankton-eating animals, Basking sharks may be an indicator species for warming of the oceans, which has resulted in significant declines in plankton populations in some regions where Basking sharks have been found in large numbers (Roemmich and McGowan 1995).

Two other examples that demonstrate the unique sensitivity of some species to greenhouse gas emissions and climate change need to be mentioned, as they illustrate the complexity of species responses to changes in the environment, and the potentially serious implications of these changes for ecosystems. The moth (*Helicoverpa armigera*) has recently been shown to be unable to accurately detect small changes in CO_2 released by food plants (as a result of their metabolic activity), because its CO_2 receptor organ is adapted to preindustrial CO_2 concentrations (280 ppm) in the atmosphere and is thrown off by the high concentrations currently present (350 ppm) (Stange and Wong 1993). While this moth is a major agricultural pest and some may rejoice that its destructive habits may be curtailed because of its receptor insensitivity, one must ask if other interactions between insects and plants, some essential for the health of ecosystems or for the viability of food crops such as pollination, are similarly disrupted.

The other example concerns changes in amphibian reproductive cycles in Britain as a consequence of warming temperatures. Several amphibian species of frogs, toads, and newts studied from 1977 to 1994 showed earlier spawning times than usual, the changes closely correlated with the observed increases in average temperatures during the period studied (Beebee 1995). The implications of these changes for the ecosystems in which these amphibians live are not well understood.

Similarly, food crops and livestock may show a marked sensitivity to climate change, as the highly virulent potato blight during the last several years in the United States, possibly caused by the heavy rains during this time (Dao 1995, Edwards 1996), and the high mortality among cattle and poultry during the July 1995 heat wave across the United States, have demonstrated (*New York Times* 7/14/95). The implications for food production from global climate change, both for agricultural crops and for livestock, is a critically important area that needs much greater attention.

Habitat Destruction

All of the global environmental changes mentioned may contribute to a loss of species and biodiversity, but it is habitat destruction by human activity that is the greatest destroyer. This is especially the case with the destruction of tropical rainforests, which we are cutting or burning at a rate of about 140,000 square kilometers per year, approximately 2% of the amount still standing, an area the size of Switzerland and the Netherlands combined (Wilson 1993).

The diversity of life in these forests is extraordinary. As an example, one investigator found an average of 300 distinct tree species in a series of 1-hectare plots in Peru (Gentry 1988); the total number of native tree species in all of North America is only 700. The same wealth of species diversity has been found for other organisms, and has led many to conclude that tropical rainforests, which comprise only 6% of the world's land mass, contain perhaps 50% of its species. Other species-rich habitats include coral reefs, and the deep oceans.

Only about 1.5 million species have been recorded and given scientific names (May 1988), but new species are being discovered all the time, and it is clear that we have only the barest knowledge of microscopic species like bacteria, fungi, and mites, and of species that have been largely inaccessible, such as in the tropical rain forest canopy, in the deep oceans (Grassle and Maciolek 1992, Hilbig 1994), and even far beneath the Earth's surface (Fredrickson and Onstott 1996). One recent report even described the finding of large numbers of previously undiscovered anaerobic bacterial species at depths of up to 1 mile below the surface of the earth, presumably living there undisturbed since the Triassic epoch (Parkes et al. 1994).

With habitat destruction, we are destroying countless species before we know their identities, and for those that we do know, before we have characterized their chemistries and physiologies.

Implications for Human Health

What are the potential human health consequences of this wholesale destruction?

Loss of Potential Medicines

For one, we are losing plants, animals, and microorganisms that may contain valuable medicines to treat human diseases that are presently untreatable and that cause enormous human suffering. At the present time, a significant proportion of the total pharmaceutical armamentarium is derived from these natural sources. For example, of all the prescriptions dispensed from community pharmacies in the United States from 1959 to 1980, some 25% contained active ingredients extracted from higher plants (Farnsworth 1990), and of the 150 drugs most commonly prescribed in the United States, a significant proportion are derived wholly or in part from natural sources (see Chapter 7). The proportion is even higher in the developing world, where people are more apt to rely on traditional medicines using natural substances.

Many of today's most useful medicines come from tropical rainforests. These include quinine and quinidine from the Cinchona tree (*Cinchona officianalis*), the latter, still one of the most important drugs in treating cardiac arrhythmias; D-tubocurarine, which has revolutionized general surgery by promoting deep muscle relaxation without high doses of general anesthetics, from the Condrodendron vine (*Chondrodendron tomentosum*); streptomycin, neomycin, amphotericin, and erythromycin, widely used antibiotics derived from tropical soil fungi; and vinblastine and vincristine from the Rosy periwinkle plant (*Vinca rosea*), the former extremely effective in treating Hodgkin's disease—achieving marked improvement in 50 to 90% of cases (Calabresi and Chabner 1990)—the latter the most effective agent known against acute childhood leukemias (Calabresi and Chabner 1990).

But it is not only in the tropics that new medicines may be found—temperate regions and the oceans are also vast storehouses of biodiversity, about which extremely little is known. Some examples of drugs from temperate zones include: aspirin, originally extracted from the willow tree (*Salix alba*); digitalis in the forms digoxin and digitoxin, still the main treatment for increasing cardiac contractility in heart failure, from the foxglove plant (*Digitalis purpurea*); and taxol from the Pacific Yew tree (*Taxus brevifolia*). Taxol provides an important example of the potential consequences of our squandering species in our own backyard, as it was discovered through the National Cancer Institute's screening program for new drugs with anticancer activity in the bark of the Pacific yew, a tree that had originally been discarded during commercial logging in old growth forests of the Pacific Northwest. Original reports showed taxol able to induce remission in 30% of advanced ovarian cancer cases that had been unresponsive to other treatment (McGuire et al. 1989). While some of the initial excitement about taxol as the new cancer wonder drug has waned, it remains an important new advance in cancer chemotherapy and may still be the most promising medication known for the treatment of ovarian and breast cancer (Nicolaou et al. 1996).

Other potential new medicines from temperate organisms that hold great promise include a substance, betulinic acid, from the bark of the white birch, which in laboratory tests has demonstrated marked activity in reducing malignant melanomas (*New York Times* 3/28/95), and salivary anticoagulants from leeches, ticks, blackflies, and other organisms (Bang 1991), which may lead to powerful new anticoagulants.

The oceans, the original home for life on this planet, have barely been explored for new drugs. One on the market, known as cytarabine or ara-C, comes from the Caribbean sponge (*Tethya crypta*), and is one of the most effective agents known for inducing remissions in cases of acute myelocytic leukemia in children and adults (Calabresi and Chabner 1990). Another, bryostatin l, from the bryozoan *Bugula neritina,* has shown a great deal of promise in Phase 1 trials for inhibiting tumor growth (Prendiville et al. 1993, Philip et al. 1993). Finally, a group of substances called pseudopterosins, extracted from the Caribbean Sea Whip (*Pseudopterogorgia elisbethae*), have been demonstrated to possess anti-inflammatory and analgesic properties more potent than currently available nonsteroidal anti-inflammatory agents such as indomethacin, through mechanisms that have yet to be defined (Look et al. 1986).

pretty dated section

Loss of Medical Models

With a loss of species, we are also losing valuable medical models that may help us understand human physiology and disease. Several examples follow.

Dart-poison frogs, the *Dendrobatidae,* from the lowland tropical forests in Central and South America are fast disappearing as they are particularly vulnerable to deforestation (Myers and Daly 1983). Each of the 100 or so species occupies a very specific location in the rainforest and cannot live naturally anywhere else. They are enormously useful to medicine. These frogs produce some of the deadliest natural toxins known, which have been used for centuries by native Indians to poison arrows and blowgun darts. The active ingredients are alkaloids, and it has been discovered that the reason for their extreme toxicity is that they bind selectively to sodium and potassium channels, calcium pumps, and acetylcholine receptors in nerve and muscle membranes, resulting in paralysis (Albuquerque et al. 1988). One of the toxins, batrachotoxin, for example, binds to a specific site in sodium channels and has been used increasingly by neurophysiologists to understand the structure and function of this most fundamental cellular unit, and to comprehend how nerves and muscles transmit electrical impulses (Albuquerque et al. 1971). If rainforest destruction continues at its present rate, these extremely valuable frogs could be lost.

In many parts of the world bears are also endangered, because of hunting and destruction of their habitats, but in particular, because of the very high prices their body parts bring on the Asian black market for use in traditional medicines and as gourmet food. Bear gallbladders are worth 18 times their weight in gold. The irony is that in medical terms, living bears are worth far more than the sum of all their body parts.

Hibernating, or more accurately denning, bears are biological wonders. Though largely immobile for 4 to 5 months or more, they do not lose bone mass (Floyd et al. 1990). A similar period of immobility or lack of weight bearing results in humans losing one-quarter or more of their bone mass—seen in bedridden and paralyzed patients, in astronauts, and in the inactive elderly. Understanding how bears prevent bone loss during immobility (the only vertebrate that can do this) could lead to ways of preventing or treating osteoporosis, a largely untreatable condition afflicting 25 million Americans, resulting in 1.5 million bone fractures and 50,000 deaths annually (New York Times 11/20/94), and costing the U.S. economy $10 billion dollars in direct health care costs and lost productivity each year (National Osteoporosis Foundation 1993).

Bears also do not urinate for the months of denning, as they are somehow able to recycle their urea to make new proteins (Nelson 1987, 1989). Humans, by contrast, unable to excrete their urinary wastes, die after a few days. If we understood how bears accomplished this feat, we might be able to find effective long-term treatments for those with chronic renal failure, who are now totally reliant on dialysis. Renal failure in the United States costs another $7 billion annually, a figure expected to increase to $10 billion by the year 2000 (New York Times 11/4/93).

Sharks are also being lost in record numbers from overfishing, and as a result we may never discover why sharks rarely develop tumors or infections, presumably the culmination of 400 million years of successful evolution (Stevens 1992). Some clues have been discovered—MIT researchers have isolated a substance from one shark species, the Basking shark (see page 19), that strongly inhibits the growth of new blood vessels toward solid tissue tumors, thereby preventing their growth (Lee and Langer 1983). This and other work on angiogenesis-inhibiting factors offers one of the best new hopes for developing effective chemotherapeutic agents for some malignant tumors.

[handwritten: prescient]

Another group of researchers has recently isolated a compound they named squalamine from the tissues of the dogfish shark. Squalamine has demonstrated extremely potent activity in lab tests against a variety of bacteria, fungi, and parasites through mechanisms not yet understood, but that seem entirely different from those of all other known antibiotics (Moore et al. 1993). As infectious agents are developing growing resistance to currently used antibiotics, the search for new drugs working through different pathways becomes increasingly important.

Predatory cone snails of the genus *Conus,* comprising approximately 500 species and found in tropical reef communities, produce a wide variety of toxins to capture their prey. These toxins are small peptides that act on voltage-sensitive calcium channels, sodium channels, N-methyl-D-aspartate (NMDA) receptors, nicotinic acetylcholine receptors, and vasopressin receptors in neuromuscular systems (Olivera et al. 1990). They have been increasingly used to understand the molecular structure of these channels and receptors and how they work, to provide clues for the mechanism of action of various human disease conditions, and to find new treatments for these conditions. Calcium channel antagonists derived from *Conus* toxins, for example, have been shown to help alleviate the exaggerated pain responses in some peripheral nerve lesions (Chaplan et al. 1994) and to differentiate the paraneoplastic neuromuscular transmission disorder, Lambert-Eaton syndrome, associated with small-cell lung carcinomas (Lennon et al. 1995) from myasthenia gravis (Miljanich and Ramachandran 1995). The laboratory findings that calcium channel toxins from cone snails also have the potential to prevent neuronal degeneration following focal (Takizawa et al. 1995) and global (Valentino et al. 1993) cerebral ischemia have generated the greatest interest, as these compounds may someday be used to limit nerve cell death in cases of bypass surgery, head trauma, cardiac arrest, and strokes (Miljanich and Ramachandran 1995).

Pit vipers (family *Crotalidae,* which includes rattlesnakes, cobras, and water moccasins) have also become invaluable to medical science. In the 1960s, it was discovered that pit viper venom contained compounds that were lethal to prey because they intensified the animal's response to bradykinin (Ferreira 1965), a potent vasodilating substance, and thereby caused fatal hypotension. These bradykinin-potentiating factors were shown to be a family of peptides that inhibited kininase II, an enzyme that inactivates bradykinin. Kininase II was subsequently demonstrated to be the same enzyme as angiotensin converting enzyme (ACE), which catalyzes the conversion of angiotensin I, a relatively inactive substance, to angiotensin II, a powerful vasopressor (Garrison and Peach 1990). This understanding of the renin–angiotensin system and of the role of bradykinin gleaned from understanding the action of pit viper venom as an ACE inhibitor led to the development of a whole class of synthetic ACE inhibitors that have become some of the most effective agents known for treating hypertension.

The sea squirt, the only animal besides human beings to form stones naturally in their kidney-like organs, has been used to understand the mechanisms of how kidney stones and gout develop in humans. By studying the formation of both uric acid and calcium oxalate stones in one type of sea squirt, the sea grape (*Molgula manhattensis*), Mary Beth Saffo of the University of California at Santa Cruz is trying to develop strategies for preventing and treating these conditions (Yoon 1994).

The discovery of a new bacterial species, *Thermus aquaticus* (from thermal springs in Yellowstone National Park), which lives at extremely high temperatures (70–75°C) (Brock and Freeze 1969), led to the isolation of polymerase enzymes that have revolutionized DNA and RNA research (Air and Harris 1974) and to countless other applications, including the identification of criminals through their DNA.

Finally, the horseshoe crab (*Limulus polyphemus*), an animal which first evolved 350 million years ago, has proved invaluable in two areas of biomedical research. The first concerns work on vision. Because the horseshoe crab possesses one of the largest and most accessible optic nerves in the animal kingdom, and has the largest photoreceptors of any animal, it has been extensively studied to help understand the fundamental mechanisms of animal vision, and of the way the brain interacts with sensory organs (Hartline et al. 1956, Barlow 1990), insights that may lead to a greater understanding of human vision. The second concerns diagnostic bacteriology. As a result of early discoveries by Frederik Bang on the *Limulus'* clotting response to injections of gram negative bacteria (Bang 1956), the processed blood from the crab, *Limulus* amoebocyte lysate, has developed into one of the most sensitive tests available to detect the presence of gram negative endotoxins. It was because of this test that batches of swine flu vaccine were first shown to be contaminated in the mid 1970s (Sargent 1987).

There are countless other examples of unique microorganisms, plants, and animals holding the secrets of hundreds of millions of years of evolution and hundreds of billions of biological experiments, which, because of species extinctions, are in danger of being lost forever to medical science.

Infectious Diseases

It is important to discuss how alterations in biodiversity can disrupt ecosystem equilibria resulting in the emergence or reemergence of infectious diseases in man. This may be one of the most serious public health consequences of biodiversity loss; it is also perhaps one of the most neglected and poorly understood.

The case of the hantavirus outbreak in the United States serves as a valuable model illustrating the potential effects of such alterations, as well as our relative lack of knowledge about infectious agents in the environment, and the processes that uncover them. Other cases will be discussed more fully in Chapter 2.

Six years of drought in the Four Corners Area of the American Southwest, where Utah, Colorado, Arizona, and New Mexico meet, ended in the late winter and spring of 1993 with unusually heavy snows and rains (Wenzel 1994). These dramatic weather patterns, seen in other parts of the world as well and thought to result directly from an intense and prolonged El Niño (and perhaps indirectly from greenhouse gas-induced warming of the seas), resulted in a superabundance of piñon nuts and grasshoppers, food for the native deer mouse. At the same time, the drought had led to a reduction in the mouse's predators—foxes, snakes, and owls, and as a result, the mouse population exploded, increasing tenfold by May of 1993 over levels of the year before (Wenzel 1994). Deer mice, it was subsequently learned, carry a hantavirus that they shed through their saliva and excreta, and as a consequence of greatly increased exposure by people in the area to the huge mouse population, an epidemic of a rapidly fatal respiratory syndrome developed—17 previously healthy people, most of them Native Americans, initially got the disease; 13 died (Duchin et al. 1994). The number of confirmed cases of hantavirus pulmonary syndrome in the United States has reached 131 in 24 states, mostly in the West, of which almost one-half have been fatal (Freeman 1996).

Other examples of changes in habitats and biodiversity associated with infectious disease spread include Kyasanur Forest Disease, a tickborne hemorrhagic viral disease identified in Mysore, India and linked to deforestation in that region (Morse 1991); Argentine Hemorrhagic Fever, a viral ill-ness carried by a mouse which proliferated and spread with the clearing of the pampas grasslands in Argentina for the purpose of growing corn (Morse 1990); and malaria in the Amazon which has reached epidemic proportions in the past 20 years (Kingman 1989), as a result of forest clear-ing and the creation of large numbers of still pools where *Anopheles dar-lingi,* the most important malaria mosquito vector in the area, can multi-ply unchecked (Walsh et al. 1993). It is not clear whether the recent cases of locally acquired malaria in the United States (Brook et al. 1994, Bell et al. 1995) (which hadn't been seen in 50 years) are also the result of habi-tat changes, secondary to the unusually warm summers of the past several years which favored the proliferation of native mosquitoes and the trans-mission of malaria, originally acquired abroad.

These cases illustrate how fragile and complex the equilibria among species in ecosystems are, how little we understand these equilibria, and how a loss of biodiversity (in some cases through changes in weather, but in other cases, perhaps through habitat destruction, the direct elimination of predators, stratospheric ozone depletion, or other means) may result in potentially catastrophic consequences for human beings. No one knows how many viruses or other infectious agents in the environment, poten-tially harmful to man, are now being held in check by the natural equilib-ria afforded by biodiversity.

Conclusion

Global climate change, stratospheric ozone depletion, toxic substances re-leased into the environment, and habitat destruction all have the capacity to lead to species extinction and biodiversity loss. Current rates of species loss from human activity rival the great geologic extinctions of the past. With a loss of biodiversity, we risk losing potential new medicines to treat human illnesses that are presently untreatable, foreclosing the possibility of discovering valuable medical models that may help us understand human physiology and disease, and disrupting ecosystem balances on which all life, including human life, depends.

The fundamental relationship between biodiversity and human health is generally unappreciated by policymakers and the public, and as a result, the preservation of habitats and of species is given a low priority. In addition, there is little understanding that the study of species and of biodiversity may offer the best means we have for predicting future dangers to human beings from environmental degradation, as some species may be affected at lower thresholds to environmental hazards than we are and serve as our "canaries," or so-called "indicator species."

How can the importance of biodiversity become central to the concerns of policymakers and the public? There is a lesson to be learned from the history of chlorofluorocarbons and the ozone layer. Once the magnitude of the danger was clear, and the consequences to human life were obvious, then the world community was able to take effective action to protect the ozone layer. The same must occur with the issue of biodiversity—it is essential that people understand that their health and lives, and those of their children, depend so completely on the health of other species and of global ecosystems. Only then will they develop the motivation to support personal behaviors and public policies that protect biodiversity. There is no task more important for physicians, other health professionals, scientists, and concerned citizens than helping to promote this understanding.

References

Air, G.M., and Harris, J.I. (1974). DNA-Dependent RNA Polymerase from the Thermophilic Bacterium *Thermus aquaticus*. *FEBS Letters* 38(3):277–281.

Albuquerque, E.X., et al. (1971). Batrachotoxin: Chemistry and Physiology. *Science* 172:995–1002.

Albuquerque, E.X., et al. (1988). Macro-molecular Sites for Specific Neurotoxins and Drugs on Chemosensitive Synapses and Electrical Excitation in Biological Membranes. In *Ion Channels,* Vol. 1 (T. Narahashi, ed.), Plenum Publishing Co.

Altman, L.K. "Sharks Yield Possible Weapon Against Infection," *New York Times,* 2/15/93.

Altshuller, A.P., and Linhurst, R.A. (eds.) (1983). *The Acidic Deposition Phenomenon and its Effects: Critical Assessment Review Papers,* EPA Office of Research and Development, Washington, DC.

Arnold, S.F., et al. (1996). Synergistic Activation of Estrogen Receptor with Combinations of Environmental Chemicals. *Science* 272:1489–1492.

Auger, J., et al. (1995). Decline in Semen Quality among Fertile Men in Paris during the Past 20 Years. *New England Journal of Medicine* 332(5):281–285.

Bang, F.B. (1956). A Bacterial Disease of *Limulus polyphemus. Bulletin of Johns Hopkins Hospital* 325–336.

Bang, N.U. (1991). Leeches, Snakes, Ticks and Vampire Bats in Today's Cardio-vascular Drug Development. *Circulation* 84(1):436–437.

Barlow, R.B. (1990). What the Brain Tells the Eye. *Scientific American,* April, pp. 90–95.

Barry, J.P., et al. (1995). Climate-Related, Long-Term Faunal Changes in a California Rocky Intertidal Community. *Science* 267:672–675.

Beebee, T.J.C. (1995). Amphibian Breeding and Climate. *Nature* 374:219–220.

Beland, P., et al. (1993). Toxic Compounds and Health and Reproductive Effects in St. Lawrence Beluga Whales. *Journal of Great Lakes Research* 19(4):766–775.

Bell, R., et al. (1995). Local Transmission of *Plasmodium vivax* Malaria—Houston, Texas 1994. *Morbidity and Mortality Weekly Report* 44(5):295, 301–303.

Blaustein, A.R., and Wake, D.B. (1990). *Trends in Ecology and Evolution* 5:203.

Blaustein, A.R., and Wake, D.B. (1995). The Puzzle of Declining Amphibian Populations. *Scientific American* 272(4):52–57.

Blaustein, A.R., et al. (1994). Amphibian Declines: Judging Stability, Persistence and Susceptibility of Population to Local and Global Extinction. *Conservation Biology* 8(1):60–71.

Blaustein, A.R., et al. (1994). UV Repair and Resistance to Solar UV-B in Amphibian Eggs: A Link to Population Declines? *Proceedings of the National Academy of Sciences* 91:1791–1795.

Bridigare, R.R. (1989). Potential Effects of UV-B on Marine Organisms of the Southern Ocean: Distribution of Phytoplankton and Krill during Austral Spring. *Photochemistry and Photobiology* 50:469–478.

Brock, T.D., and Freeze, H. (1969). *Thermus aquaticus* gen. n. and sp. n., a Nonsporulating Extreme Thermophile. *Journal of Bacteriology* 98(1):289–297.

Broecker, W.S. (1987). Unpleasant Surprises in the Greenhouse? *Nature* 328:123–126.

Brook, J.H., et al. (1994). Brief Report: Malaria Probably Locally Acquired in New Jersey. *New England Journal of Medicine* 331(1):22–23.

Brown, J., et al. (1988). Lag from Exposure to Appearance of Melanoma in World War II Veterans Serving in the South Pacific. *International Journal of Dermatology* 14:31.

Brown, L., and Young, G.E. (1990). Feeding the World in the Nineties. In *State of the World, 1990* (L. Brown, ed.), Norton.

Calabresi, B.A., and Chabner, B.A. (1990). Antineoplastic Agents. In *The Pharmacologic Basic of Therapeutics,* 8th ed. (A.G. Gilman et al., eds.), Pergamon.

Carlsen, E., et al. (1992). Evidence for Decreasing Quality of Semen during Past 50 Years. *British Medical Journal* 305:609–613.

Chaplan, S., et al. (1994). Role of Voltage-Dependent Calcium Channel Subtypes in Experimental Tactile Allodynia. *Journal of Pharmacology and Experimental Therapeutics* 269(3):1117–1123.

Chilvers, C., et al. (1984). Apparent Doubling of Frequency of Undescended Testes in England and Wales in 1962–1981. *The Lancet* ii:330–332.

Christiani, D.C. (1993). Urban and Transboundary Air Pollution: Human Health Consequences. In *Critical Condition: Human Health and the Environment* (E. Chivian, M. McCally, H. Hu, and A. Haines, eds.), MIT Press.

Colborn, T., and Clement, C. (eds.) (1992). *Chemically-Induced Alterations in Sexual and Functional Development: The Wildlife/Human Connection,* Princeton Scientific Publishing.

Colborn, T., et al. (1993). Developmental Effects of Endocrine-Disrupting Chemicals in Wildlife and Humans. *Environmental Health Perspectives* 101:378–384.

Colborn, T., et al. (1996). *Our Stolen Future,* Dutton.

Coleman, M., et al., (1993). *Trends in Cancer Incidence and Mortality,* International Agency for Research on Cancer, Lyon.

Crichton, M. (1990). *Jurassic Park,* Ballantine Books, New York.

Dao, J. "Worst Blight Since Ireland's is Chilling Potato Farmers," *New York Times,* 7/30/95, p. 33.

Davis, D.L., et al. (1993). Medical Hypothesis: Xenoestrogens as Preventable Causes of Breast Cancer. *Environmental Health Perspectives* 101:372–377.

Davis, D.L., and Bradlow, H.L. (1995). Can Environmental Estrogens Cause Breast Cancer? *Scientific American* 273(4):166–172.

De Guise, S., et al. (1995). Possible Mechanisms of Action of Environmental Contaminants on St. Lawrence Beluga Whales (*Delphinapterus leucas*). *Environmental Health Perspectives* 103(Suppl. 4):73–77.

Descotes, J. (1988). *Immunotoxicology of Drugs and Chemicals,* Elsevier.

Dold, C. "Toxic Agents Found to be Killing Off Whales," *New York Times,* 6/16/92, p. C4.

Domingo, M., et al. (1990). Morbillivirus in Dolphins. *Nature* 348:21.

Duchin, J.S., et al. (1994). Hantavirus Pulmonary Syndrome: A Clinical Description of 17 Patients with a Newly Recognized Disease. *New England Journal of Medicine* 330:949–955.

Edwards, R. (1996). Tomorrow's Bitter Harvest. *New Scientist,* August 17, pp. 14–15.

Ehrlich, P.R., and Wilson, E.O. (1991). Biodiversity Studies: Science and Policy. *Science* 253:758–762.

Eldredge, N. (1991). *The Miner's Canary,* Princeton University Press.

Epstein, P.R. (1995). Emerging Diseases and Ecosystem Instability: New Threats to Public Health. *American Journal of Public Health* 85(2):168–172.

Epstein, P.R., and Colwell, R.R. (1993). Marine Ecosystems. *The Lancet* 342:1216–1219.

Erwin, T.L. (1983). *Bulletin of the Entomological Society of America* 30:14.

Farman, J.C., et al. (1985). Large Losses of Total Ozone in Antarctica Reveal Seasonal ClO_x/NO_x Interaction. *Nature* 315:207–210.

Farnsworth, N.R. (1990). The Role of Ethnopharmacology in Drug Development. In *Bioactive Compounds from Plants,* Ciba Foundation Symposium.

Ferreira, S.H. (1965). A Bradykinin-Potentiating Factor (BPF) Present in the Venom of *Bothrops jararaca*. *British Journal of Pharmacology* 24:163–169.

Fisch, H., et al. (1996). Semen Analysis in 1283 Men from the United States Over a 25-Year Period: No Decline in Quality. *Fertility and Sterility* 65(5):1009–1014.

Floyd, T., Nelson, R.A., and Wynne, G.F. (1990). Calcium and Bone Metabolic Homeostasis in Active and Denning Black Bears. *Clinical Orthopaedics and Related Research* 255:301–309.

Fredrickson, J.K., and Onstott, T.C. (1996). Microbes Deep Inside the Earth. *Scientific American* 275(4):68–73.

Freeman, K. "Cases of Fatal Virus Are Edging Up, Mostly in the West, Agency Says," *New York Times,* 4/12/96.

Garrett, L. (1994). *The Coming Plague.* Farrar, Strauss & Giroux, New York.

Garrison, J.C., and Peach, M.J. (1990). Renin and Angiotensin. In *The Pharmacologic Basis of Therapeutics,* 8th ed. (A.G. Gilman et al., eds.), Pergamon.

Gawel, J., et al. (1996). Role for Heavy Metals in Forest Decline Indicated by Phytochelatin Measurements. *Nature* 381:64–65.

Gentry, A.H. (1988). Tree Species Richness of Upper Amazonian Forests. *Proceedings of the National Academy of Sciences* 85:156–159.

Grabherr, G., Gottfried, M., and Pauli, H. (1994). Climate Effects on Mountain Plants. *Nature* 364:448.

Grachev, M.A., et al. (1989). Distemper Virus in Baikal Seals. *Nature* 338:209.

Grassle, J.F., and Maciolek, N.J. (1992). Deep-Sea Species Richness: Regional and Local Diversity Estimates from Quantitative Bottom Samples. *American Naturalist* 139:313–314.

Grigg, R.W., and Epp, D. (1989). Critical Depth for the Survival of Coral Islands: Effects on the Hawaiian Archipelago. *Science* 243:638–641.

Gross, R.E. (1956). *The Surgery of Infancy and Childhood,* W.B. Saunders.

Group, J.R.H.C.S. (1986). Cryptorchidism: An Apparent Substantial Increase Since 1960. *British Medical Journal* 293:1401–1404.

Hamburg, S.P., and Cogbill, C.V. (1988). Historical Decline of Red Spruce Populations and Climatic Warming. *Nature* 331:428–431.

Hartline, H.K., et al. (1956). Inhibition in the Eye of *Limulus. Journal of General Physiology* 39(5):651–673.

Harwood, J. (1989). Lessons from the Seal Epidemic. *New Scientist,* Feb. 18, pp. 38–42.

Hilbig, B. (1994). Faunistic and Zoogeographical Characterization of the Benthic Infauna on the Carolina Continental Slope. *Deep Sea Research II* 41(4–6):929–950.

Hilchey, T. "Rise in CO_2 May Shift Key Biochemical Cycles," *New York Times,* 7/27/93, p. C4.

Hoel, D.G., et al. (1992). Trends in Cancer Mortality in 15 Industrialized Countries, 1969–1986. *Journal of the National Cancer Institute* 84:313–320.

Hoffman, J.S., and Longstreth, J. (1987). *Ultraviolet Radiation and Melanoma with a Special Focus on Assessing the Risks of Stratospheric Ozone Depletion.* EPA 400/1-87/001D, Washington, DC, December.

Hutchinson, T.C., and Havas, M. (eds.) (1980). *Effects of Acid Precipitation on Terrestrial Ecosystems,* Plenum Press.

Intergovernmental Panel on Climate Change (IPCC). (1995). Working Group II Second Assessment Report, "Summary for Policymakers: Impacts, Adaptation and Mitigation Options," World Meteorological Organization and United Nations Environment Programme.

Jeevan, A., and Kripke, M.L. (1993). Ozone Depletion and the Immune System. *Lancet* 342:1159–1160.

Kalkstein, L.S. (1993). Health and Climate Change—Direct Impacts in Cities. *Lancet* 342:1159–1160.

Kellermann, A.L., and Todd, K.H. (1996). Killing Heat. *New England Journal of Medicine* 385(2):126–127.

Kennedy, S., et al. (1988). Viral Distemper Now Found in Porpoises. *Nature* 336:21.

Kerr, J.B., and McElroy, C.T. (1993). Evidence for Large Upward Trends of Ultraviolet-B Radiation Linked to Ozone Depletion. *Science* 262:1032–1034.

Kilbourne, E.M. (1992). Illness Due to Thermal Extremes. In *Public Health and*

Preventive Medicine, 13th ed. (J.M. Last and R.B. Wallace, eds.), Maxcy-Rosenau-Last. Appleton Lange, Norwalk.

Kingman, S. (1989). Malaria Runs Riot on Brazil's Wild Frontier. *New Scientist* 123:24–25.

Korner, C., and Arnone, J.A. (1992). Responses to Elevated Carbon Dioxide in Artificial Tropical Ecosystems. *Science* 257:1672–1675.

Kricker, A., et al. (1994). Skin Cancer and Ultraviolet. *Nature* 368:594.

Kripke, M.L. (1990). Effects of UV Radiation on Tumor Immunity. *Journal of the National Cancer Institute* 82:1392–1396.

Leaf, A. (1993). Loss of Stratospheric Ozone and Health Effects of Increased Ultraviolet Radiation. In *Critical Condition: Human Health and the Environment* (E. Chivian, M. McCally, H. Hu, and A. Haines, eds.), MIT Press.

Lee, A., and Langer, R. (1983). Shark Cartilage Contains Inhibitors of Tumor Angiogenesis. *Science* 221:1185–1187.

Leggett, J. (1990). The Nature of the Greenhouse Threat. In *Global Warming: The Greenpeace Report* (J. Leggett, ed.), Oxford University Press.

Lennon, V.A., et al. (1995). Calcium-Channel Antibodies in the Lambert-Eaton Syndrome and Other Paraneoplastic Syndromes. *New England Journal of Medicine* 332:1467–1474.

Line, L. "Acid Rain Leading to Moose Deaths," *New York Times,* 3/12/96, p. C4.

Look, S.A., et al. (1986). The Pseudopterosins: Anti-inflammatory and Analgesic Natural Products from the Sea Whip *Pseudopterogorgia elisbethae. Proceedings of the National Academy of Sciences* 83:6238–6240.

Martineau, D., et al. (1987). Levels of Organochlorine Chemicals in Tissues of Beluga Whales (*Delphinapterus leucas*) from the St. Lawrence Estuary, Quebec, Canada. *Archives of Environmental Contamination and Toxicology* 16:137–147.

Martineau, D., et al. (1988). Pathology of Stranded Beluga Whales from the St. Lawrence Estuary, Quebec, Canada. *Journal of Comparative Pathology* 98:287–311.

Maskell, K., et al. (1993). Basic Science of Climate Change. *Lancet,* October 23.

Matlai, P., and Beral, V. (1985). Trends in Congenital Malformations of External Genitalia. *Lancet,* January 12, p. 108.

May, R.M. (1988). How Many Species are There? *Science* 241:1441–1449.

McGuire, W.P., et al. (1989). Taxol: A Unique Antineoplastic Agent with Significant Activity in Advanced Ovarian Epithelial Neoplasms. *Annals of Internal Medicine* 111(4):273–279.

McLaughlin, S.B., and Downing, D.J. (1995). Interactive Effects of Ambient Ozone and Climate Measured on Growth of Mature Forest Trees. *Nature* 374:252–254.

Miljanich, G.P., and Ramachandran, J. (1995). Antagonists of Neuronal Calcium Channels: Structure, Function and Therapeutic Implications. *Annual Review of Pharmacology and Toxicology* 35:707–734.

Molina, M.J., and Rowland, F.S. (1974). *Nature* 249:810.

Moore, K.S., et al. (1993). Squalamine: An Aminosterol Antibiotic from the Shark. *Proceedings of the National Academy of Science* 90:1354–1358.

Morbidity and Mortality Weekly Reports, Heat-Related Deaths—United States, 1993. *MMWR* 42(28):558–560, 7/23/93.

Morbidity and Mortality Weekly Reports. Heat-Related Mortality—Chicago. *MMWR* 95(44):577–579, July 1995.

Morse, S.S. (1990). Stirring Up Trouble: Environmental Disruption Can Divert Animal Viruses into People. *The Sciences* 30:16–21, The New York Academy of Sciences.

Morse, S.S. (1991). The Origins of New Viral Diseases. *Environmental Carcinogens and Ecotoxicology Reviews* C9(2):207–228.

Murray, K. (1995). A Morbillivirus that Caused Fatal Disease in Horses and Humans. *Science* 268:94–97.

Myers, C.W., and Daly, J.W. (1983). Dart-Poison Frogs. *Scientific American* 248(2):120–133.

National Geographic (1995). "Huge, Gentle Basking Sharks are Vanishing," *National Geographic* 188(5).

National Osteoporosis Foundation. (1993). *Fast Facts on Osteoporosis,* Washington, DC.

National Research Council. (1984). *Toxicity Testing—Strategies to Determine Needs and Priorities,* National Academy Press, Washington, DC.

National Research Council. (1992). *Biological Markers in Immunotoxicology,* National Academy Press, Washington, DC.

Nederlof, K.P., et al. (1990). Ectopic Pregnancy Surveillance, United States 1970–1987. *Morbidity and Mortality Weekly Report* 39:9–17.

Nelson, R.A. (1987). Black Bears and Polar Bears—Still Metabolic Marvels. *Mayo Clinic Proceedings* 62:850–853.

Nelson, R.A. (1989). Nitrogen Turnover and its Conservation in Hibernation. In *Living in the Cold II* (A. Malan and B. Canguilhem, eds.), Colloque Inserm, John Libbey Eurotext Ltd., pp. 299–307.

New York Times, "Northern Hemisphere Ozone at 14-Year Low," 4/23/93, p. A26.

New York Times, "A Bleak U.S. Report on Kidney-Failure Patients," 11/4/93, p. A17.

New York Times, "FDA Panel Votes Against Drug to Treat Osteoporosis," 11/20/94, p. 33.

New York Times, "Birch Used to Shrink Melanoma in Mice," 3/28/95, p. C11.

New York Times, "Reasons Sought for Decline of Basking Shark," 6/27/95, p. C4.

New York Times, "Preview of the Heat Wave," 7/14/95, p. B2.

New York Times, "A Wave of Manatee Deaths Puzzles Scientists," 3/19/96, p. C7.

New York Times, "Red Tide Is a Suspect in the Deaths of Manatee Off Florida Coasts," 4/12/96.

Nicolaou, K.C., et al. (1996). Taxoids: New Weapons Against Cancer. *Scientific American* 274(6):94–98.

Olivera, B.M., et al. (1990). Diversity of *Conus* Neuropeptides. *Science* 249: 257–263.

Osterhaus, A.D.M.E., et al. (1988). Canine Distemper Virus in Seals. *Nature* 335:403–404.

Parkes, R.J., et al. (1994). Deep Bacterial Biosphere in Pacific Ocean Sediments. *Nature* 371:410–413.

Parmesan, C. (1996). Climate and Species Range. *Nature* 382:765–766.

Parrilla, G., et al. (1994). Rising Temperatures in the Subtropical North Atlantic Ocean over the Past 35 Years. *Nature* 369:48–51.

Paulsen, C.A., et al. (1996). Data from Men in Greater Seattle Area Reveals No Downward Trend in Semen Quality: Further Evidence That Deterioration of Semen Quality Is Not Geographically Uniform. *Fertility and Sterility* 65(5):1015–1020.

Perring, F.H. (1965). The Advance and Retreat of the British Flora. In *The Biological Significance of Climatic Changes in Britain* (C.J. Johnson and L.P. Smith, eds.), Academic Press.

Peters, R., and Darling, J.D.S. (1985). The Greenhouse Effect and Nature Reserves. *BioScience* 35(11):707–717.

Philip, A.P., et al. (1993). Phase I Study of Bryostatin 1: Assessment of Interleukin 6 and Tumor Necrosis Factor Induction *In Vivo. Journal of the National Cancer Institute* 85(22):1812–1818.

Phillips, O.L., and Gentry, A.H.(1994). Increasing Turnover Through Time in Tropical Forests. *Science* 263:954–957.

Pierce, B.A. (1985). Acid Tolerance in Amphibians. *BioScience* 35(4):239–243.

Pimm, S.L., and Sugden, A.M. (1994). Tropical Diversity and Global Change. *Science* 263:933–934.

Ponting, C. (1991). *A Green History of the World: The Environment and the Collapse of Great Civilizations,* Penguin Books.

Prendiville, J., et al. (1993). A Phase I Study of Intravenous Bryostatin 1 in Patients with Advanced Cancer. *British Journal of Cancer* 68:418–425.

Prinz, B. (1987). *Environment* 29:10–37.

Reijnders, R.J.H. (1986). Reproductive Failure in Common Seals Feeding on Fish from Polluted Coastal Waters. *Nature* 324:456–457.

Rigel, D.S. (1994). Environmental Atmospheric Issues and Their Effects on Skin Cancer. In *Proceedings of the National Conference on Environmental Hazards to the Skin,* American Academy of Dermatology.

Roberts, L. (1989). How Fast Can Trees Migrate? *Science* 243:735–737.

Roemmich, D., and McGowan, J. (1995). Climatic Warming and the Decline of Zooplankton in the California Current. *Science* 267:1324–1326.

Rogot, E., and Padgett, S.J. (1976). Associations of Coronary and Stroke Mortality with Temperature and Snowfall in Selected Areas of the United States, 1962–1966. *American Journal of Epidemiology* 103:565–575.

Sargent, W., (1987). *The Year of the Crab,* W.W. Norton & Company, New York.

Schindler, D.W., et al. (1996). Consequences of Climate Warming and Lake Acidification for UV-B Penetration in North American Boreal Lakes. *Nature* 379:705–708.

Seddon, B. (1971). *Introduction to Biogeography,* Barnes & Noble.

Semenza, J.C., et al. (1996). Heat-Related Deaths During the July 1995 Heat Wave in Chicago. *New England Journal of Medicine* 335(2):84–90.

Shope, R. (1991). Global Climate Change and Infectious Diseases. *Environmental Health Perspectives* 96:171–174.

Sims, C. "A Hole in the Heavens (Chicken Little Below?)," *New York Times,* 3/3/95.

Stange, G., and Wong, C. (1993). Moth Response to Climate. *Nature* 365:699.

Stanley, S.M. (1987). *Extinction,* Scientific American Library.

Stevens, W.K. "Terror of Deep Faces Harsher Predator," *New York Times,* 12/8/92, p. C1 and C12.

Stevens, W.K., "Emissions Must be Cut to Avert Shift in Climate, Panel Says," *New York Times,* 9/20/94, p. C4.

Stevens, W.K. "Data Give Tangled Picture of World Climate Between Glaciers," *New York Times,* 11/1/94, p. C4.

Stevens, W.K. "A Global Warming Resumed in 1994, Climate Data Show," *New York Times,* 1/27/95, p. A1.

Stevens, W.K. " '95 the Hottest Year on Record As the Global Trend Keeps Up," *New York Times,* 1/4/96, p. A1.

Stork, N.E. (1988). *Biological Journal of the Linnaean Society* 35:321.

Suarez-Varela, M.M., et al. (1992). Non-Melanoma Skin Cancer: An Evaluation of Risk in Terms of Ultraviolet Exposure. *European Journal of Epidemiology* 8:838–844.

Sudetic, C. "Roving Manatee's Whale of a Trail," *New York Times,* 8/8/95, p. B5.

Taylor, H.R., et al. (1988). The Effect of Ultraviolet Radiation on Cataract Formation. *New England Journal of Medicine* 319:1411–1415.

Takizawa, S., et al. (1995). A Selective *N*-Type Calcium Channel Antagonist Reduces Extracellular Glutamate Release and Infarct Volume in Focal Cerebral Ischemia. *Journal of Cerebral Blood Flow and Metabolism* 15:611–618.

Terry, D., "Heat Death Toll Rises to 436 in Chicago," *New York Times,* 7/20/95, p. A16.

U.S. Environmental Protection Agency. (1990). *Toxics in the Community, National and Local Perspectives, The 1988 Toxic Release Inventory National Report,* EPA 560/4-90-017, U.S. Government Printing Office, Washington, DC.

Valentino, K., et al. (1993). A Selective *N*-Type Calcium Channel Antagonist Protects Against Neuronal Loss after Global Cerebral Ischemia. *Proceedings of the National Academy of Sciences* 90:7894–7897.

Wake, D.B. (1991). Declining Amphibian Populations. *Science* 253:860.

Walsh, J.F., et al. (1993). Deforestation: Effects on Vector-Borne Disease. *Parasitology* 106:S55–S75.

Warrick, R.A., and Oerlemans, J. (1990). Sea Level Rise. In *Climate Change: The IPCC Scientific Assessment* (J.T. Houghton et al., eds.), Cambridge University Press, Cambridge.

Webb, T. (1992). Past Changes in Vegetation and Climate: Lessons for the Future. In *Global Warming and Biological Diversity* (R.L. Peters and T.E. Lovejoy, eds.), Yale University Press.

Wenzel, R.P. (1994). A New Hantavirus Infection in North America. *New England Journal of Medicine* 330:1004–1005.

Wilford, J.N. "Antarctic Ozone Hits Record Low," *New York Times,* 10/19/93, p. A23.

Wilson, E.O. (1993). *The Diversity of Life,* Harvard University Press.

Worrest, R.S., and Caldwell, M.M. (eds.). (1986). *Stratospheric Ozone Reduction, Solar Ultraviolet Radiation and Plant Life,* Springer-Verlag.

Yoon, C.K. "Odd Biology: Sea Squirt is a Three-in-One Creature," *New York Times,* 3/15/94, p. C1.

Zurer, P. (1995). Record Low Ozone Levels Observed Over Arctic. *Chemical and Engineering News,* April 10, pp. 8–9.

CHAPTER 2

Earthly Dominion: Population Growth, Biodiversity, and Health

ROBERT ENGELMAN

Introduction

Human activities, especially habitat alteration, are the decisive factor in the present acceleration of species extinctions. What is occurring can be characterized as a global assault on the nonhuman living world from multiple directions, as the scale of humanity's needs for space and natural resources continues to expand. Human population pressures are a critical factor in this relationship, but surprisingly little effort has been made to explore or articulate the influence of population on biological diversity more precisely. Most efforts have been more conceptual than empirical, reflecting obstacles to quantification and precision that are commonly encountered in the complex interdisciplinary field of population and the environment (see Figure 2.1).[1]

Biologists are confident that since the middle of this century the pace of species extinctions has increased markedly, and some predict a rate of extinctions over the coming century that may compare in scale to those that marked the end of the Cretaceous period, as forests are burned and cut down and human settlement covers increasing portions of the earth's land surface.[2] This acceleration in species loss comes as world population has moved from less than 3 billion to close to 6 billion people. In the same time period, the technology involved in massive alteration of natural habitats has become mature, inexpensive, and widespread. Clearly these trends are linked, but we cannot say precisely how. We need to identify new ways of looking at—and researching further—the impacts of population dy-

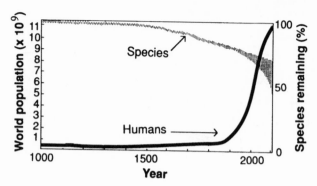

FIGURE 2.1

Expected inverse correlation between human population size and the survival of species worldwide.

namics on biodiversity and, in turn, on human health. And we need to consider what the implications of such a population–biodiversity–health continuum might be. What, if anything, can we and should we *do* about population's impact on biodiversity? Or is the continuing growth of human population merely an unstoppable force we can at best only prepare for, rather than influence?

The Current and Projected Future State of the World's Human Population

The planet is now home to 5.7 billion people, having gained more since 1950 than it had in all of time to that point. By one estimate, 105 billion human beings have lived on earth to date, but for the vast majority of the tens of thousands of years human beings have walked on earth, world population was in the low millions, not the low billions.[3]

Today four-fifths of the world's population lives in developing countries. And more than half of all the world's population growth is occurring in countries with tropical forests, such as Zaire, Brazil, Indonesia, the Philippines, Mexico, and the nations of Central America.

Yet the one-fifth of the world's people who live in industrialized countries have an impact on biodiversity disproportionate to their numbers. Each person in these countries consumes far more fuel, more water, more soil, more wood and minerals, more habitation and work space on average than the residents of developing countries.

Three United Nations projections suggest that world population in 2050 will be as low as 7.6 billion or as high as 9.8 billion, with 8.3 billion a medium figure.[4] These are *not*, however, predictions, but scenarios based on assumptions about future birth, death, and migration rates that may or may not be close to the reality. United Nations projections suggest future population size as far as 2150 (Figure 2.2), when human numbers could be either still growing rapidly (high projection), roughly stable (medium projection) at around 12 million, or declining gradually (low projection) based on low global average fertility rates.[5] The implications of each of these projections for the survival of biodiversity over the next two centuries are obvious, in that species conservation is certain to be a far more successful endeavor if world population stabilizes earlier and at relatively low levels.

The year 2150 is admittedly a rather distant-time to contemplate. The loss of biodiversity, however, beckons us to a much longer view than might be appropriate for, say, economic planning. The question relative to biodiversity is this: What will be the ultimate maximum size of the human population, and when will it be reached? The magnitude and date of the maximum size of the human population will matter greatly to the number and identities of living species surviving (with humanity, we hope and presume) into future geological epochs.

Another trend important to biodiversity beyond population size and

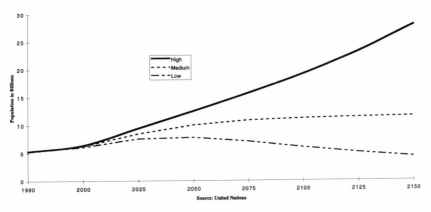

FIGURE 2.2
Three projections for world population, 1990–2150.

growth is urbanization. More than half of the world's inhabitants are projected to be living in urban areas by the year 2005. Especially significant to
biodiversity is the growing concentration of population in dozens of
"megacities" of 10 million or more.[6] This is a scale of urban agglomera-
tion far beyond any the world has ever known.

At first glance, it may seem that urbanization could protect biodiversity
by relieving population pressure in thinly settled regions where there is
much more scope for the survival of nonhuman organisms. But this benefit is an illusion. Population growth continues in most rural areas, albeit at
levels below those of nearby urban areas. And city dwellers have needs for
food, fuel, and materials that force the extension of supply networks
around the globe. The demands of wealthy urban dwellers for tropical
woods and hamburgers have contributed in past years to considerable destruction of rain forests. Finally, large cities produce air, water, and land pollution that can have remote impacts on species survival thousands of miles
away.

Since most megacities are on or near coasts, the world's population is
also moving toward coastal areas, which inevitably mark meeting places of
marine, estuarine, riverine, and terrestrial ecosystems, with all the diversity
of species such ecotones imply. Such "coastalization" of population, while
not well quantified, is undoubtedly taking an increasing toll on species
health and survival in these areas.

Population momentum, resulting from a the bulge of young people now
approaching or traversing their childbearing years, means that unless global
catastrophe intervenes world population will continue to grow for some
time. The force of population momentum can be reduced somewhat by
policies that successfully encourage later marriages and greater spacing of
childbirths. But unless birthrates go into an unexpected tailspin or death
rates rise, or both, it is unrealistic to expect world population to stop growing before the year 2020.

Positive demographic trends include the growing desire for smaller families and personal regulation of fertility and recent international consensus
on the key elements of enlightened population policy. Both fertility and
desired family size have been declining in most regions since the 1960s,
and average fertility rates in developing countries have dropped from
roughly six children per woman to fewer than four—halfway to the approximately two children per couple that constitutes "replacement fertility."[7] More than half of all married couples in developing countries now
use contraception, up from about 10 percent in 1960. And this trend has
recently become apparent even in some sub-Saharan countries, such as
Kenya, Zimbabwe, Nigeria, and Botswana.[8] As more people move to cities,

seek education for their children and encounter electronic media and the consumer culture, the desire for large families often weakens quickly.

From the standpoint of government policy, the 1994 International Conference on Population and Development, held in Cairo, offered real hope that population issues will be addressed in the future out of a shared commitment to human development and individual freedom to determine the timing of childbearing. The Cairo conference was historically unprecedented in achieving near consensus among 179 nations for this strategy. The longstanding debate over the importance of contraception was resolved in favor of broad agreement that quality choices in family planning services should be universally available. So, too, should the educational and economic opportunities that allow women to make informed, effective and safe choices about when, with whom, and how often to bear children in good health.

Surprisingly, almost 45 percent of the world's people now live in countries in which average fertility rates have reached or fallen below "replacement"—including almost all industrialized countries as well as China and a number of other developing countries in Asia. If fertility rates do not rebound significantly and if net immigration is not a significant demographic factor (as is the case in the United States, Canada, and Australia), population size of these countries will stabilize by the middle of the next century—and may then begin to decline gradually.

Nonetheless, uneven access to family planning and reproductive health care services worldwide still condemns many couples and individuals to unintended pregnancy and childbearing. And in the months following the Cairo conference, it remains uncertain whether developed and developing countries will follow through effectively on their commitments to make these services available to all. While we cannot predict precisely how much of a problem future human population growth will present for biodiversity or human health, we can state confidently that the actions governments take today in the population arena will influence profoundly the size and date at which human population will stabilize in the next 150 years or beyond.

Scientific Uncertainty and the Multiple Factors Interacting in Population–Environment Relationships

The existing population–environment research base tends to support the idea that human population itself—the number of human beings in any unit of land—does not directly and linearly alter environments. Rather, population acts with and through such variables as social organization, government policy, dominant technology, and personal consumption behavior. The connections are complex, indirect, and variable by location and time scales. Often a rather long sequence of causation leads from population growth to changes in the environment, and this is especially true of biological diversity.

The complexity of these interactions is partly captured in the equation $I = PAT$ (environmental Impact = Population $\times$ Affluence, or consumption $\times$ Technology), which applies to biodiversity as to other environmental issues. This equation sometimes can usefully capture the dynamics of change in the various actors to assign relative responsibility for environmental degradation, at least in small time frames. But it does not adequately capture the role of long-term population growth during one period on environmental problems that emerge in a later one, when growth rates may have declined and even reversed. The magnitude of today's greenhouse-gas emissions, for example, owes much to the rapid growth of population in the United States and Europe during the 18th and 19th centuries—when the industrial revolution was not yet far enough advanced to result in a rapidly changing atmosphere.[9] Along the same lines, even a slowly declining population of human beings a century or more hence could witness multiple species extinctions related to the rapid population growth and large population size of the second half of the 20th century.

Biologists, especially zooarcheologists, can now state with confidence that human beings played an important role in the extinction of certain mammalian and avian species in prehistoric times. This was certainly the case in some Polynesian Islands[10] and was probably true as well in Australia and perhaps even in the Americas. This suggests that neither large population size nor rapid population growth is necessarily a prerequisite for human-caused species extinctions and ecosystem disruption. Yet even in these cases, population size and growth did play roles as multipliers of individual human impacts on species survival. And, it is important to recall, these impacts on biological diversity were localized and extended across thousands of years in time. This presents a contrast to today's human threat to biodiversity, which is global in nature and appears to be gathering overwhelming force in a time span of mere decades.

Plausible Hypotheses for Population's Role in Biodiversity Loss

Population acts as a "scale multiplier" on biological diversity through chains of causation and influence that vary by time, place, and ecosystem.

This may be as simple as the expansion of settlement, which typically includes the elimination of human and livestock predators. It may involve overhunting and overfishing, as the numbers of meat and fish consumers (on the demand side) and hunters and fishers (on the supply side) increase. This is evident in ocean fisheries today, where *Homo sapiens* is still essentially a hunter–gathering species.

The population–biodiversity linkage likewise may involve the introduction of previously absent species—including disease vectors—as human networks of settlement, travel, and trade expand and proliferate. Unforeseen biological thresholds can be crossed as a result of these introductions, or those of previously absent competing species. At some combination of timing and magnitude of both the introduction and environmental conditions, epidemics result or species composition is altered.

Often the linkage involves other aspects of global environmental change, such as greenhouse-related climate change. In such cases, one must consider whether population growth contributes to the magnitude of these changes (Figure 2.3). In the case of climate change, to the extent it can be demonstrated that population growth is a contributing factor, the link is also made to any associated species loss. As one example, recent research documents the decline of zooplankton in the waters off of California and assigns likely responsible to temperature increases in the water which may or may not be related to greenhouse-driven global warming.[11] A similar case could potentially be made in the case of ozone depletion and its proposed impacts not only on marine phytoplankton but on amphibian species.[12] Although much less studied, it seems likely that the pollution of air, soil, and water has multiple impacts on nonhuman species survival, and population growth clearly plays an important role in the magnitude and geographic spread of these activities as well.

Population growth, often but not always combined with growth in per capita resource consumption, can be the critical factor pushing a human-caused environmental impact past a natural threshold of ecosystem tolerance. In the case of biodiversity, this threshold may involve some minimum area or species composition of habitat in which a specific species can survive indefinitely. Ecologists have found evidence that in many ecosystems extinction rates—including "commitment" to future extinctions caused by present environmental impacts—are anything but linear. The first 60 percent of a habitat that is destroyed or degraded may cause few extinctions, while the next 20 to 30 percent loss will cause devastating species loss.[13] In such cases, a curve of exponential growth (in the extinction of species) may be superimposed atop another similarly shaped curve described by the growth of human numbers. The fact that world population took until 1950 to reach 2.5 billion people and then climbed to 5.7 billion in less than 50

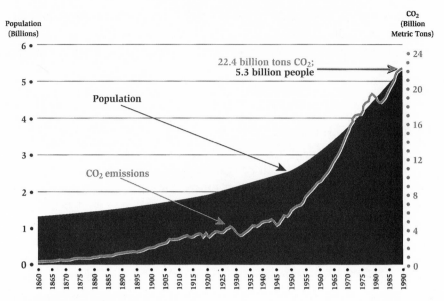

FIGURE 2.3

World population and carbon dioxide emissions, 1860–1990. As the industrial revolution progressed from 1860 to 1970, global industrial emissions of carbon dioxide grew much more rapidly than world population. Since 1970, however, both emissions and population have grown at similar rates. At first glance, these trends might suggest that consumption growth drove the global emissions increases before 1970 and population growth thereafter. In reality, however, no such generalizations can be made without a more detailed examination of historic population and consumption trends in each region of the world.

years may go a long way toward explaining the current wave of extinctions if we consider ecologist David Tilman's thesis that the rate of extinctions increases as the last remaining portions of a habitat are degraded. By one recent estimate, nearly three-quarters of the inhabited portion of the world today consists of ecosystems that have been disturbed in some significant way by human activity.[14]

Critically for policy, population growth has the effect of limiting our choices on biological diversity, forcing us into the kind of "lose–lose" policy choices that are best illustrated by the ongoing debate on how to gain public support for the reauthorization of the Endangered Species Act. In this case, landowners and developers are being asked to make financial sacrifices because their use of the environment happens to come at a critical threshold point in a long history of human–ecosystem interactions. Their predecessors didn't face these kind of choices and sacrifices, and today's landowners quite naturally resent the fact that they do. As population increases and habitat decreases, the choices will become even starker and the search for what could be called "near-miraculous" equitable solutions more challenging.

Population Dynamics in Three Species-Rich Arenas

It may be easier to understand the role of human population dynamics in species extinctions by considering how the two sets of dynamics operate in specific ecosystems known to be home to many species, genera, and families. In tropical forests, often cited as the threatened heart of the earth's biodiversity, population growth has been variously estimated as somewhat less than half to as much as 80 percent of the root cause of the loss of forest cover.[15] In reality, such precision is difficult to support empirically. But certainly in recent times the role of land-poor farmers in the overwhelming transformation of forest has been paramount in Africa, southeast Asia, and parts of Latin America. A critical factor in the migration of farmers into tropical forests—though by no means the only factor—has been the increasing subdivision of family farms (combined with exhaustion of soil and water resources) to the point that they no longer can support a family, forcing a search for new livelihoods.

These settlers, along with others who have few resources and options, further deplete forests by using biomass for fuel beyond the capacities of forest plants to regenerate. This activity is completely sustainable if the magnitude of extraction is small enough, but when population growth pushes fuelwood use above key thresholds of forest regrowth, the forest simply retreats.

Other factors contributing to deforestation are the demand in industrialized countries for tropical wood, spurred by growth of wood-consuming populations and changes in consumption patterns, in concert with evolving global trading patterns. Population growth, in wealthy as well as less wealthy countries, plays a role in these developments as well, although it acts remotely rather than locally and is much harder to isolate as a causative factor.

Within and beyond tropical forests, the earth's waters are home to uncounted millions of the world's species. And these waters are subject to human population pressure from mountain streams to the ocean depths. Although the ocean covers more than two-thirds of the world's surface, many or most of its fisheries are in decline as the result of overexploitation for humanity's growing appetite for seafood. Overfishing does not appear to have produced outright extinctions to date, but certainly the possibility of eventual extinctions looms for long-targeted species like the bluefin tuna and several varieties of whales.

Closer to land, human activities are likely to be responsible for the recent decline in sea corals, and coastal ecosystems are suffering from the growing loads of sediment, toxic chemicals, plastic, human sewage, and other sources of nitrogen and phosphorus. "Coastalization" is leaving few stretches of the earth's hundreds of thousands of miles of coastline undeveloped. Continental shelf areas increasingly are intended and unintended receptacles of human wastes of all kinds. These disruptions are exacerbated by more direct habitat alteration, such as the destruction of wetlands and the paving of agricultural land for buildings, housing developments, and roads.

A major cause of extinctions in riverine and estuarine habitats is withdrawal of freshwater beyond critical thresholds. Because renewable water—that which is provided by the hydrological cycle as rain and snow falling on land—is finite, the amount of this water available to each person declines as population grows. Although renewable water can and must be used more efficiently, it seems all but inevitable over the long term that population growth will result in increased human withdrawals from streams and rivers, with all that implies for the health and survival of aquatic species.

If we use the common hydrological benchmark of 1,000 cubic meters of renewable water per person per year as the upper limit of water scarcity in a country, it is possible to quantify how many people are projected to live in conditions of scarcity or water stress (between 1,000 and about 1,700 cubic meters of water) and in which countries. Data for both population and total annual renewable water are available for 149 countries. As can be seen in Figure 2.4, not only the number but the proportion of people living in water-short countries will depend greatly on which trajectory of population growth the world follows in the coming century.

A similar relationship between population growth and per capita resource availability applies to the use of farmland, which is not often thought of in terms of biodiversity beyond crop and livestock species and strains. Yet the soils under mature and old-growth forests have proven to be among the richest ecosystems in species diversity. And the spread of farming has undoubtedly depleted this species diversity, especially as land is worked more intensively and is subjected to fertilizers and pesticides.

The global area of arable land has grown in rough correspondence with human population since the dawn of agriculture—until the past few decades, when the expansion of arable land slowed down. World popula-

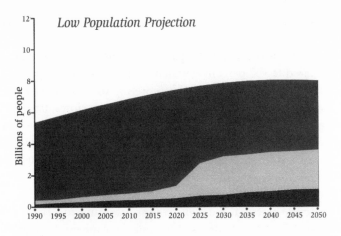

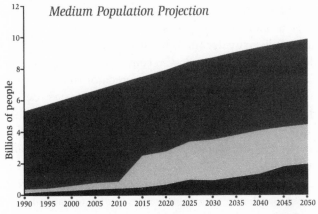

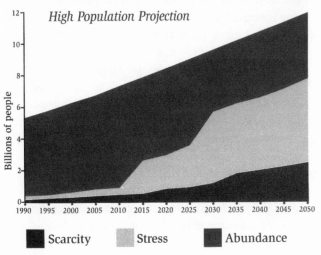

FIGURE 2.4

Population experiencing fresh water scarcity, 1990–2050. "Scarcity" is the lower field in each graph; "stress," the middle field; and "abundance," the upper field.

tion now grows at eight times the pace of total area of arable land—1.6 percent compared to 0.2 percent—which guarantees that the world's available arable land must be worked much more intensively to bridge the gap between the two rates.[16] As the agricultural frontier closes, land intensification is replacing land extensification.

Without special effort, this results in a depletion of soil-based carbon, which is the food on which vast and diverse populations of arthropods and microorganisms depend. There is very little research examining the dynamics of soil biodiversity, but there is increasing evidence that a wealth of species helps cycle essential nutrients—essential either to plant or human health or both—between bedrock, soil, and atmosphere. Some microorganisms protect food plants from pathogens and disease.[17]

Moreover, as monoculture expands in farming areas, both food-related and associated biodiversity decline, often necessitating greater use of chemical pesticides. The loss of a meadow, recent research suggests, can remove species that in a healthy ecosystem act as a predator to crop pests.[18] In developing countries, and even in some developed ones, overuse of land can result in a degradation of soil structure and biological activity so complete it is called "desertification," which can result in a near absence of species above the level of microorganism. Although the phenomenon is not well monitored or quantified, logic suggests that we are altering species composition and quite possibly causing significant species extinctions in soils.

Because arable land, like renewable water, is a finite natural resource (albeit much more flexible than renewable water), similar calculations can be made about the future of per capita arable land availability. No societies have managed to feed their populations with less than 0.07 hectares of arable land per person, unless they have used the kind of intensive, chemical-based agriculture common in the western Europe, Japan, and the American Midwest. Based on this benchmark, if we assume that global arable land extent will remain constant at 1990 levels (it is not expected to grow by much and could easily decline), the rate of population growth over the next 150 years will determine whether 1.6 billion people, 5.5 billion, or some number in between live in countries with scarcities of arable land in 2050 (see Figure 2.5).

What are the implications of these examples of threats to biodiversity? Tilman's research on "species survival" thresholds suggests the likelihood in some ecosystems of eventual extinctions even if the decline isn't evident today. As habitats shrink or are otherwise degraded, survival for individual species members becomes more unlikely, and the probability of reproduction lower, until the genetic viability of the species is weakened to the point of immediate or eventual extinction. We cannot predict with any certainty or detail the impacts of demographic trends and patterns of human behavior on this mechanism. The linkages are far too complex and not well enough understood. Nor should we expect ever to achieve perfect understanding of these interactions. The need is rather to recognize

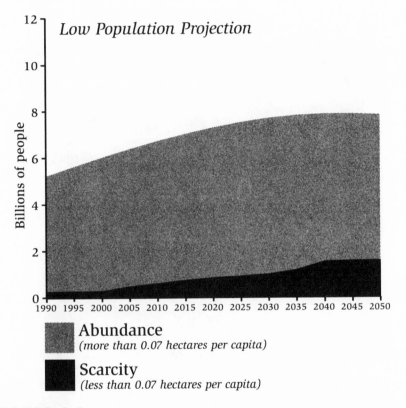

Abundance
(more than 0.07 hectares per capita)

Scarcity
(less than 0.07 hectares per capita)

FIGURE 2.5

Three projections for world population and arable land availability, 1990–2050. "Abundance" (*denoted by the upper field in each graph*) represents more than 0.07 hectares per capita; "scarcity" (*lower field in each graph*) represents less than 0.07 hectares per capita. The speed of population growth largely determines the number of people living in countries with less than 0.07 hectares of arable land per person, a benchmark of arable land scarcity. In 2050, the number of people living with arable land scarcity could range from 1.6 billion under the United Nations' low population projection to 5.5 billion under the high projection. (*Figure continues.*)

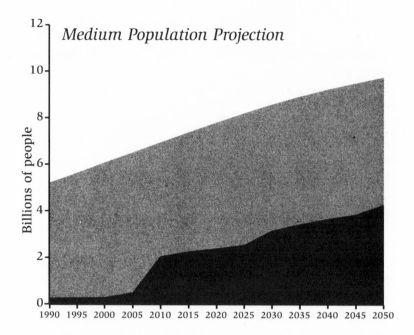

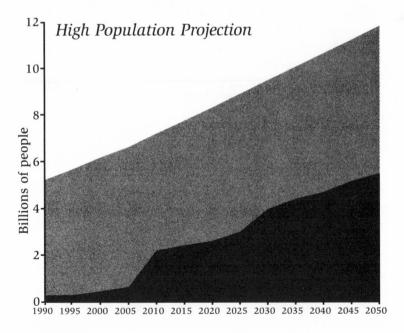

FIGURE 2.5

(*Continued*)

that the risk of extinctions rises with the growing magnitude of human intervention in natural systems, and that this growth in magnitude is coupled
closely—though not in direct or linear fashion—with population growth.

The implications for human health are significant. There may in fact be
a kind of "negative feedback," in which population growth contributes to
biodiversity loss, which in turn contributes to threats to human health and
well-being. The spread of pathogens, for example, can be spurred by the
extinction of natural predators as well as by changes in climate that expand
hospitable ecological conditions for the pathogens. It is possible that biodiversity influences climate change as well as the other way around. The
work of ecologists at the Imperial College's Center for Population Biology at Silwood Park west of London, for example, suggests that biologically diverse forest and soil ecosystems absorb more atmospheric carbon
than less diverse ones.[19]

The loss of diverse sources of food presents another hazard to human
health. The declining diversity of crop and livestock species and strains has
reached the attention of the news media and public. Less often remarked
upon, however, is the ongoing loss of wild foods and medicinal plants.
Africans, for example, are estimated to rely on 500 species of wild plants
for food, and wild animals are in many tropical countries a major source of
meat. These are important supplements to diets that increasingly lack variety and micronutrients—in large part because land pressure has forced
greater cultivation of high-yield crops like cassava that are rich in calories
but poor in nutrients.[20] The importance of wild food in the diets of the
poor is an underutilized argument for the conservation of forests and other
unmanaged ecosystems. The argument is all the more compelling in that it
relates not to the needs of the world at large but to those of the people
who live close to such biologically rich areas. It also captures neatly the cir-
cular nature of the population–biodiversity–health dynamic, since population growth is a major factor threatening the biodiversity that provides the
basis of nutrient-rich diets, which in turn supports good health. The same
argument applies to medicinal plants.

On a more global scale we see similar trends threatening the security of
food supplies in the ocean, as marine fisheries decline precipitously, and on
land, as soil degradation reduces the ecological diversity of the living
geomembrane that makes the growth of food plants possible. Indeed, as a
general principle we can hypothesize that once population growth has
"filled out" most of the available ecosystems (a situation that probably applies to most countries today), further growth limits the very biological diversity needed to make large populations sustainable. There is thus a "cir-

cle of risk" that runs from human population growth through biodiversity to human health, and back to population growth again.

There is little doubt that human presence on earth has reached a magnitude that has put this circle of risk in motion. On land, research calculations suggest that human beings use, waste, or otherwise appropriate between 35 and 40 percent of net photosynthetic production.[21] In marine environments outside of the deep oceans, a recently derived comparable estimate of "aquatic primary production" appropriated by human beings is from 24 to 35 percent.[22] If even roughly accurate, these findings indicate that we are hitting limits of what ecosystems will provide sustainably. As human populations increase, the risk of species loss grows, and it tends over time to grow much faster than population itself.

Implications for Policy

In recent decades and in wealthy countries, there is strong evidence that good policy and creative conservation can accomplish much to mitigate the effects of population on habitat and diversity. Sometimes mere good fortune and technological change are responsible—as in the return of forests to previously deforested land in the northeastern United States.

But even the best policies and most fortunate circumstances are imperiled by the continuation of population growth. Recently, U.S. government officials proposed changing the legal status of the American bald eagle from endangered to threatened under the Endangered Species Act. But as one newspaper article noted, such a change can be at most a temporary reprieve if long-term economic and demographic trends threaten future degradations of habitat. In the case of the bald eagle, it remains an open question how the species would survive the near doubling of U.S. population growth projected by 2050.[23]

Ultimately, the need for near-miraculous policy solutions to the loss of biodiversity will be reduced, and the effects of good policies and programs can only be enduring, if population itself is addressed as a critical factor in a matrix of several factors.

At local level, much more should be learned about the needs of human communities influencing and influenced by changes in biodiversity. Researchers, nongovernmental organizations, and others working in species-rich areas can and should address the demographic aspects of community and habitat change, and especially the potential need and desire for family planning services and reproductive health care. All activities affecting these communities should grow out of genuine participatory decisionmaking by the communities themselves.

The global scope of biodiversity loss, however, requires that the influence of population growth be addressed globally. The accelerating loss of species adds to the many existing arguments for stabilizing global population size within the next few decades.

Demographic research demonstrates that while some population growth is virtually inevitable, there is broad scope for action that will make a real difference at the zenith of the predicted biodiversity crisis. In the middle of the next century, world population is projected to amount to anywhere from 7.9 billion people, at the low end, to 11.9 billion, at the high end. This difference of 4 billion people is roughly the size of world population in the mid 1970s.[24] For some biodiversity-rich regions and countries, the high projections of population growth make it hard to imagine the survival of any intact, reasonably undisturbed ecosystems. Achieving or improving upon the UN's low projection clearly would offer more hope to the long-term preservation of the estimated tens of millions of nonhuman species.

Fortunately, policymakers have already reached a global consensus on the kinds of policies most effective and most needed to slow the growth of world population. The representatives of national governments who attended the International Conference on Population and Development in Cairo stressed the basic human right to regulate one's own reproduction. The Cairo agreement recognized that these services enabling couples, and especially women, to do this in good health should be voluntary and should offer a range of safe and effective contraceptive options. And they should be available and affordable to all who need them. Where possible, they should be provided in the context of services that include a broader range of maternal and child health care as well as reproductive health care services.

The survival of children also contributes over the long term to an easing of population growth rates—despite the fact that the initial influence may for obvious reasons be in the direction of higher population growth. There are two important aspects, however, beyond the obvious ethical ones, of enabling infants and children to survive to their own reproductive years. One is that when first and second pregnancies are successful and children survive their early years, parents are much less likely to insist on multiple pregnancies beyond the number of children they ultimately want in their families. The other aspect is that one of the most important health interventions in aiding child survival is the use of contraceptives, which enables women to delay births beyond adolescence and then to space births at least two years apart. By attenuating each woman's reproduction (and men's too, on average), postponement of first childbearing and the spacing of subsequent pregnancies can have an impressive impact on birth and population growth rates—as well as on the health and lives of women and children—even if they had no impact on fertility.[25] This adds to the arguments for universal access to voluntary family planning services.

As the Cairo conferees also agreed, it is also critical to provide girls and young women with at least the same schooling that boys and young men receive. Educated young women are far less likely to get pregnant while in their teen years, less likely to have large families, and more likely to use contraception—not to mention more likely to make a contribution to their societies and economies beyond that of motherhood. Ultimately, the synergistic effect of access to family planning, education, and economic opportunities is so powerful, it is quite likely that something like the UN's low population projection could be achieved if all three could be provided to all the world's women. The low population trajectory would be the product not of successful "population control" or of global demographic targets, but of the collective actions of hundreds of millions of men and women with the capacity to make and put into effect their own decisions about childbearing.

The effect of such policies—while immediate in the lives of women, couples, and communities—will be modest in the near term as far as global population growth and biodiversity preservation are concerned. Over several decades, however, the impact of population policies will help determine how many other species—our only certain companions in the universe—we carry with us into future millennia.

At least one analyst has applied the term "demographic winter" to the biodiversity crisis.[26] What is needed is a strategy for emerging on the other side of a much larger human population with as many habitats and species as possible, and it is here that population policies can help—only help, as there is much else that must be done. But the help of good population policies is absolutely essential to success.

Pushing the stark metaphor of demographic winter one step further, stabilization of human population can be compared to a solstice, in which the coldest weeks of winter lie ahead but the sun actually is already returning. The stabilization of human population size will mark the beginning of the end of massive loss of nonhuman species. This is not because no more species will be lost. The worst could still lie ahead. But with wise policies, a stable or even gradually declining population of human beings will be much more capable of managing its affairs in ways that do not rob from natural habitats, and may even supplement and enrich them. We will be better off if this demographic winter solstice is reached earlier rather than later and at lower rather than higher levels of global population.

not

recently

Population policies centered on human development and individual freedom would merit support even if they had no impact on population growth rates, because they improve human development and well-being in multiple ways that have nothing to do with demographic change. Such policies are all the more urgent because of the close interdependence among human population dynamics, the survival of biological diversity, and the health of humanity.

Notes

1. Michael E. Soulé. (1991). Conservation: Tactics for a Constant Crisis. *Science* 253:744-750. See also Norman Myers. (1994). Tropical Deforestation: Rates and Patterns. In *The Causes of Tropical Deforestation* (K. Brown and D.W. Pearce, eds.), pp. 27–40. London: University College Press.

2. Edward O. Wilson. (1992). *The Diversity of Life.* Cambridge: Harvard University Press.

3. Carl Haub. (1995). How Many People Have Ever Lived on Earth? *Population Today,* February. Population Reference Bureau.

4. United Nations. (1995). *World Population Prospects: The 1994 Revision.* New York: The United Nations.

5. United Nations. (1992). *Long-Range World Population Projections: Two Centuries of Population Growth, 1950–2150.* New York: United Nations.

6. United Nations. (1994). *Urban Agglomerations 1994.* New York: United Nations.

7. Steven W. Sinding. (1992). "Getting to Replacement: Bridging the Gap between Individual Rights and Demographic Goals." Paper delivered at International Planned Parenthood Federation Family Planning Congress. October 23–25. Delhi, India.

8. John C. Caldwell, I.O. Orubuloye, and Pat Caldwell. (1992). Fertility Decline in Africa: A New Type of Transition? *Population and Development Review* 18(2), June.

9. Robert Engelman. (1994). *Stabilizing the Atmosphere: Population, Consumption and Greenhouse Gases.* Washington, DC: Population Action International.

10. David W. Steadman. (1995). Prehistoric Extinctions of Pacific Island Birds: Biodiversity Meets Zooarcheology. *Science* 267:1123–1131.

11. Dean Roemmich and John McGowan. (1995). Climatic Warming and the Decline of Zooplankton in the California Current. *Science* 267:1324–1326.

12. Andrew R. Blaustein and David B. Wake. (1995). The Puzzle of Declining Amphibian Populations. *Science* 269:52–57.

13. David Tilman et al. (1994). Habitat Destruction and the Extinction Debt. *Nature* 371:65–66.

14. Lee Hannah et al. (1994). A Preliminary Inventory of Human Disturbance of World Ecosystems. *Ambio* 23(4–5), July.

15. Norman Myers. (1994). Tropical Deforestation: Rates and Patterns. In *The Causes of Tropical Deforestation* (K. Brown and D.W. Pearce, eds.), pp. 27–40. London: University College Press. See also Paul Harrison. (1993). *The Third Revolution: Population, Environment and a Sustainable World,* 2nd ed. London: Penguin.

16. Robert Engelman and Pamela LeRoy. (1995). *Conserving Land: Population and Sustainable Food Production.* Washington, DC: Population Action International.

17. Committee on International Soil and Water Research and Development. (1991). *Toward Sustainability: Soil and Water Research Priorities for Developing Countries.* Washington, DC: National Academy Press.

18. Andreas Kruess and Teja Tscharntke. (1994). Habitat Fragmentation, Species Loss and Biological Control. *Science* 264 (June 10).

19. "Biodiversity: There's a Reason for It." *Science* 262:1511 (December 3, 1993).

20. Vaclav Smil. (1994). How Many People Can the Earth Feed? *Population and Development Review* 20(2), June.

21. Peter M. Vitousek et al. (1986). Human Appropriation of the Products of Photosynthesis. *BioScience* 36(6):368–373.

22. D. Pauly and V. Chrisensen. (1995). Primary Production Required to Sustain Global Fisheries. *Nature* 374:255–257.

23. Timothy B. Wheeler. (1994). "No Longer Endangered, Bald Eagles Must Now Fight Development of Bay Habitat." *The Baltimore Sun,* July 3.

24. United Nations. (1995). *World Population Prospects: The 1994 Revision.* New York: United Nations, pp. 217 and 219.

25. John Bongaarts. (1994). Population Policy Options in the Developing World. *Science* 263:771–776.

26. Colin Tudge. (1992). *Last Animals at the Zoo: How Mass Extinction Can Be Stopped.* Washington, DC: Island Press.

CHAPTER 3

Biodiversity and Emerging Infectious Diseases: Integrating Health and Ecosystem Monitoring

PAUL R. EPSTEIN, ANDREW DOBSON, AND JOHN VANDERMEER

" . . . a small percentage [of insects] have come into conflict with human welfare . . . as competitors for the food supply and as carriers of human disease. . . . eliminating the coyote has resulted in plagues of field mice . . . [Pest] populations are kept in check . . . by the resistance of the environment [and] their internecine warfare. . . . neglected is the truly explosive power of a species to reproduce once the resistance of the environment has been weakened . . ."
—Rachel Carson, *Silent Spring,* 1962

Introduction

The role of biodiversity in controlling pests, pathogens, and human parasites has received little attention to date. This protective "function" provided by diversity of defensive responses, acting at all scales, is essential for the preservation of healthy living systems. What do losses of species key to ecosystem-level processes (Likens 1992), such as large predators, mean for natural biological controls? Declining biodiversity may play a significant role in the current emergence, resurgence, and redistribution of infectious diseases of animals and plants. Mechanisms that remove most harmful mutations on the cellular level are mirrored by interspecific actions on the macro level that prevent opportunistic species from overexploiting and dominating many undegraded natural systems. Although each individual organism works to maximize its own lifetime reproductive success, the numerous interactions between exploiters and victims (predator and prey, parasite and host, herbivore and plant), the diversity of which characterize the biodiversity of a region, also often act to regulate the abundance of any individual species and preserve the health and dynamic integrity of the ecosystem.

Predator/prey relationships are central to biological control. Owls, coyotes, and snakes help regulate populations of rodents—opportunistic species involved in the transmission of Lyme disease, hantaviruses, arenaviruses, leptospirosis, and human plague. Freshwater fish, reptiles, and bats help limit the abundance of mosquitoes, some carrying malaria, yellow fever, dengue fever, and many encephalitides. Finfish, shellfish, and sea mammals affect the dynamics of marine algal populations—some toxic, others anoxic, still others transporters of cholera bacteria.

Ecosystems are complex, dynamic, adaptive systems (Lewin 1992), with many internal, self-regulating mechanisms. Successional states with relative stability among functional groups (competitors, scavengers, decomposers, etc.) may be viewed as healthy, vigorous, and productive.

Population explosions of nuisance organisms and disease carriers may be seen as emergent phenomena, signs of ecosystem disturbance, with reduced resilience and resistance. Rodents, insects, and algae represent key biological indicators, rapidly responding to environmental change.

The current rate of extinctions (Line 1996) assumes additional significance in this respect: historically periods of mass extinctions—punctuations in prolonged phases of evolutionary equilibrium—herald the emergence of multiple levels of new biological variation. Will the current extinction period initially favor opportunistic organisms?

This chapter explores mechanisms and provides examples of the role of biodiversity in controlling disease "vectors" and microorganisms. Those affecting terrestrial and marine food sources are highlighted as nutrition is fundamental to human health. The analysis expands the classic disease epidemiological framework of host–environment–agent interactions taking a systems-based approach to examine the compounding influences of ecological change, global environmental trends, and social practices on species dynamics in controlling disease emergence; it also suggests future directions for selecting biological indicators to improve disease surveillance and ecosystem monitoring.

Ecology, Pests, and Disease: Terrestrial Ecosystems

Competitive and cooperative mechanisms act within nuclei, cells, and organisms and at community levels to regulate anomalies, harmful mutations, cancerous cells, and opportunistic organisms. Fitness for survival frequently reflects coevolution of species and their resulting role in creating and maintaining well-functioning, self-regulating systems.

Explicit intranuclear regulatory systems have coevolved to control mutations: $p53$ gene products (exonucleases) "proofread" DNA errors by excising mismatched base pairs (Harris and Holstein 1993, Holland 1993, Modrich 1994, Sancar 1994, Hanwalt 1994). In the cytoplasm, peroxidases and other anti-oxidants detoxify harmful oxygen-derived free radicals (de Duve 1996).

Within organisms, cytokines and antibodies tender a multifarious orchestra of defensive responses against parasitic invasion. For example, persons with depressed immune systems, such as those with HIV/AIDS, fall prey to opportunistic infections. At the ecosystem level, birds and parasitic wasps often limit population explosions of opportunistic insect herbivores (Blaustein and Wake 1995).

Regulatory systems can involve multiple species and a variety of chemicals and mechanisms. Just as cytokines summon immune cells at the organismic level, semiochemicals (e.g., terpenes)—released only when herbivores feed on leaves (and not from each alone)—attract parasitic wasps that infest and reduce the population of plant pests (Tumlinson et al. 1993). In this system of three species (plant, herbivore, and wasp) a regulatory system has evolved affording resistance to pest overgrowth.

However, perturbations can act at several points to disrupt regulatory dynamics. On the genetic level, toxins, ultraviolet-B and X-radiation harm both informational and regulatory genes, a defect in the latter removing control over mutations throughout the cell's lifetime. In nature, excessive pesticides select for hardy, resistant insects *and* harm predators (birds, ladybugs, and lacewings) that limit insect overgrowth. Meanwhile, on the macro level, cooperative and regulatory mechanisms may help determine the shape of communities. A colony of fire ants behaves and develops in relation to environmental constraints and competition (Gordon 1995, Holldobler and Wilson 1994), as does a stand of trees.

What is termed "redundancy," often mediated by spatial heterogeneity, allows large levels of species diversity to accumulate; thus complex systems may be more stable than simpler, more degraded ones under some conditions. A variety of organisms perform similar tasks (predation, protein breakdown, nutrient recycling, etc.), and the often intense competition between species performing similar functions means the independent decline of one usually leads to an increase in the abundance of another (Tilman and Downing 1994, Tilman et al. 1994, Dobson 1996).

At each level, the diversity of responses is key to the resilience and resistance of the systems, reducing vulnerability to the colonization, invasion, and dissemination of infections and nuisance organisms. Stresses acting at multiple points can be destabilizing to ecosystem function. Simplified ecosystems further stressed by habitat fragmentation, overuse of pesticides, and a changing climate can experience disruption of many functions, leading to collapse of defenses and an invasion of opportunistic infections, much as one sees with HIV/AIDS patients.

Opportunists and Competitors: A Complex Balance

MacArthur and Wilson (1967) proposed that the most desirable attributes ("r-selected") for a successful invader and good colonizer are good dispersal ability, high reproductive rates, and hermaphroditism or asexual reproduction. The first attribute allows species to cross barriers to dispersal easily, the second allows invaders to build up a population rapidly after introduction, and the third allows them to found a new population from a few propagules. Most helminths, arthropods, protozoa, bacteria, and viruses excel at these attributes. Parasites, in particular, have much higher rates of reproduction than free-living species. Data comparing the fecundity of free-living and parasitic helminths suggest that parasitic species produce two to three orders of magnitude more eggs than free-living species of similar size. Part of the massive increase in reproduction reflects increases in the number of reproductive stages needed to ensure that at least one parasite infective stage locates a suitable host; part of it must also reflect the huge energetic savings achieved by living a parasitic life style (Calow 1979, 1983).

Species exhibiting what is called "K-selection" tend to be larger, reproduce later in life, and are slower to develop, but are superior competitors in a stable environment (Ehrlich 1986). Indeed, r-strategists would proliferate exponentially if not kept in check by the K-strategists through predator/prey relationships and competition—the systems of biological control (the "immune," regulatory systems of the environment). However, as "specialists," frequently dependent on localized niches and often more limited diets, predator K-strategists are ultimately more fragile than their opportunistic prey (Davis and Zabinski 1990).

Weeds, rodents, insects, and many microorganisms are opportunistic species; they tend to grow rapidly, are small in body size, and have huge broods and good dispersal mechanisms. As competitors, small mammals consume insects; thus the interplay and emergence of pests is complex. Environmental conditions (such as climate and nutritional state) are also critical components in this delicate balance and crises tend to favor opportunists. (See the section titled "Nutrition and Pathogen Evolution" later in this chapter.) Extreme weather events (floods and droughts) can be conducive to insect population explosions and cause rodents to flee inundated burrows (Epstein et al. 1995). Overall, opportunists (r-strategists) are good colonizers and achieve dominance in disturbed environments (just as HIV infection and poor nutrition allow opportunistic infections to proliferate in weakened hosts).

The dominance of scavengers and "generalists" (such as crows, gulls, and Canadian geese) over "specialists" (who lose circumscribed niches) in rural and urban settings may impact disease transmission. For example, the transmission in New England of the viral disease Eastern equine encephalitis involves starlings and robins (generalists) in the bird–mosquito–human–equine life cycle (A. Spielman, personal communication, 1995). Hardy generalists might be apt to more easily acquire, maintain, and circulate pathogens than those with isolated ranges and restricted diets.

The Volterra predator–prey relationship (described by the Lotka-Volterra equation) (Ehrlich 1986) is an ecological principle fundamental for understanding disease emergence and resurgence. When predator and prey are both reduced as a result of habitat fragmentation, monoculture, overuse of pesticides, or climate extremes, the prey—more rapidly reproducing and evolving (resistance)—can rebound with punishing ferocity (the *Silent Spring* of no birds to consume insects, as Rachel Carson eloquently depicted).

Models of population dynamics of predators and prey that exclude refuges, camouflages, and migration among systems are not sustainable. Predator and prey are eliminated. One result of this theoretical prediction may be that habitat reserves are necessary to preserve refugia for raptors (like owls) that control rodent populations, for example. Corridors can be crucial in maintaining metapopulations of predators and competitors (Levins 1970).

So-called "redundant" species may provide insurance against invasion in the face of reduced populations of one or more predators, or with severe environmental stresses. If coyote populations fall, their competitors— snakes and owls—may compensate in regulating rodent populations.

Biological Invasions

Both on islands and on continents introduced species that manage to successfully colonize often grow rapidly to high population densities.

Another feature that may allow invaders to achieve this is the absence of natural enemies such as pathogens, parasites, and infectious diseases. For example, the house sparrow (*Passer domesticus*), which has successfully colonized the United States, New Zealand, Australia and many oceanic islands from its native Europe, has an unusually low burden of specific microparasites (Manwell 1957, Brown and Wilson 1975, Dobson and May 1986). Indeed, only one of the species of parasite known to infect the sparrow in Europe has been recorded in the United States; other sparrow pathogens can generally infect other host species as well.

Introduced species can frequently turn into pest species. In particular, where they have been introduced onto oceanic islands, rats, cats, goats, and dogs have been responsible for a majority of documented extinctions that have occurred to date (Diamond 1989, Atkinson 1985). When compared to populations on the mainland, populations of rats and mice on oceanic islands have considerably lower burdens of parasitic helminths and a depauperate community of parasites (Dobson 1988). This presumably reflects a process which is analogous to the founder effect: the small number of individuals that colonize an island bring with them only a subsample of parasites with which to colonize the host population as it grows. It is more likely that the successfully invading hosts contain only benign parasites rather than virulent ones, which require larger host populations to sustain them and are likely to kill colonizers in the early stages after their arrival.

The relative absence of parasites and pathogens from introduced island populations of rats, cats, dogs, and goats on oceanic islands suggests the possible use of pathogens as biological control agents for these pest species (Dobson 1988). A number of theoretical studies suggest that this approach could be quite successful (Anderson 1980, Hochberg 1989). Both microparasites and macroparasites could be considered as possible biological control agents. When the invasion by an introduced species is of a limited number of individuals, or closely related individuals, genetic diversity of the resulting population can be quite low. This can be true for host organisms as well as their pathogens. If this genetic founder effect has produced low levels of genetic variability in the introduced populations, it is likely that a parasite that can successfully infect one host individual will rapidly spread through the population and quickly reduce its density.

A number of theoretical studies indicate that the potential control achieved using a parasitic helminth will be considerably reduced if the parasites are aggregated in their distribution in the host population (Anderson 1980). Two major factors determine the frequency of parasites in the host population: spatial heterogeneity in the distribution of hosts, compounded by the transmission process, and genetic heterogeneity in the susceptibility of different individuals to infection (Anderson and Gordon 1982). If genetic variability in the host population is reduced because of genetic founder effects, then only spatial heterogeneity will affect the frequency of the macroparasites of the host populations. When potentially reduced levels of aggregation are combined with high levels of transmission, as might occur in a high density introduced species, parasite-induced deaths and reductions in fecundity may rapidly produce declines in pest population numbers.

Obviously, total eradication of the host will not be achieved by a pathogen, but the pest population will be reduced to a density that roughly corresponds to the threshold for establishment of the parasite. At this point the last few individuals may be removed by alternative more labor intensive techniques. Above all it is crucially important that any attempt to introduce a parasite pathogen to control an introduced pest species has to carefully ensure that the endemic species on the island or on surrounding islands are not susceptible to the pathogen used as a biological control agent.

Biological Control of Pests

The use of microorganisms for pest control was first envisioned by 19th century zoologist Elie Metchnikoff (Carson 1962). *Bacillus thuringiensis,* discovered in 1911—and appreciated in the 1930s as the cause of milky disease in Japanese beetles—has proved helpful and environmentally friendly in controlling *Aedes* species involved in Eastern equine encephalitis (Edman et al. 1993).

Predator/prey, competitor, and cooperative dynamics have also been employed—with varying degrees of success—to institute biological control. Guppies, goldfish, and *Gambusia affinis* have been introduced into ponds to limit mosquito larvae. Birdhouses can improve peridomestic mosquito control. Failures at biological control may in part be attributable to oversimplified approaches, and the failure to appreciate the multiple ("redundant") and structural interventions required to improve generalized defenses and ensure healthy ecosystem function. Trees surrounding fields of crops, and flowers with nectar, may do more than individual introductions to sustain the diverse community of birds, parasitic wasps, spiders, lace-wings, and ladybugs needed to preserve the dynamic ratios of r- and K-strategists.

Negative Synergies and Outbreaks

Synergistic effects among perturbations can be important in precipitating pest population explosions. Regarding hantavirus emergence in the U.S. Southwest, land-use changes and climate factors (prolonged drought) may have reduced rodent predators, while heavy rains in 1993 (providing piñon nuts and grasshoppers) helped precipitate a tenfold increase in rodent pop-ulations. Thus a "new" disease (hantavirus pulmonary syndrome) was spread through rodent feces and urine (Levins et al. 1993).

Similarly, an experimental study of Canadian snowshoe hares depicts such a synergy. Here, reduced predation caused a twofold population in-crease; new food sources increased hare populations threefold. But both to-gether ultimately conveyed a tenfold population explosion (Krebs et al. 1995, Stenseth 1995). Such synergies may often be critical in population explosions and discontinuities and in nonlinear changes in population dy-namics.

Thus some stresses may increase the sensitivity and vulnerability to the impacts of other stresses. Extensive land clearing (especially that which al-ters species composition) is one example. As humans enter forested areas, deforestation (Molyneaux 1992) can lead to a shift from forest to field mice, while the overuse of pesticides and herbicides can reduce predators. This combination of events occurred with the sudden appearance of Machupo virus in eastern Bolivia in 1962, and with Junin (Argentine he-morrhagic fever) in 1953 (Garrett 1994) that killed corn harvesters. In the former, pesticides harmed predators directly when cats were killed from the heavy amounts of DDT used to control *Anopheline* mosquitoes carry-ing malaria. When cats were reintroduced, the epidemic of Bolivian hem-orrhagic fever abated. In Argentina, tall grasses where predators take their refuge were razed and replaced by cornfields. Similar mechanisms have been involved with the emergence of rodent-borne Guaranito (Venezuela) and Sabia (Brazil) arenavirus (Coimbra et al. 1994).

In the South Pacific on Nissan Island, pesticide use was intense during World War II. Soon after the war, swarms of malaria-carrying mosquito populations reinvaded in the absence of natural predators that had been killed off by the spraying (Carson 1962).

In the southern United States, excessive use of pesticides to control boll weevil has reduced wasps and spiders that subdue other pests; and the explosion of other pests is causing a backlash among Texas and Alabama farmers (Verhovek 1996). Echoes of *Silent Spring*—wherein pesticide-resistant insects were left to flourish in the absence of the chorus of birds to consume them.

Current land-use practices, overuse of chemicals, and habitat fragmentation may significantly augment the chances for such "nasty synergies" (Peters and Lovejoy 1992). A disturbance in one factor can be destabilizing; multiple perturbations can affect the resistance and the resilience of systems.

Anthropogenic Forces: Effects on Terrestrial Systems

Various human activities affect biodiversity and the control of pests and disease. Monocultures diminish genetic and species diversity and are sometimes more vulnerable to invasions; habitat fragmentation, edge effects (Skole and Tucker 1993). The excessive use of pesticides harm predators and favor pests (Carson 1962, French-Constant 1995); and change in climate norms and its variability (extreme events) can disrupt demographic relationships among species, releasing opportunist r-strategists from control by predator K-strategists and competitors.

Periods of ecological and climate change may be associated with extinctions of some species and the emergence of new ones (Wilson 1992, Pimm et al. 1995). Pests and pathogens may be among the first to emerge during some of these transition periods. An altered balance of species, combined with a stressful environment, could initially provide selective advantage to pioneering and opportunistic organisms and contribute to a redistribution of disease vectors, animal reservoirs, and microorganisms as old systems collapse.

Ecology, Pests, and Disease: Marine Ecosystems

Each year 30,000 to 60,000 U.S. residents suffer from contaminated seafood. New diseases are emerging from marine systems that threaten human health, food safety, and food security. In the marine environment, ecologists portray complex microecosystems involving algae, bacteria, and viruses associated with the emergence of new infections and biotoxic events. These include vibrio (*Vibrio cholerae* [Colwell et al. 1980, Byrd et al.

1991, Epstein et al. 1993] and *V. vulnificus* [CDC 1993b]); viral contamination of oysters in the Gulf of Mexico (CDC 1993c); domoic acid induced amnesic shellfish poisoning on both coasts of North America (Todd 1993); the emergence of *Pfeisteria piscicida* (a highly toxic dinoflagellate) as a cause of large fish mortalities on the East Coast (Burkholder et al. 1993); multiple marine mammal die-offs; a green turtle fibropapilloma epizootic (Williams 1987); and a brown tide off Long Island first crippling the scallop industry (Cosper et al. 1987), then turning toxic, with a viral agent later implicated (Milligan and Cosper, 1994).

Biodiversity in marine ecosystems is important in controlling diseases in marine flora and fauna, and those species dependent upon them (e.g., shorebirds, and humans). For example, among the factors influencing the proliferation and persistence of coastal phytoplankton populations are eutrophication (excess nitrogen and phosphorous); the removal of wetlands that filter excess nutrients; and warm water events (with stratification in the water column); and diminished grazing due to overfishing. The widespread epidemic of harmful algal blooms—the increase in their extent, duration, and appearance of novel toxic species—has had multiple health effects. Included are (1) an increase in shellfish and finfish poisonings, (2) anoxia in seagrass beds (shellfish nurseries), and (3) increased human health hazards related to inland and seashore swimming (gastroenteritis, ear and eye infections).

The "Taura syndrome," plaguing aqua/mariculture in Ecuador, Panama, Honduras, Venezuela, China, Japan, and elsewhere is illustrative. Marine monocultures are typically (1) confined in coastal ponds carved from surrounding mangroves and vegetation, (2) contaminated with pesticides from surrounding agriculture, and (3) flooded with broad-spectrum antibiotics (e.g., chloramphenicol). Shrimp in such farms have become immunocompromised and fall prey to opportunistic infections that include several viruses (one baculo- or insect-borne virus) and vibrios. Oysters have been infected with rickettsia. Such semi-enclosed systems are themselves vulnerable, devoid of refugia and in- and out-migration, and may not be a sustainable answer to the worldwide decline in ocean fisheries.

Global Change, Biodiversity, and Marine-Related Disease

Changes in climate dynamics and variability (CLIVAR 1992) can have profound impacts on the distribution of flora and fauna, previously acclimatized to particular regimes (Bakun 1993, Barry et al. 1995, Tester et al. 1991). Ocean warming may also critically alter functional relationships within the marine food web, beginning with plankton biomass (Roem-mich and McGowan 1995) and eventually affecting shorebirds and fish (Hill 1995). Paralytic, diarrheal, neurologic, and amnesic shellfish poisoning, Ciguatera, Pufferfish, and Scromboid fish poisoning—all related to algal biotoxins (Epstein 1993b)—appear to be spreading in an apparent "global epidemic" of coastal algal blooms (Anderson 1992, Smayda and Shimizu 1993), with direct consequences for human health and nutrition.

Timing is also critical. A species with yearly rhythms may be able to tolerate specific temperatures at one time during the year, for example, but it—or its predators or food sources—may be unable to tolerate the same temperatures at another. In a warm El Niño–Southern Oscillation (ENSO) event, salmon move north along the North American Pacific coast. But it is the northward shift of California mackerel—consumers of young salmon exiting Northwest rivers—that decreases adult salmon stocks two years hence. Thus the timing, intensity, duration, and frequency of anomalies can alter population dynamics.

Temperature thresholds are important in species as diverse as humans and mosquitoes. Surpassing temperature thresholds affects the health of coral reefs and other marine life, habitats important for human nutrition. A 1996 report (Kushmaro et al. 1996) implicates a *Vibrio* bacterium in coral bleaching, with warm sea surface temperatures (SSTs) a contributing factor, possibly increasing bacterial growth and lowering the resistance of the coral. The cumulative ecological impacts of the unusually prolonged El Niño (5 years, 8 months) that ended in August, 1995 and transitioned into a cold ENSO ("La Niña") event (1995/96) (Trenberth and Hoar 1996) have yet to be evaluated. In 1995, warming in the Caribbean led to coral bleaching for the first time in Belize—as SSTs surpassed the 29°C threshold (Hayes and Goreau 1996). The die-off of manatees (and cormorants) from a red tide (*Gymnodium brere*) may be another example of the cumulative impacts of the "longest [El Niño period] on record" (Trenberth and Hoar 1996).

When the cold upwelling phase followed in 1995/96, many regions of the globe that had previously been subjected to prolonged drought now experienced intense rains and flooding. Conditions in Colombia and Central America, for example, were optimal for insect breeding (resulting in dengue fever and Venezuelan equine encephalitis). At the same time rodents fled flooded areas (Epstein et al. 1995), spreading leptospirosis (a treatable, but sometimes fatal bacterial disease). Unusually warm temperatures in June 1995 also precipitated toxic and anoxic algal blooms in a northern Colombia lagoon, killing 350 tons of fish. Thus extreme climatic conditions may in some instances favor clusters of disease outbreaks.

These and other impacts of climate change (warming trends and patterns of variability) on biodiversity (Peters and Lovejoy 1992) and disease effects deserve greater attention. Mass and chronic mortalities of marine biota frequently result from an altered or contaminated food web and can lead to substantial economic losses. Changes in climate (mean temperature and precipitation and the frequency of anomalies) can affect the ranges and dynamics of toxic phytoplankters, and this may affect the overall health of coral reefs, seagrasses, shell and finfish, sea mammals, seabirds, sea turtles, invertebrates, and humans.

Anthropogenic Forces: Effects on Marine Environments

Industrial and agricultural activities in watersheds, coastal development, and fishing practices are collectively contributing to the extent, duration, and toxicity of harmful algal blooms worldwide. While nutrient-rich, cool upwelling currents are associated with blooms and healthy fishing grounds, warm SSTs—given adequate nutrients—may play a significant role in stimulating blooms, toxic and otherwise.

The key anthropogenic elements are (1) eutrophication—from sewage, fertilizers, and acid precipitation that contributes over 1/2 the nitrogen along the U.S. Atlantic coast (Howarth et al., in press); (2) chemical pollutants, chiefly chlorinated hydrocarbons; (3) the previously undetected abundance of marine bacteria and viruses (Paul 1993) (some that transfer genetic material as bacteriophages); (4) overharvesting of shell- and finfish (plankton consumers); (5) loss of wetlands ("nature's kidneys" that filter nutrients and toxins); (6) increased UV-B radiation; and (7) warmer SSTs.

Eutrophication in coastal waters sets the stage for toxic algal blooms; increased phosphates augment ATP (adenosine triphosphate) and energy levels, while warmer SSTs increase metabolic rates and photosynthesis (Valiela 1984). Warm SSTs, accompanied by stratification in the water column, are also associated with shifts in species composition toward the picophytoplankton (cyanobacteria and dinoflagellates) (Valiela 1984), often more toxic and less palatable to predators, resulting in further algal growth.

Biological Indicators

Bioindicators have primarily been used to monitor environmental accumulations of chemicals. But insects and rodents are also carriers of many plant and animal (including human) pathogens and some are avid herbivores. Plankton (Colwell and Spira 1992) serve as a reservoir and amplify *Vibrio cholerae* and other bacteria. Monitoring the abundance and distribution of these three key groups may facilitate impact monitoring, climate change detection, and the design of interventions to improve generalized resistance of ecosystems.

Insects

Insects, the most abundant and diverse group of animals, are particularly sensitive biological indicators in terrestrial ecosystems—many are adapted to narrow thermal conditions and have short generation times. Paleological records support a strong link between past climatic transitions and increases in insect fauna. Fossil assemblages from ecotones (the edges of ecosystems) demonstrate that insects responded rapidly to warming in North America and Europe during the last deglaciation (10,000 y.a.), redistributing in years; while grasslands, shrubs, trees, and wildlife took decades to centuries to shift. "Beetles are better climatic indicators than bears," wrote P.D. Moore (*Nature* 1968:294–385 in Elias 1994). Changes in distribution of insects were most closely correlated with changes in night-time (minimum) temperatures and winter temperatures.

While abundance and diversity may be controlled by biotic factors (predator/prey ratios, competition, and parasitism) at the center of a species' range, thresholds and optimum variances in bioclimatic conditions (temperature and humidity) limit populations near the edges of their range. Abiotic factors such as these impact a variety of phenomena, including breeding sites, maturation rates, intrinsic parasite incubation, and biting behavior (Billett 1974, Dobson and Carper 1993). Thus the persistence of insect populations at new altitudes and latitudes provides a sensitive indicator of climatic change (Peters 1991). And while climate limits the distribution of insect vector-borne diseases, weather contributes to the timing of outbreaks. Overall, the abundance and behavior of insects may integrate many aspects of ecosystem health and integrity, including diversity, productivity, vigor, invasibility, resilience, and ambient conditions.

Rodents

Rodent populations in rural and urban settings are another group which responds rapidly to environmental change (e.g., food supplies and wastes) and to altered biodiversity (reduction in predators). With catholic diets and extensive networks of underground burrows, these small mammals can rapidly proliferate with surges in food supplies or declines (or lags) in natural enemies. Rodents have been involved in crop failures and plague in Southern Africa (Epstein and Chikwenhere 1994); hantaviruses in the United States, the former Yugoslavia, and Taiwan; plague in India; and crop damage in Northern Australia (associated with prolonged drought and the 1991–95 El Niño). Some 16% of a recent sample of citizens in Baltimore, Maryland, demonstrated previous exposure (antibodies) to leptospirosis (Glass 1996). In Latin America, most of the hemorrhagic arenaviruses listed above are carried by rodents. Rodents as pests and vectors of pathogens are a growing problem today worldwide in urban and rural settings.

Algae

In coastal marine systems, algae function as key indicators of ecosystem integrity, responding rapidly to environmental changes and integrating many of the factors outlined earlier. In fresh water—lacking dissolved carbonate—CO_2 fertilization adds further stimulus to phytoplankton growth. The global increase in algal blooms and toxic phytoplankton species reported by planktonologists may reflect widespread ecological and global changes and represent adaptive responses to altered marine biodiversity and environmental overloads (providing an alternative carbon sink). Indeed, "the worldwide increase in coastal algal blooms may be one of the first biological signs of global change" (T. Smayda, personal communication, 1995).

Algae (e.g., *Anabaena variabilis* [Islam et al. 1990]) may (1) function as biological "biofilms" for bacteria; (2) harbor some bacteria internally (e.g., *Legionella* bacteria survive chlorine disinfection when inside protozoa); (3) supply nutrients for bacteria; (4) involve bacteria as endosymbionts (as demonstrated by electron microscopy); (5) be assisted by some bacteria in making biotoxins (e.g., saxitoxin) (Doucette 1993); and (6) involve bacteriophages in the transfer of genetic information, in the production of algal toxins, and thus contribute to the "crash" of algal blooms.

Algae and zooplankton harbor *Vibrio cholerae* and other human enteric pathogens. *Vibrio* and other gram-negative bacteria attach to a wide taxonomic range of plankton, assuming spore-like "quiescent" forms (Colwell and Spira 1992). When plankton blooms, vibrios return to a culturable and infectious state. Thus algal blooms and zooplankton (e.g., copepods amplify *V. cholerae* bacteria up to 106-fold in their egg sacs) can serve as reservoirs of vibrios and other bacteria, causing gastroenteritis. The bacteria can then be consumed directly (imbibed in estuarine water) or transferred to humans through a contaminated seafood chain—notably via bivalves, crustacea, and fish.

This mechanism may have been involved in the rapid and intense cholera epidemic beginning in January 1991 in the Americas (entering within weeks from three coastal ports), resulting in over 500,000 cases and 5,000 deaths over the first 18 months (Swerdlow et al. 1992). In late 1992, a new variant of cholera (O139) emerged in the Bay of Bengal and threatened to become the agent of an eighth cholera pandemic (Islam et al. 1994).

In freshwater ponds, algae bind disinfectants, forming toxic chlorinated hydrocarbons associated with hepatic cancers; they can cause anoxia in seagrass, shellfish beds, and coral reefs, increasing the growth of pathogenic anaerobes. As discussed earlier, these phytoplankton blooms are a major health hazard for humans, shorebirds, fish, and sea mammals.

Chemicals, Biodiversity, and Health: An Integrated Ecological Risk Assessment

Risk analyses of chemicals in the environment presently account only for direct effects: cancers, birth defects, premature births, etc. Recently dioxin was shown to activate HIV-1 gene expression through an oxidative stress pathway affecting cytochrome P450 CYP1A1 enzyme, an overlooked first-order effect (Yao et al. 1995). But like increased UV-Bs, chemicals like dioxin, PCBs, and pesticides may also have indirect—second- and third-order—effects on the emergence of infectious diseases.

Through disruption of hormones and through immune suppression (Colburn et al. 1996), such chemicals may directly affect bird and fish populations—not merely serving as sentinels of future human impacts. Bird populations with weakened immune systems might support an elevated burden of viruses (e.g., Eastern equine, Western equine, and St. Louis encephalitides); and, with reduced numbers of inland fish to consume insect larvae, mosquitoes may flourish.

Morbilliviruses (measles, canine distemper virus, rinderpest) have been fatal to aquatic and terrestrial mammals (seals and dolphins, Australian horses, Serengeti lions) (Kennedy et al. 1988, Ferrer and Pumarola 1990, Morell 1994, Murray et al. 1995). Here PCBs, biotoxins, warm SSTs, and altered food sources acted together or in combination to lower the immune responses of seals, for example, allowing previously innocuous *morbilliviruses* to opportunistically gain a foothold. How do these evolving epidemics across a wide taxonomic range reflect the evolution of the disease agents and/or altered environments affecting the vulnerability of hosts?

Future Directions: Integrating Research and Management

Few research or management programs integrate biological invasions and ecosystem properties (Vitousek 1990); even fewer integrate health and ecosystem surveillance (Epstein 1995). The National Science Foundation funded the Long-Term Ecological Research site (LTER) in New Mexico, which is an exception. There, collaboration with the Centers for Disease Control and Prevention (CDC) to monitor rodent ecology and hantaviruses has become established. The 18 U.S. LTERs and international Man and Biosphere (MAB) programs of the United Nations could include monitoring rodent populations and surveillance for viruses and bacteria. In instances where insects are also monitored, disease burden and incidence could be combined.

Advances in climate forecasting (Cane and Zebiak 1987, Glantz et al. 1991) can be useful in the development of health early warning systems (HEWS), providing advance notice of bioclimatic conditions conducive for disease and pest outbreaks. "HEWS" could permit timely implementation of environmentally sound interventions (e.g., community education, vaccination campaigns, water boiling, benign bacterial larvacide applications, removal of peridomestic mosquito breeding sites and the like). Critical, vulnerable (Almendares et al. 1993), and particularly sensitive regions (e.g., mountain ranges) can be targeted for monitoring.

The International Research Institute (IRI) for climate prediction plans regional application centers, first focusing on hydrology, agriculture, and health—the three basic human needs! The Global Change System for Analysis, Research and Training (START) program of the International Geosphere–Biosphere Program (IGBP), World Climate Research Program and International Human Dimensions Program is to involve 13 regional research networks, numerous regional research centers, and affiliated sites.

Large marine ecosystem (LME) monitoring of the sustainability and biomass yields of coastal waters (Sherman et al. 1993, Sherman 1994) is being organized for the Gulf of Guinea and proposals have been submitted for the Yellow Sea and the Black Sea. The intention is to expand coverage to the world's 49 LMEs. These comprehensively monitored ecosystems can serve as coastal components of a Global Ocean Observing System (GOOS). Monitoring the health of the living marine components, as well as monitoring plankton, bivalves, and finfish for biotoxins and bacteria (and coastal nations for causes of shellfish poisoning and cholera) can form an integral part of these projects. The relative contribution of the chief driving forces impacting each system (e.g., pollution/habitat loss, overfishing, and climate) will be evaluated in order to inform policies for mitigation and prevention. Integrating adverse and disease events into these larger monitoring schemes can aid in assessing the driving forces and in evaluating the costs of policies and actions.

A network of regional centers with support from the World Climate Program (Global Climate Observing System or GCOS) is being planned to monitor worldwide meteorological data. Other programs are planned to monitor ecosystems (the Global Terrestrial and World Hydrological Observing Systems or GTOS and WHYCOS), with a coordinated Integrated Global Observing System (IGOS). Chronicling of disease events and assorted biological indicators can play roles in these systems as well.

Such monitoring will be augmented by remote sensing (i.e., NASA Mission to Planet Earth and other national space programs), with data assimilated into mapped, overlapping geographic information systems. In the United States, the Climate Analysis Center (NOAA) defines four regions for weather surveillance. Coordinating ecological (EPA and National Biological Survey), agricultural (USDA), and health data (CDC) through these centers could provide the basis for integrated regional assessment and monitoring.

The CDC, the World Health Organization, and the Program for Monitoring Emerging Diseases (ProMED) are planning an international consortium of regional centers for enhanced surveillance and response through clinical, laboratory, and epidemiological means to respond to the global emergence, resurgence, and redistribution of infectious diseases. Sentinel sites and data gathering can be coordinated.

Conclusion

Biodiversity provides an essential buffer against the uncontrolled proliferation of opportunistic pests and harmful microorganisms. Disease outbreaks often reflect breakdowns in these defensive systems, and illness can affect many taxa. The impacts of epidemics can ripple through societies and economies (Todd 1993, Epstein 1993b), affecting productivity, commerce, tourism, and travel. Conversely epidemics of parasitic diseases can themselves reduce biodiversity. Dynamic equilibrium among species and functional groups is thus essential for health.

The alarming growth of pesticide (French-Constant 1995) and antibiotic resistance (Neu 1992) serve to emphasize the importance of improving our understanding of the environmental reservoirs and species dynamics involved in the emergence and spread of infectious diseases.

Combining the surveillance of biological indicators and relevant health outcomes with ecosystem monitoring can support a systems-based approach to the design of adaptive and preventive responses.

Forestry, fishery, and fossil fuel policies must all be considered in light of their effect on biodiversity and the expanding conditions conducive to the emergence, resurgence, and redistribution of infectious diseases throughout the world.

Biodiversity and Health: Some Underlying Principles

The following summarizes key underlying principles regarding biodiversity and health.

1. Infectious disease events have multiple causes and involve an agent, a vulnerable host, and a conducive environment.

2. K-strategists (including many predators) are frequently more vulnerable to environmental challenges than r-strategists (including many prey, pests, and pathogens). Species that are specialists (i.e., with limited ecological niches) are most sensitive to changes in the environment and in climate variability. Thus generalists may become dominant where habitat loss and widespread environmental stress occurs.

3. Monocultures—in terrestrial and marine systems—may be more vulnerable to opportunistic species invasion, colonization, and spread.

4. Sensitivity to stress depends on the current state and on the presence of cofactors (e.g., the effect of temperature on growth depends on adequate moisture and nutrients).

5. Vulnerability to infection is compounded when multiple perturbations occur over a short period of time.

6. The ability of an ecosystem to recover from one or more stresses (its resilience) depends on the health and flexibility of other components.

7.   The complexity of a food chain affects stability and resilience in the face of extreme events, and epidemics may be viewed as symptoms of ecosystem dysfunction.

8.   Inputs (e.g., nutrients) at the bottom of food chains will reverberate throughout trophic levels, while the impacts of those entering at the top (e.g., overfishing) or at the middle (e.g., copepod damage from excess UV-B radiation) are more variable depending on the strength of linkages between levels.

9.   Dispersion of disease agents can occur contiguously or hierarchically (e.g., through air or sea [ballast] transport), with amplification and persistence of exotic species dependent on local bioclimatographic factors.

References

Almendares, J., Sierra, M., Anderson, P.K., and P.R. Epstein. (1993). Critical Regions, a Profile of Honduras. *Lancet* 342:1400–1402.

Anderson, D.M. (1992). The Fifth International Conference on Toxic Marine Phytoplankton: A Personal Perspective. *Harmful Algae News, Supplement to International Marine Science (UNESCO)* 62:6–7.

Anderson, I. (1995). Is a Virus Wiping Out Frogs? *New Scientist* 7 (January 7).

Anderson, R.M. (1980). Depression of Host Population Abundance by Direct Life Cycle Macroparasites. *J. Theor. Biol.* 82:283–311.

Anderson, R.M., and D.M. Gordon. (1982). Processes Influencing the Distribution of Parasite Numbers within Host Populations with Special Emphasis on Parasite-Induced Host Mortalities. *Parasitology* 85:373–398.

Atkinson, I.A.E. (1985). The Spread of Commensal Species of *Rattus* to Oceanic Islands and Their Effects on Island Avifauna. *ICBP Technical Publications* 3:35–81.

Bakun, A. (1993). The California Current, Benguela Current, and Southwestern Atlantic Shelf Ecosystems: A Comparative Approach to Identifying Factors Regulating Biomass Yields. In *Stress, Mitigation, and Sustainability of Large Marine Ecosystems* (K. Sherman, L.M. Alexander, and B.D. Gold, eds.), pp. 99–221. Washington, DC: AAAS Press.

Barry, J.P., Baxter, C.H., Sagarin, R.D., and S.E. Gilman. (1995). Climate-Related, Long-Term Faunal Changes in a California Rocky Intertidal Community. *Science* 267:672–675.

Billett, J.D. (1974). Direct and Indirect Influences of Temperature on the Transmission of Parasites from Insects to Man. In *The Effects of Meteorolog-*

ical Factors Upon Parasites (A.E.R. Taylor and R. Muller, eds.), pp. 79–95. Oxford: Blackwell Scientific Publications.

Blaustein, A.W., and D.B. Wake. (1995). The Puzzle of Declining Amphibian Populations. *Scientific American* April:52–57.

Bothwell, M.L., Sherbot, M.J., and C.M. Pollock. (1994). Ecosystem Response to Ultraviolet-B Radiation: Influence of Trophic-Level Interactions. *Science* 265:97–100.

Brown, N.S., and G.I. Wilson. (1975). A Comparison of the Ectoparasites of the House Sparrow (*Passer domesticus*) from North America and Europe. *Am. Midl. Nat.* 94:154–165.

Buchmann, S.L., and G.P. Nabhan. (1996). *The Forgotten Pollinators.* Washington, DC: Island Press.

Burkholder, J.M., Glasgow, H.B., and K.A. Steidinger. (1993). Unraveling Environmental and Trophic Controls on Stage Transformations in the Complex Life Cycle of An Ichthyotoxic "Ambush Predator" Dinoflagellate. Sixth International Conference on Toxic Marine Phytoplankton, Nantes, France, 18–22 Oct. (Abstr.).

Byrd, J.J., Huai-shu, X.U., and R.R. Colwell. (1991). Viable but Non-culturable Bacteria in Drinking Water. *Appl. Environ. Microbiol.* 57:875–878.

Calow, P. (1979). Costs of Reproduction—A Physiological Approach. *Biol. Rev.* 54:23–40.

Calow, P. (1983). Pattern and Paradox in Parasite Reproduction. *Parasitology* 86:197–207.

Cane, M., and S.E. Zebiak. (1987). Prediction of El Niño Events Using a Physical Model. In *Atmospheric and Oceanic Variability* (H. Cattle, ed.), pp. 153–182. London: Royal Meteorological Society.

Carson, R. (1962). *Silent Spring.* Boston: Houghton Mifflin.

Centers for Disease Control and Prevention (CDC). (1993a). Isolation of *Vibrio cholerae* O1 from Oysters: Mobile Bay, 1991–1992. *MMWR* 42:91–93.

CDC. (1993b). *Vibrio vulnificus* Infections Associated with Raw Oyster Consumption. *MMWR* 42:405–407.

CDC. (1993c). Multistate Outbreak of Viral Gastroenteritis Related to Consumption of Oysters—Louisiana, Maryland, Mississippi, and North Carolina, 1993. *MMWR* 42:945–948.

CLIVAR. (1992). A study of climate variability and predictability, World Climate Research Program. WMO, Geneva.

Coimbra, T.L.M., Nassas, E.S., Burattini, M.N., de Souza, L.T., Ferreira, I.B., Rocco, I.M., da Rosa, A.P.A.T., Vasconcales, P.F., Pinheiro, F.P., LeDuc, J.W.,

Rico-Hesse, R., Gonzalez, J-P., Jahrling, P.B., and R.B. Tesh. (1994). New Arenavirus Isolated in Brazil. *Lancet* 343:391–392.

Colburn, T., Dumanoski, D., and J.P. Meyers. (1996). *Our Stolen Future.* New York: Dutton.

Colwell, R.R., Belas, M.R., and A. Zachary. (1980). Attachment of Microorganisms to Surfaces in the Aquatic Environment. *Dev. Ind. Microbiol.* 21:169–178.

Colwell, R.R., and W.M. Spira. (1992). The Ecology of *Vibrio cholerae.* In *Cholera: Current Topics in Infectious Disease.* (D. Barua and W.P. Greenrough III, eds.), pp. 107–127. New York: Plenum Medical Book Company.

Cosper, E.M., Dennison, W.C., Carpenter, E.J., Bricelj, M.V., Mitchell, J.G., Kuenstner, S.H., Colflesh, D., and M. Dewey. (1987). Coastal Marine Ecosystems. *Estuaries* 10:284–290.

Davis, M.B., and C. Zabinski. (1990). Changes in Geographical Range Resulting from Greenhouse Warming on Biodiversity in Forests. In *Proceedings of the World Wildlife Fund Conference on Consequences of Greenhouse Effect for Biological Diversity* (R.L. Peters and T.E. Lovejoy, eds.), New Haven, CT: Yale University Press.

de Duve, C. (1996). The Birth of Complex Cells. *Scientific American* April: 50–58.

Diamond, J.M. (1989). The Present, Past and Future of Human–Caused Extinctions. *Phil. Trans. R. Soc. Lond. B.* 325:469–477.

Dobson, A., and R. Carper. (1993). Climate Change and Human Health: Biodiversity. *Lancet* 342:1096–1131.

Dobson, A.P. (1988). Restoring Island Ecosystems: The Potential of Parasites to Control Introduced Mammals. *Cons. Biol.* 2:31–319.

Dobson, A.P. (1996). *Conservation and Biodiversity.* Scientific American Books.

Dobson, A.P., and R.M. May. (1986). Patterns of Invasions of Pathogens and Parasites. In *Ecology of Biological Invasions of North America and Hawaii* (H.A. Mooney and J.A. Drake, eds.), pp. 58–76. New York: Springer-Verlag.

Doucette, G.J. (1993). Assessment of the Role of Procaryotic Cells in Harmful Algal Bloom Development. In *Proceedings of Sixth International Conference on Toxic Marine Phytoplankton* (P. Lassus, ed.), Nantes, France, October.

Edman, J.D., Timperi, R., and D. Werner. (1993). Epidemiology of Eastern Equine Encephalitis in Massachusetts. *J. Fla. Mosquito Control Assoc.* 64:84–96.

Ehrlich, P.R. (1986). *The Machinery of Nature.* New York: Simon and Schuster.

Elias, S.A. (1994). *Quaternary Insects and Their Environments.* Washington, DC: Smithsonian Institution Press [based on work of R. Coope and others].

Epstein, P.E. (1994). Framework for an Integrated Assessment of Climate Change and Ecosystem Vulnerability. In *Disease in Evolution: Global Changes and Emergence of Infectious Diseases* (M.E. Wilson, R. Levins, and A. Spielman, eds.), pp. 423–435. New York: NYAS.

Epstein, P.R. (1993a). The Role of Algal Blooms in the Spread and Persistence of Human Cholera. *BioSystems* 31:209–221.

Epstein, P.R. (1993b). The Costs of Not Achieving Climate Stabilization. *Ecological Economics* 8:307–308.

Epstein, P.R. (1995). Emerging Infections and Global Change: Integrating Health Surveillance and Environmental Monitoring. *Current Issues in Public Health* 1:224.

Epstein, P.R., and G. Chikwenhere. (1994). Biodiversity Questions. *Science* 265:1510–1511.

Epstein, P.R., Ford, T.E., and R.R. Colwell. (1993). Climate Change and Human Health: Marine Ecosystems. *Lancet* 342:1216–1219.

Epstein, P.R., Pena, O.C., and J.B. Racedo. (1995). Climate and Disease in Colombia. *Lancet* 346:1243–1244.

Ferrer, M., and N. Pumarola. (1990). Morbillivirus in Dolphins. *Nature* 348:21.

French-Constant, R.H. (1995). The Zap Trap. *The Sciences* March/April 31: 31–35.

Garrett, L. (1994). *The Coming Plague: Newly Emerging Diseases in a World Out of Balance.* New York: Farrar, Strauss and Giroux.

Glantz, M.H., Katz, R.W., and N. Nicholls (eds.). (1991). *Teleconnections Linking Worldwide Climate Anomalies.* Cambridge: Cambridge University Press.

Glass, G.E. (1996). Rodent Ecology and Infectious Disease Risk. Presentation at AAAS, Baltimore, MD, Feb. 8–13.

Gordon, D.M. (1995). The Development of Organization in an Ant Colony. *American Scientist* 83:50–57.

Hanwalt, P.C. (1994). Transcription-Coupled Repair and Human Disease. *Science* 266:1957–1958.

Harris, C.C., and M. Hollstein. (1993). Clinical Implications of the *p*53 Tumor-Suppressor Gene. *New Engl. J. Med.* 329:1318–1327.

Hayes, R.L., and T.J. Goreau. (1996). Mass Coral Reef Bleaching: Climate, Temperature and Ultraviolet Effects. Abstract, AAAS, Baltimore, MD, Feb. 8–13.

Hill, D.K. (1995). Pacific Warming Unsettles Ecosystems. *Science* 267:1911–1912.

Hochberg, M.E. (1989). The Potential Role of Pathogens in Biological Control. *Nature* 337:262–265.

Holland, J. (1993). Replication Error, Quasispecies Populations, and Extreme Evolution Rates of RNA Viruses. In *Emerging Viruses* (S.S. Morse, ed.), pp. 203–218. Oxford: Oxford University Press.

Holldobler, B., and E.O. Wilson. (1994). *Journey to the Ants: A Story of Scientific Exploration.* Cambridge, MA: Belknap Press of Harvard University Press.

Howarth, R.W., Billen, G., Swaney, D., Townsend, A., Jaworski, N., Lajtha, K., Downing, J.A., Elmgren, R., Caraco, N., Jordan, T., Berenose, K., Freney, J., Kudeyarov, V., Murdoch, P., and Z. Zhao-Liang. (in press). Regional Nitrogen Budgets and Riverine N & P Fluxes for the Drainages to the North Atlantic Ocean: Natural and Human Influences. *Biogeochemistry.*

Islam, M.S., Drasar, B.S., and D.J. Bradley. (1990). Long-Term Persistence of Toxigenic *Vibrio cholerae* O139 in the Mucilaginous Sheath of a Blue-Green Algae, *Anabaena variabilis. J. Trop. Med. Hyg.* 93:133–139.

Islam, M.S., Hasam, M.K., Miah, M.A., Yunus, M, and M.J. Albert. (1994). Isolation of *Vibrio cholerae* O139 Synonym Bengal: Implications for Disease Transmission. *Appl. Environ. Microbiol.* 60:1684–1686.

Jeveen, A., and M.L. Kripke. (1993). Ozone Depletion and the Immune System. *Lancet* 342:1158–1159.

Kennedy, S., Smyth, J.A., McCullough, S.J., Allan, G.M., and F. McNeilly. (1988). Confirmation of Recent Seal Deaths. *Nature* 335:404.

Krebs, C.J., Boutin, S., Boonstra, R., Sinclair, H.R.E., Smith, J.N.M., Dale, M.R.T., Martin, K., and R. Turkington. (1995). Impact of Food and Predation on the Snowshoe Hare Cycle. *Science* 269:1112–1115.

Kruess, A., and T. Tschamtke. (1994). Habitat Fragmentation, Species Loss and Biological Control. *Science* 264:1581–1584.

Kushmaro, A., Loya, Y., Fine, M., and E. Rosenberg. (1996). Bacterial Infection and Coral Bleaching. *Nature* 380:396.

Levins, R. (1970). Extinction. *Lectures on Mathematics in the Life Sciences* 2:75–107.

Levins, R., Epstein, P.R., Wilson M.E., Morse, S.S., Slooff, R., and I. Eckardt. (1993). Hantavirus Disease Emerging. *Lancet* 342:1292.

Lewin, R. (1992). *Complexity: Life at the Edge of Chaos.* New York: Collier Books, MacMillan.

Likens, G.E. (1992). *The Ecosystem Approach: Its Use and Abuse.* Ecology Institute, Germany.

Likens, G.E., Driscoll, C.T., and D.C. Buso. (1996). Long-Term Effects of Acid Rain: Response and Recovery of a Forest Ecosystem. *Science* 272:244–246.

Line, L. (1996). "1,096 Mammals and 1,108 Bird Species Threatened," *New York Times,* p. C4. [Based on "The Red List," Species Survival Commission of the World Conservation Union.]

MacArthur, R.H., and E.O. Wilson (1967). *The Theory of Island Biogeography.* Princeton: Princeton University Press.

Manwell, R.D. (1957). Blood Parasitism in the English Sparrow with Certain Biological Implications. *J. Parasit.* 43:428–433.

Milligan, K.L.D., and E.M. Cosper. (1994). Isolation of Virus Capable of Lysing the Brown Tide Microalga *Aureococcis efferens. Science* 266:805–807.

Modrich, P. (1994). Mismatch Repair, Genetic Stability, and Cancer. *Science* 266:1959–1960.

Molyneaux, D.H. (1992). Deforestation: Effect of Deforestation on Vector Borne Disease. Paper presented at conference: The Impact of Global Change on Disease. Joint meeting of the British Society for Parasitology, The Linnean Society of London, and the Royal Society of Tropical Medicine and Hygiene, 30 Sept.

Morell, V. (1994). Mystery Ailment Strikes Serengeti Lions. *Science* 264:1404.

Murray, K., Rogers, R., Selvey, L., Selleck, P., Hyatt, A., Gould, A., Gleeson, L., Hooper, P., and H. Westbury. (1995). A Novel *Morbillivirus* Pneumonia of Horses and Its Transmission to Humans. *Emerg. Infect. Dis.* 1:31–33.

Neu, H.C. (1992). The Crisis in Antibiotic Resistance. *Science* 257:1064–1073.

Paul, J.H. (1993). The Advances and Limitations of Methodology. In *Aquatic Microbiology: An Ecological Approach* (T.E. Ford, ed.), pp. 483–511. Boston: Blackwell Scientific Publications.

Peters, R.L. (1991). Consequences of Global Warming for Biological Diversity. In *Global Change and Life on Earth* (R.L. Wyman, ed.), pp. 99–118. New York: Chapman and Hall.

Peters, R.L., and T.E. Lovejoy (eds.). (1992). *Global Warming and Biodiversity.* New Haven, CT: Yale University Press.

Pimm, S. L., Russell, G.J., Gittleman, J.L., and T.M. Brooks. (1995). The Future of Biodiversity. *Science* 269:347–350.

Roemmich, D., and J. McGowan. (1995). Climatic Warming and the Decline of Zooplankton in the California Current. *Science* 267:1324–1326.

Sancar, A. (1994). Mechanisms of DNA Excision Repair. *Science* 266:1954–1956.

Schindler, D.W., Curtle, P.J., Parker, B.R., and M.P. Stainton. (1996). Consequences of Climate Warming and Lake Acidification for UV-B Penetration in North American Boreal Lakes. *Nature* 379:705–708.

Schoener, T.W., and D.A. Spiller. (1995). Effect of Predators and Area on Invasions: An Experiment with Island Spiders. *Science* 267:1181–1183.

Sherman, K. (1994). Sustainability, Biomass Yields, and Health of Coastal Ecosystems: An Ecological Perspective. *Marine Ecology Progress Series* 112: 277–301.

Sherman, K., Alexander, L.M., and B.D. Gold (eds.). (1993). *Large Marine Ecosystems: Stress, Mitigation, and Sustainability*, pp. 301–319. Washington, DC: AAAS Press.

Skole, D., and C. Tucker. (1993). Tropical Deforestation and Habitat Fragmentation in the Amazon: Satellite Data from 1978 to 1988. *Science* 260:1905–1910.

Smayda, T.J., and Y. Shimizu. (1993). *Toxic Phytoplankton Blooms in the Sea*. London: Elsevier.

Stenseth, N.H. (1995). Snowshoe Hare Populations Squeezed from Below and Above. *Science* 269:1061–1062.

Stevens, W.K. (1994). "Extinction of the Fittest May Be the Legacy of Lost Habitats," *New York Times*, 9/27/94, p. B8.

Summich, J.L. (1984). *An Introduction to the Biology of Marine Life*, 3rd ed. Dubuque, IA: W.C. Brown.

Swerdlow, D.L., Mintz, E.D., Rodriguez, M., Tejada, E., Ocampo, C., Espejo, L., Greene, K.D., Saldana, W., Seminario, L., Tauxe, R.V., Wells, J.G., Bean, N.H., Ries, A.A., Pollack, M., Vertiz, B., and P.A. Blake. (1992). Waterborne Transmission of Epidemic Cholera in Trujillo, Peru: Lessons for a Continent at Risk. *Lancet* 340:28–33.

Tester, P.A., Stumpf, R.P., Vukovich, F.M., Fowler, P.K., and J.T. Turner. (1991). An Expatriate Red Tide Bloom, Transport, Distribution, and Persistence. *Limnol. Oceanogr.* 36(5):1053–1061.

Tilman, D.R., and J.A. Downing. (1994). Biodiversity and Stability in Grasslands. *Nature* 367:363–365.

Tilman, D.R., May, R.M., Lehman, C.L., and M.A. Nowak. (1994). Habitat Destruction and the Extinction Debt. *Nature* 371:65–66.

Todd, E.C.D. (1993). Seafood-Associated Diseases in Canada. *J. Assoc. Food and Drug Officials* 56(4):45–52.

Trenberth, K.E., and T.J. Hoar. (1996). The 1990–1995 El Niño–Southern Os-cillation Event: Longest on Record. *Geophysical Research Letters* 23:57–60.

Tumlinson, J.H., Lewis, W.J., and L.E.M. Vet. (1993). How Parasitic Wasps Find Their Hosts. *Scientific American* March: 100–106.

Valiela, I. (1984). *Marine Ecological Processes.* New York: Springer-Verlag.

Verhovek, S.H. (1996). "In Texas, an Attempt to Sweat an Old Pest Stirs a Re-volt," *New York Times,* 1/24/96, p. A10.

Vitousek, P.M. (1990). Biological Invasions and Ecosystem Properties: Towards an Integration of Population Biology and Ecosystem Studies. *Oikos* 57:7–13.

Williams, E.H., Jr. (1987). Caribbean Mass Mortalities: A Problem with the Solution. *Oceanus* 30:69–75.

Wilson, E.O. (1992). *The Diversity of Life.* Cambridge, MA: Harvard Univer-sity Press.

Wyman, R.L. (1991). Multiple Threats to Wildlife: Climate Change, Acid Pre-cipitation, and Habitat Fragmentation. In *Global Change and Life on Earth* (R.L. Wyman, ed.), pp. 134–155. New York: Chapman and Hall.

Yao, Y., Hoffer, A, Chang, C., and A. Puga. (1995). Dioxin Activates HIV Gene Expression by an Oxidative Stress Pathway Requiring a Functional Cy-tochrome *P*450 CYP1A1 Enzyme. *Envir. Health Perspec.* 103:366–371.

Suggested Further Reading

CDC. (1994). *Addressing Emerging Infectious Disease Threats: A Prevention Strategy for the United States.* Atlanta, Georgia.

Edwards, P.J., May, R.M., and N.R. Webb. (1994). *Large-Scale Ecology and Con-servation Biology.* Oxford: Blackwell Scientific Publications.

Epstein, P.R. (1995). Emerging Diseases and Ecosystem Instabilities: New Threats to Public Health. *Am. J. Pub. Health* 85:168–172.

Haines, A., Epstein, P.R., and A.J. McMichaels. (1993). Global Health Watch: Monitoring Impacts of Environmental Change. *Lancet* 342:4464–4469.

Kingsolver, J., Kareiva, P., and R.B. Huey (eds.). (1993). *Biotic Interactions and Global Change.* Sunderland, MA: Sinauer.

Krause, R.M. (1992). The Origin of Plagues: Old and New. *Science* 257:1073–1078.

Levins, R., Auerbach, T., Brinkmann, U., Eckardt, I., Epstein, P., Makhonl, N., dePossas, C.A., Puccio, C., Spielman, A., and M.E. Wilson. (1994). The Emergence of New Diseases. *American Scientist* 82:52–60.

Morse, S.S. (1995). Factors in the Emergence of Infectious Diseases. *Emerging Infectious Diseases* 1:7–15.

Root, T., and S.H. Schneider. (1995). Ecology and Climate: Research Strategies and Implications. *Science* 269:334–341.

Walker, B.H. (1994). Landscape to Regional-Scale Responses of Terrestrial Ecosystems to Global Change. *Ambio* 23:67–73.

CHAPTER 4

Fatal Synergisms: Interactions between Infectious Diseases, Human Population Growth, and Loss of Biodiversity

ANDREW DOBSON, MARY S. CAMPBELL, AND JENSA BELL

Introduction

The recent outbreak of Ebola virus in Zaire and the current HIV/AIDS epidemic have focused increasing attention on the threat that emerging diseases present to the human population. Infectious disease is caused by an interaction between a pathogen and its host. Examining the ecology of pathogens and hosts can give us insights into the emergence and mainte-nance of disease in human populations. We focus on two main themes throughout: the first of these examines the emergence of new diseases from an ecological perspective. Here we illustrate that it is the relationship between human population growth, urbanization, malnutrition, and poverty which permits the establishment of novel pathogens in human populations. We conclude that pathogens can more readily establish in large populations, particularly those aggregated into large urban concentrations. If the human population continues to increase in size, then newly emer-gent pathogens, and drug-resistant forms of older scourges, will find it pro-gressively easier to become established.

The second major theme of the chapter is that the "pulp science fiction" image of tropical forests as a source of new diseases is a distortion. In many ways this distortion is as inaccurate as Hollywood's portrayal in the movie *Outbreak* of the U.S. military as the most effective organization available to deal with any emergent disease outbreak. Both ideas are dangerous fan-tasies that are ill-supported by the available data. The most significant fu-ture disease threats to humans, wildlife, and domestic livestock are likely to why not mutation ?

be well-established pathogens whose range expands due to the evolution of drug resistance, or whose geographic range expands due to climate change. Because pathogens evolve much more rapidly than their hosts, they are likely to capitalize on the new opportunities that global climate change will provide for them. This may well lead to pronounced increases in the incidence of some human diseases and the disruption of ecosystem function in habitats where a pathogen's impact is focused upon "keystone" species. Furthermore, tropical forests, instead of acting as a source of new pathogens, are more likely to serve as a source of new drugs and medicines that can be used in the fight against disease. This may provide one of the strongest arguments for the protection of tropical forests. They are likely to contain the solution to the mounting disease problems that pathogens will cause if the human population continues to increase and poverty and malnutrition come to define the dominant lifestyle of the twenty-first century.

Classification of Pathogens

Epidemiologists have used life history characteristics to broadly classify pathogens into two groups—microparasites and macroparasites. Microparasites are the relatively small organisms, such as viruses, bacteria, protozoa, and fungi, which undergo within-host reproduction, usually infecting hosts for short periods of time, and generating immune responses in infected and recovered individuals. Conversely, macroparasites—predominantly intestinal worms, ticks, and fleas—cannot reproduce directly within individual hosts. These pathogens often have complex life cycles with infective stages that must pass out of the host in order to infect other individuals. The adult stages of macroparasites, though persistent, fail to produce strong immune responses in infected hosts. Whereas the severity of microparasitic infections is not directly related to the actual number of individual pathogens within the host, host macroparasitic burden is important to consider when looking at the disease on the population level. In both classes of parasitism, the parasite's most important resource is its host population; in the absence of the host, the parasite will decline to extinction; however, as the host increases in abundance, the resources available for the parasite increase and it is likely to prosper.

Host and Pathogen Life Histories

It is also useful to examine and compare host and pathogen life histories when studying disease. Among other traits, an organism's life history in-

cludes its size, life expectancy, and reproductive rate. The differences between pathogens and hosts are obvious when it comes to size. Although
the sizes of pathogens differ by orders of magnitude—from Kuru-causing
prions, the smallest known pathogens, at less than 50 nm to 20- to 30-cm
long guinea worms.

The life expectancies of pathogens also vary widely, but they are all
shorter than that of their hosts. The long human life span has made us
more vulnerable to the effects of pathogens. A vivid example of this can
be found in HIV. Rarely have AIDS-like syndromes been observed in
green monkeys, the main primate reservoirs of HIV-1 in Africa. However,
with few exceptions, nearly every human infected with HIV eventually
succumbs to AIDS. One possible explanation for the difference in pathology and impact on infected hosts is that the incubation period for AIDS
averages around ten years. Green monkeys, with a life expectancies of 8 to
10 years, may not live long enough to develop AIDS (Anderson 1991).

A final life history trait to consider is the reproductive rate of hosts and
pathogens. From the evidence above, it follows that hosts and pathogens are
operating on different time scales. Because pathogens live for such short
periods of time, relative to humans and other hosts, their reproductive rates
naturally must be much higher. These high rates of reproduction in
pathogens have consequences on diseases in hosts. A high reproductive rate
on the part of the pathogen means that random mutations in DNA and
RNA will occur frequently during in-host replication. These mutations
can lead to increased virulence as the pathogen "adapts" to the genome of
the host. Conversely, the host is powerless to adapt its genome to the
pathogen during the course of infection, although the specialized cells of
the immune system are able to recognize and adapt to new infective strains
of microparasites. It is this particular interaction that may lead to the onset
of AIDS in HIV-infected individuals. A study of the dynamics of HIV in
single-infected individuals revealed that the number of genetically distinct
strains of HIV gradually rises in the years following infection (Nowak
1991). As the diversity of HIV strains challenging the host's immune system increases, the immune system's inability to fight off the barrage of new
strains may precipitate its collapse and the development of AIDS. Several
later papers corroborated this theory and provided more evidence for the
high rate of reproduction in HIV. Wei et al. (1995) found that the half-life
of an individual virus particle in the human body was about 2 days, with
10^8–10^9 new virus particles being produced each day. The dynamics of the
human immune system in response to HIV infection were discovered to
be no less impressive. In combating HIV, the immune system may produce
a daily average of 2×10^9 CD4 T cells. Losses in CD4 cells, although small
(20–200×10^6 per day), account for the gradual onset of AIDS as the im-

mune system wanes (Ho et al. 1995). The HIV virus wins the race against its host partly because no infected individual is able to change its genome in response to selection pressures for a more efficient immune system.

Host Switching

Another consequence of the high reproductive rate and mutation rate of pathogens is their ability to exploit new hosts. Host-switching has frequently led to the emergence of new diseases in humans. Examples of human diseases caused by pathogens whose genomes diverge only slightly from pathogens that infect other animals are malaria, measles, smallpox, and influenza. One instance of a small genetic change in a virus leading to a new disease in a new host is that of canine parvovirus. In the mid to late 1970s, epidemics of a previously unknown disease in dogs were reported worldwide. By 1980, all populations of domestic and wild dogs had tested positive for exposure to this new virus. Isolation of the virus from infected tissues revealed that it was closely related to the feline panleukopenia virus. Although domestic dogs were widely vaccinated after its accidental introduction, CPV on Isle Royale in Lake Superior decimated the wolf population, disrupting one of the longest-running ecological studies of a natural predator–prey system. Without these predators, the island moose population has increased, which in turn has had a negative impact on the island's vegetation (Mech and Goyal 1993; Peterson and Page 1988).

Once again, HIV serves as an example of an emerging disease that most likely resulted from a host switch. In the search for the origin of the pathogen that causes AIDS, epidemiologists identified several diseases in animals that resemble human AIDS. The earliest known AIDS-like immunodeficiency disease observed in animals was *visna* (Brown 1990), a syndrome which occurred in Icelandic sheep after the introduction of a hardier German breed in the 1940s (Brown 1990). When this virus was sequenced, along with similar ones from goats and horses, molecular biologists found that it fit into the retrovirus subgroup of RNA viruses. As time passed, immunodeficiency viruses were also detected in cats, cows, and monkeys. The simian immunodeficiency viruses (SIV) are particularly fascinating because not all of them elicit symptoms in their hosts. Although SIVmac often leads to death in macaques within twelve months, SIVsm and SIVagm, which infect sooty mangabeys and African green monkeys, respectively, do not give rise to disease. Studies of the relationship of HIV viruses to the SIV viruses of nonhuman primates have shown that HIV-2, the West African strain of HIV, has a genetic sequence that matches SIVsm at 82% of the loci on a key gene (Brown 1990). HIV-1, the virus that infects most AIDS patients in the United States and East Africa, does not show a high degree of homology to any of the SIV viruses. This is not surprising because the HIV virus evolves faster than any other known organism. Strains found within a single patient may vary by as much as 1 to 2%. Between-patient genetic variation ranges from 6 to 9% in North America and up to 20% in Africa (Brown 1990). Attempts to quantify the mutation rate of HIV have estimated that of one in every 1,000 nucleotides changes each year. The RNA of HIV has a total of 5,000 to 10,000 bases. The wide variation in HIV is also reflected in SIVagm viruses. The strains that infect the many species of African green monkeys show a genetic similarity of only 83%. With such differences in the viral RNA from wild animal populations, the chance that a mutant SIV strain leapt from the simian genera to that of humans seems quite possible.

Morbillivirus Epidemics

In September 1994, a previously unknown virus fatally infected mammals in two different orders. On a ranch in Brisbane, Australia 21 horses and two men came down with high fevers, flu-like symptoms, and extreme breathing difficulties. Fourteen of the horses and one man died during the two-week outbreak. Virologists have since identified and sequenced the pathogen, a member of the *Morbillivirus* genus, the group that contains canine distemper virus, rinderpest, and measles (Murray 1995). From the genetic code, it has been hypothesized that the equine *Morbillivirus* is not just a slight variant of one of the known morbilliviruses; instead, it is a pathogen that normally infects another host, but suddenly was able to exploit a new niche because of ecological changes (Murray 1995).

The *Morbillivirus* outbreak in Australian race horses has been disconcertingly coincident with outbreaks of morbilliviruses in seals in the North Sea (Heide-Jorgensen et al. 1992) and of distemper in lions in the Serengeti region of Tanzania (Roelke-Parker et al. 1996). There are several historical records of major disease epidemics when morbilliviruses successfully transferred between hosts, particularly the rinderpest pandemic that destroyed perhaps a third of sub-Saharan Africa's game at the end of the nineteenth century.

Epidemiological Concepts and Human Demography

Despite the relatively low reproductive rate of our species (from the pathogen's point of view), our population has been growing at an astounding rate. It has increased from less than 4 billion to close to 6 billion in the last twenty-five years (Daily and Ehrlich 1992; Cohen 1995). This population growth has fueled the spread of infectious disease. As our numbers have been growing, we have gradually crossed the thresholds for establishment of many diseases. The continued presence of pathogens in a population of hosts is entirely dependent upon the rate at which susceptible hosts are added to the host population. This means that most pathogens can only establish once their host population exceeds a certain density which is usually referred to as the threshold for establishment, H_T, or critical community size. For highly contagious diseases such as measles, the minimum population of people needed for the disease to be continually present is a city of around 300,000 (Black 1966). Less contagious and more severe diseases, such as polio and smallpox, have higher thresholds for establishment.

Closely related to the threshold for establishment is the the basic reproductive ratio of the pathogen, R_0, this is an estimate of the rate of spread of the pathogen between susceptible hosts. More specifically, R_0 is the number of hosts an infected individual would infect in a population consisting purely of susceptibles. When R_0 equals one, each infection results in one new case of the disease, when R_0 is less than one, the pathogen will die out in the host population. Epidemiologists estimate thresholds for establishment by setting R_0, the basic reproductive rate of the disease, equal to one. The thresholds for establishment are used to determine the level of vaccination coverage needed for eradication; a disease with an R_0 value of less than one dies out, so immunizing the population to set the number of susceptibles below the threshold for establishment, in theory, will eradicate the disease.

Human population growth has led to the establishment of infectious diseases, but it may also be true that infectious disease has curtailed further increases in population. During the first 5,000 years of the Agricultural Revolution, which began approximately 10,000 years ago, human population grew by a factor of 20, from 5 million to 100 million. But during the next 5,000 years, the population grew only by a factor of five (Anderson and May 1979). Anderson and May (1991) have suggested that this depression in the population growth rate was a consequence of the establishment of infectious disease.

Spatial Distribution

Another important feature of human demography relates to the distribution of our species on the planet. The human population is not homogeneously distributed but tends to be aggregated into villages, towns, and cities. Figure 4.1 illustrates the size of different towns and cities in human history for the last 3,000 years. Not only has the total size of human populations increased but the proportion of people living in larger conglomerations has also significantly increased from less than 0.1% of people before 1400 A.D. and less than 2% at the beginning of the nineteenth century to greater than 20% of people at the end of the twentieth century.

Subgroups

The huge human population is also divided by genetics, behavior, and geography into subgroups, each of which may show a different susceptibility and response to a pathogen. In some subgroups, certain diseases are more prevalent, while in others, infections are less likely. The most intrinsic way in which organisms can be isolated from one another is through their genes. The biological concept of a species rests upon this phenomenon— no matter how closely two populations' geographical ranges overlap, they remain separate unless their genetic makeup is such that they produce fertile offspring upon mating. However, through inbreeding and strong selection pressures, populations can diverge genetically and become less susceptible to diseases that affect other members of their species. One example of this relates to a recent distemper outbreak in East Africa. This disease has killed up to one-third of the lions in two national parks in Kenya and Tanzania, the Masai Mara, and the Serengeti, but it has had little effect on the inbred lions in Ngorongoro Crater. Here, the population's geographic isolation that has caused their genetic isolation has perhaps also protected them from exposure to the distemper strain passing through the rest of the lion population. Instances of genetic isolation can also be found in our own species. People who carry the gene for sickle cell anemia but do not suffer from the disorder are at least partially protected from malaria. Similarly, it has been hypothesized that people who are heterozygous for the genetic disorder of cystic fibrosis are less susceptible to cholera.

A second way in which populations are isolated is through their behavior. Some diseases are predominant within small enclaves, but are nearly absent from the rest of society. Sexually transmitted and intravenous drug use–associated diseases are spread by people who frequently engage in the particular risky behaviors required for transmission. Unlike other infectious diseases, which tend to increase in prevalence as a direct consequence of human population growth, these diseases are maintained by people in behavioral subgroups. The long persistence times of these pathogens and

YEAR

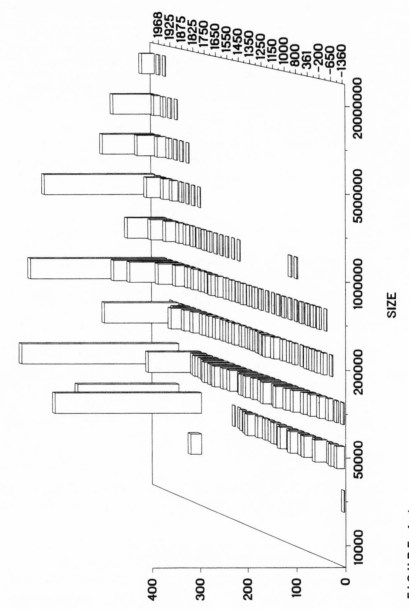

FIGURE 4.1
City size: 1360 B.C. to 1992 A.D.

their failure to produce immunity allow them to be continually present, even in populations of low density (Anderson and May 1991). When people in high-risk behavioral subgroups have intimate contact with individuals in other sections of the population, the pathogens can spread into these sections of the population. Here, they either cause a low level epidemic that is maintained by occasional contacts with high-risk groups, or, they spread at a slower rate through the lower risk groups, as seems to be the case with HIV/AIDS.

While the subdivision of the human population with regard to disease susceptibility varies with social activities, there is probably also variability in disease susceptibility associated with nearly every human activity—their water sources, their pets, or their occupations. For example, seemingly similar tribes in Tanzania have vastly different rates of infection with schistosomiasis, a water-borne protozoan that enters the body through the skin. The members of the Bantu tribe have severe schistosomiasis infections, but this disease is almost unknown in the Hazda. Digging holes in dry riverbeds to collect rainwater, instead of coming into contact with cercariae-infested rivers, accounts for the discordant incidences of this disease between the two tribes (Bennett et al. 1970). Interestingly, in classifying human behaviors into metabolic, reproductive, protective, comfort-providing, territorial, and information gathering, the behaviors that put people at the greatest risk for acquisition of parasitic diseases were metabolic in nature—breathing, coughing, sneezing, food cultivation and gathering, drinking, eating, storage, defecation, and urination (Croll 1983), none of which could be considered morally objectionable by even the most fervant religious zealot.

Another behavior that puts people at risk for disease is close association with animals. McNeill (1976) tallied the number of diseases of domesticated animals that also infect humans to be 296. Of these, the largest number, 65, are diseases of dogs. The common practice of keeping dogs for work and companionship has certainly impacted health. Religious taboos, such as that in Islam, against keeping dogs as pets probably emerged from experiences with canine diseases (Croll 1983). Parasitic worms from dogs commonly infect children in the United States as children come into contact with their dog's muzzle and paws while playing (Schantz 1983). Even people without pets are at risk for canine and feline worm infections as *Toxicara* eggs contaminate the soil of 10 to 32% of public parks (Schantz 1983).

Occupational exposure to pathogens is a third behavior that can lead to the emergence of new diseases. The incidence of equine *Morbillivirus* in the horse trainer and stable hand in Australia, mentioned earlier, is a vivid illustration of how close contact with animals can lead to infections in humans. Throughout history, people have been infected because of the peculiar hazards of their employment situations. Anthrax, which leads to septicemia and often death if left untreated, is also known as woolsorter's disease because of its prevalence among workers in the wool and hide industries (Acha and Szyfres 1980). *Erythema migrans,* a skin lesion which sometimes causes arthritis but is rarely fatal, occurs mainly among fish and poultry processing personnel (Acha and Szyfres 1980). Human occupational diseases do not always have negative outcomes; had milkmaids in the nineteenth century not contracted cowpox, the discovery of vaccination and the eradication of smallpox might continue to elude us today.

Emergence of New Pathogens and Global Change

Now that we have discussed how the diseases most familiar to humans have established in the human population, it is interesting to look at how current environmental changes will further affect pathogen prevalence and distribution. The media has attempted to present us with an image of tropical forests as a source of mysterious and devastating diseases. As with any media image, this one is distorted. Far greater threats are presented by changes in the geographical distribution of known pathogens, movement of pathogens from animal to human hosts, and uncontrolled vector population growth and adaptation. The next section will therefore discuss ecological changes that provide the basis for these concerns.

Population Movement

We established earlier the correlation between human population growth and the emergence and establishment of new diseases. In thinking about disease distribution, we must consider human population growth not only in terms of sheer numbers, but also in terms of the frequency and diversity of our interactions. That is to say that both the tendency to aggregate into towns and cities and the increasing potential for long distance movement will affect pathogen transmission. The classic and most spectacular example of human movement leading to the outbreak of infectious diseases occurred when Europeans invaded the New World in the fifteenth and sixteenth centuries. The introduction of smallpox and measles by Spanish soldiers and sailors caused huge declines in the native populations of Central and South America. The population of Mexico fell from 20 million to about 3 million in the 50 years from 1518 to 1568 and then to 1.6 million in the next 50 years (McNeill, 1976). The generous gift of smallpox from the colonizers to the American population was not left un-reciprocated: The exchange was completed with the transmission of syphilis from Native Americans to the explorers and later with its intro-duction into Europe. The first European cases of syphilis were reported in the late fifteenth century following the French invasion of Naples in 1490. Both the French and Italian troops suffered heavy losses from the disease and—as could only be expected—blamed their opponents for the origin of the epidemic. Although it was called the Italian pestilence and the French pox by respective sides, it was most likely spread through Colum-bus's crew to the Spanish army (Quetel, 1990).

In what is now known as the Columbian Exchange, the epidemics of smallpox were the result of the introduction of infected humans into a highly susceptible population. Although there are now few truly isolated populations, increased long-distance air travel creates the potential for the carriage of arthropod vectors or infected humans whose pathogens may be vectored by a species native to the new location. Similarly, movement of domestic animals and livestock brings with it the potential for carriage of viruses, bacteria, and their arthropod vectors.

A recent example of host movement and a resulting disease outbreak oc-curred with the spread of raccoon rabies in the mid-Atlantic regions of the United States. This epidemic began when raccoon hunters introduced large numbers of raccoons from the southern United States into North and South Carolina in the late 1970s. The outbreak has now spread from Virginia through Pennsylvania, New Jersey, New York, and Connecticut to Massachusetts and New Hampshire. Because the rabies strains responsible are identical to those found in the raccoon population in the southern United States, we are led to believe that the change in pathogen distribu-tion was the result of illegal raccoon movers. The ensuing epidemic is likely to cost millions of dollars to control and compensate (Torrence et al. 1992; Smith et al. 1986).

Seal mortality in the North Sea may have similarly resulted from the nat-ural movement of Harp seals across the Atlantic. Phocine distemper, a dis-tant relative of canine distemper and measles, produced widespread mor-tality in the population because it had no prior exposure to distemper-like viruses (Carter et al. 1990).

Ecological Change

Increases in world population size, long-distance travel, and the settlement of a greater portion of the earth's surface have both direct and indirect effects on pathogen distribution. As we discussed above, the introduction of new diseases into susceptible populations is the most obvious effect. Less clear-cut is the connection between human manipulation of the environment and changes in pathogen ecology. With colonization comes deforestation, irrigation, and the introduction of new species. This section discusses how "human-induced" ecological changes affect disease transmission and establishment.

Human penetration of the rainforest can have an immediate impact on pathogen ecology. Even before we account for the inevitable effects of logging, mining, agriculture, collection of fuelwood, and the development of transport routes and hydropower, human movement into the rainforest has all the predicted effects that accompany the introduction of a new species. These effects are illustrated by the persistence of yellow fever in West Africa, despite the availability of a vaccine (Morse and Schluederberg 1990). Although the disease was originally transmitted from monkey to monkey by *Aedes africanus,* it was spread to human communities when logging was initiated in Uganda, and later became an urban problem transmitted by *Ae. bromiliae* and *Ae. aegypti* (Haddow et al. 1947). Although yellow fever was first transmitted to humans before active settlement and clearance of forested areas, there is little doubt that later deforestation exacerbated the problem (Fiennes 1978).

Forest clearance for agricultural purposes is one of the primary ecological impacts of human settlement. Indeed, increasing population pressures have accelerated the rate at which tropical forests are destroyed for use as farmland. In West Africa, for example, the land is largely used for the production of cash crops such as cocoa, coffee, pineapple, and bananas. Forest destruction creates habitats that are most readily colonized by invading species, particularly mosquitoes, ticks, and fleas—species that often act as vectors for pathogens. Introduction of cattle and other domestic animals increases host abundance and provides further impetus for the explosion of the vector population. The impact of deforestation and colonization by invasive flora and fauna is exemplified by Kyasanur forest disease. Discovered in 1957 in Karnataka State, south India, KFD represents the first finding of a tick-borne arbovirus in the tropics. In 1983, at its peak prevalence, 1,555 cases and 180 deaths were recorded (Bannerjee 1988). After extensive investigation, researchers were able to trace the sudden upsurge of KFD to deforestation, invasion of a thick brush species, and introduction of cattle. Deforestation of south India's Shimoga district in the 1950s led to invasion by *Lantana camara,* a species that forms a thick undergrowth in abandoned clearings and provides a habitat for small mammals and birds. Tick density increased in the area because adult ticks, especially *Haemaphysalis spinigera,* could feed on the cattle while the abundant small mammals acted as reservoirs for the immature stages, the stages responsible for human infection. The infective nymphs then fed avidly on humans collecting wood and other forest produce (Walsh et al. 1994).

Vector control has become an increasing problem as a result of the ecological changes associated with introduction of livestock and domestic animals, deforestation, and construction of primitive irrigation systems. As the previous example shows, cattle provide additional vertebrate hosts for such vectors as ticks and hence help to maintain vector populations at high levels. In addition, irrigated savannas where rice fields have been developed and areas cleared by deforestation lend themselves to the creation of temporary freshwater pools. Surface water collection provides adequate breeding sites for anopheline mosquitos, the carriers of the malaria plasmodia. The reduction of anopheline breeding sites near human settlements will be a primary factor in decreasing malaria's annual mortality of 1 to 2 million (WHO 1990).

Some of the characteristic activities that accompany colonization and settlement amplify the problem of vector control. Overhunting of vertebrate species can lead to a reduction in the density of hosts for the many blood-sucking vectors of infectious diseases. In the absence of their preferred hosts, these species will switch to humans, producing outbreaks of pathogens which have not previously occurred in human populations. The classic example of this occurred in East Africa following the rinderpest epidemic that caused a huge reduction in the density of wild game species. The tsetse flies (*Glossina* spp.) that usually feed on wild game species experienced a massive reduction in the size of their host population and switched to feeding on humans in some areas. As tsetse flies are the vectors of sleeping sickness this caused a large outbreak of the disease throughout East Africa (Simon 1962).

Deforestation and Disease

In an effort to outline the effects of ecological change on disease distributions, we've focussed on deforestation and the associated human impacts on the environment. This is not to say that reforestation doesn't have similar consequences on pathogen prevalence. For example, the expansion of the New England white-tailed deer population came at a time when the abandoned farms of the northeastern United States had succeeded to eastern deciduous forest. After the extensive deforestation of the eighteenth and nineteenth centuries, much of the deer population had been eliminated. One of the few exceptions was Long Island, New York where both the white-tailed deer and the deer tick *Ixodes scapularis* were documented fifty years ago (Barbour and Fish 1993). Reemergence of *I. scapularis,* which transmits the Lyme disease bacterium, *Borrelia burgdorferi,* initiated the current epidemic of Lyme disease in the Northeast. In 1991, some 9,645 cases of Lyme disease were formally reported, making it now the most common arthropod-borne disease in the United States (MMWR 1992).

Agricultural practices have also contributed to the emergence of new pathogens. For example, bovine sponge encephalopathy, a scrapie-like disease in cattle, had been present in the domestic animal population for many years, particularly in British sheep, but because of changes in agricultural practices, it later produced an epidemic in the British cattle population (Nathanson et al. 1993; Anderson et al. 1996). Interestingly, BSE (or "mad cow disease" as the British press like to call it) is produced by a new type of pathogen called a prion. This very simple protein essentially has no DNA or RNA but replicates as an image of itself by some more sophisticated process. The disease outbreaks occurred due to the elimination of a solvent extraction process in the meat and bone meal fed to cattle. Essentially, the cattle were fed with limited protein from sheep which caused increasing incidences of BSE in dairy herds. Fortunately, there were no examples of infected milk being fed to human populations. The solvent extraction procedure was later reinstated after the cause of the outbreak had been determined. It is now clear that BSE had been established in the cattle population by 1992 when there was a major epidemic of the disease in Britain. It remains to be seen whether the pathogen has established in the human population as the best estimates of the incubation period of the pathogen suggest that this is around four years. However, the European Economic Community has banned the import of British beef and dairy products and these have been boycotted by both the public and several major fast-food chains. This has led to the collapse of the British beef industry and calls for compensation for farmers that may eventually lead to the collapse of the current British government.

Global Climate Change and Disease

In 1988, strong evidence for increasing temperatures on a global scale startled the public. Although several scientists at the dawn of the industrial era had predicted that the earth's climate would change in response to greater atmospheric concentrations of heat-absorbing carbon dioxide and methane, their ominous message was not heeded until 1958 when monitoring of atmospheric carbon dioxide concentrations first began (Houghton and Woodwell 1989). Since this prediction, temperatures have risen by half a degree and scientific consensus warns of another two to three degree rise by the year 2030 (Houghton and Woodwell 1989). At local and regional scales, the temperature increases may be more dramatic; for instance, winters at middle and high latitude locations may be four to six degrees warmer, with smaller changes in summer temperatures (Houghton and Woodwell 1989).

But how will changes by just a few degrees cause such catastrophe? To understand the consequences of temperature change, we need to understand that these predictions refer to rises in average temperatures. The fluctuations in temperature during one summer day in the temperate zone often lie within a twenty to thirty degree range (Stommel and Stommel 1979). After the eruption of Mount Tambora in Indonesia in 1815, global temperatures fell by about half a degree, causing snow and frosts in New England in the summer of 1816 (Stommel and Stommel 1979). Scientists estimate that temperature changes of one degree effectively change regional climate in the same way that would a move 100 to 150 kilometers north or south. In response to global warming, many temperate plant species will die out in the southernmost areas of their ranges and broad vegetation patterns will move northward.

Other climatological patterns may change as well. The centers of continental land masses may get significantly less rain (Peters and Lovejoy 1992), but equatorial regions will have more precipitation (Myers 1992). This may lead to floods and soil erosion, causing droughts during the dry season. Extreme events such as fires and hurricanes may also become more prevalent (Peters and Lovejoy 1992).

Climate: Hosts and Pathogens

Hosts

Climate and weather have always been important determinants in the patterns of disease. As we did previously, let's look at disease in terms of hosts and pathogens to understand the effects of climate. Pathogens need hosts for survival and reproduction, so when susceptible hosts are absent, diseases die out. Climate-induced changes in vegetation patterns may cause hosts to disperse into more favorable locations. Some pathogens may be able to migrate with their hosts, but some may remain in their original ranges where they may be able to exploit any new host species that arrives. The dispersing populations will have lower densities overall, making it more difficult for diseases to establish.

While the development rates of warm-blooded hosts will be largely unaffected by climate change, the time required for development in invertebrate hosts is strongly temperature dependent (Dobson and Carper 1992). An increase in ambient temperature of 10 degrees can double the rate of development in invertebrate species (Dobson and Carper 1992). The decrease in development time can lead to higher population densities in warm regions, increasing the likelihood that the threshold for establishment of various pathogens will be crossed. Because of this, it is possible that global warming will lead to more diverse and higher pathogen burdens in invertebrate hosts. One example of this occurred with snails (*Physa gyrina*) in a Canadian lake used as a cooling reservoir for a power plant. Snails in the warmer areas of the lake developed much faster and had population densities several orders of magnitude greater than in the cooler control waters. With the increase in host population density came increases in the diversity and density of its parasites (Dobson and Carper 1992; Sankurathi and Holmes 1976).

A third effect that climate change might have on hosts relates to their ability to cope with less than optimal environmental conditions. Species living in abnormally hot and dry environments may expend more energy for thermoregulation and water conservation. This stress may leave them more vulnerable to pathogens and less able to recover from infections. Evidence for high temperature and reduced water availability leading to increased susceptibility to disease has been found for plant species (Myers 1992), but it is a relatively unexplored area for animal hosts.

Pathogens

While climate change on one hand will have an indirect effect on the distribution and abundance of pathogens because of changes in host distribution and abundance, it will have a direct effect upon the regions in which some pathogens can establish. In the 1940s an Australian parasitologist devised a method that would allow livestock-raisers to forecast epidemics of parasitic worms. He plotted a graph of mean monthly rainfall and temperature in several areas of the country superimposed on isoclines of the humidity and temperature ranges needed for development of *Haemonchus contortus,* the barber's pole worm (Dobson and Carper 1993). This allowed him to determine the months during which epidemics were likely and also why animals in certain areas had never been infected. Although local variability in temperature and humidity prevents bioclimatographs from being completely reliable, they would nevertheless be extremely useful tools for targeting the crucial times for performing preventive measures. Another reason for their rarity is that they require long-term climate data and field, rather than laboratory-based, studies on the specific conditions needed for development of each species. To illustrate the specificity of conditions necessary for survival of parasitic worms, take *Necator americanus* and *Ancylostoma duodenale,* the two most common species of hookworm in humans. *N. americanus* larvae require soil between 28 and 32 degrees for development; in contrast, *A. duodenale* larvae can survive freezes and need only 20 to 28 degree soil for development (Fenner 1982). These requirements account for the prevalence of *N. americanus* in the tropics and *A. duodenale* in the temperate zone.

Temperature and humidity are also important for microparasites. Circumstantial evidence has pointed to low humidity as a factor allowing smallpox epidemics. However, the dry conditions that facilitate the spread of this disease inhibit the transmission of the common cold (Fenner 1982). Ambient temperature can even affect pathogens while they are within hosts. In the part of the plasmodium (malaria) life cycle during which the protozoan is inside the mosquito's digestive system, the air temperature is a key factor in determining whether or not the malarial parasites will complete their life cycle. Air temperatures as low as 15 degrees allow completion of the sporogenous phase in *P. vivax,* but for *P. falciparum,* the minimum required temperature is between 20 and 21 degrees, with its optimum at 30 degrees (Gillett 1974). As with hookworm, these temperature requirements for different malarial parasites limit the ranges in which they can survive.

Global Chilling and Disease

To close, let's take a final look at the global chilling pattern of 1816 that occurred after the eruption of Mount Tambora and its possible connection to disease—the first cholera pandemic. Cold-induced crop failures of 1816 precipitated a famine in India. It has been suggested that widespread malnutrition led to a local cholera epidemic in Bengal. The disease was carried by British military personnel from India to Afghanistan and Nepal, after which it gradually traveled westward to the Caspian Sea. From there, Moslem pilgrimages carried it to the Middle East and trade routes along the Volga River brought it to Baltic seaports, and eventually to western Europe. Sixteen years later, the cholera epidemic reached New York City, killing as many as 100 each day (Stommel and Stommel 1979).

The Emergence of Ebola

While cholera was the dreaded disease of the nineteenth century, a far more horrible pestilence captivates us these days. Viral hemorrhagic fevers have recently become the most talked about of emerging infections. The April 1995 outbreak of Ebola in Kikwit, Zaire hit the press with such a force that very few would now question its significance as a serious, albeit localized, threat. The stabilized death toll on May 22, 1995 totaled 101 of 136 cases isolated (WHO, 5/22/95).

VHFs are a group of diseases caused by filoviruses, arenaviruses, flaviviruses, and bunyaviruses. The Ebola virus, an RNA filovirus, was discovered in 1976 and was named for the river in Zaire, Africa, where it was first detected. The original two outbreaks occurred nearly simultaneously in western Sudan in August 1976 and in north-central Zaire in November 1976 and resulted in more than 550 cases and 340 deaths (CDC 1977). A third, and smaller, outbreak occurred in 1979 in Sudan with 34 cases and 22 fatalities. The most recent epidemic was again centered in Zaire, in Kikwit, a city of 400,000 located 400 kilometers from Kinshasa, the capital.

Usually rodents or arthropods serve as VHF hosts; however, in the case of Ebola, the natural host is unknown. Investigators from the Centers for Disease Control have tested thousands of specimens from various species captured near the outbreak sites, but all efforts to identify the viral host have been unsuccessful. Monkeys, like humans, are susceptible to the infection and, if infected, may serve as a source of the virus. A June 1977 case in the village of Tandala, Zaire, apparently unconnected to the 1976 outbreak, suggests that Ebola virus is endemic and likely enzootic in the Zaire River basin (Heymann et al. 1980; Garrett 1994).

All forms of VHFs begin with fevers and muscle aches, and then may progress to the point where the patient suffers from respiratory problems, severe bleeding, kidney problems, and shock. Symptoms begin 4 to 16 days after infection. Vomiting, diarrhea, abdominal pain, sore throat, and chest pain can follow the initial fever. The severity of any case can range from a relatively mild illness to death (CDC, 5/15/95). It is not difficult to see why the Ebola virus has captured the imagination of American authors and screen writers and secured a place alongside HIV as one of the most feared infections of our time. Any medical description of the symptoms will elicit horror—an effect that is by no means lessened by Richard Preston's interpretation of the facts in the 1994 novel, *The Hot Zone*: "The skin loosens, and is easily torn off by slight pressure. The surface of the tongue might slough off. The victim vomits "black vomit"—a mixture of blood and sloughed off tissues. The bleeding from pores and openings will not clot even as the arteries and veins become clogged with clots—the blood has been destroyed and stripped of its clotting factors" (Preston 1994).

Transmission of the disease requires close personal contact with a person who is acutely ill with the disease. The virus is spread through contact with the patient's blood or body fluids and enters its new host through skin lesions and mucous membranes. In all outbreaks seen, the CDC has attributed the spread of Ebola to hospital conditions in the developing world. Problems such as frequent reuse of needles and syringes, insufficient medical supplies, and inadequate barriers to pathogen transmission between patients and medical personnel facilitate transmission. Ebola virus can also be spread through sexual contact. Although patients who have recovered from an illness caused by Ebola do not pose a serious risk for spreading the infection, the virus may still be present in their genital secretions for a brief period after recovery, and therefore they may still be capable of infecting others through sexual contact (CDC, 5/15/95).

The Four Corners hantavirus outbreak that occurred in 1993 brings to light the importance of collaborations between ecologists and medical professionals. In the American Southwest twenty-six people died from cardiopulmonary failure four to five days after coming down with a flu-like illness (Marshall 1993, Wenzel 1994). Scientists at the CDC used molecular techniques to identify the pathogen as a hantavirus, a more virulent relative of a virus that causes kidney failure carried by rodents near the Hantaan River in Korea (Nichol et al. 1993; Murphy 1994). Once the virus had been isolated, the CDC scientists turned to mammalogist Robert Parmenter at the University of New Mexico to get leads on possible animal reservoirs for the new pathogen. The data he had been collecting on rodent populations showed that there had been an explosion in deer mouse (*Peromyscus maniculatus*) densities that year. The increase was spurred by an extraordinary abundance of piñon nuts, whose production had been enhanced by the heavy precipitation of an El Niño event. Tests on 770 deer mice from the Four Corners region revealed that 30% of the population was seropositive for the hantavirus (Stone 1993). Several other rodent species have also been identified as carriers, including four mouse species, two chipmunk species, and one squirrel and one rat species. Humans become infected with other hantaviruses when they inhale virus particles from the vapor of rodent urine and feces. Therefore, preventing future hantavirus outbreaks will require simple measures such as keeping houses free of rodents and avoiding contact with their wastes. Without the combined efforts of physicians, medical researchers, and ecologists, the origin of the hantavirus outbreak might still remain an enigma.

Conclusions

Examining illness from the ecological perspective will continue to provide insights into the origins, effects, and possible preventive measures for both established and emerging diseases. A disease inevitably results from an interaction between a host and a pathogen. This puts illness into an ecological framework, in which individual and population-level patterns have consequences for the emergence and maintenance of disease. Beyond the relatively simple two- or three-species interactions that are directly responsible for a disease, there is the vastly more complicated environment within which the interplay occurs. Even without wide scale anthropogenic disturbances, such as deforestation and global warming, diseases continue to emerge. Small perturbations to seemingly extraneous elements have led to unprecedented epidemics. The difference in the time scales over which hosts and pathogens evolve has also created great dilemmas for the treatment of disease. Although we have developed tools such as antibiotics and immunizations to fight disease, humans are unable to quickly adapt their immune systems to the ever-changing tactics of pathogens. Even as it has remedied and prevented illnesses, technology has contributed to the spread of disease. Without long-distance air travel, pathogens would remain in the places of their origin, rarely causing the outbreaks sensationalized in the recent years. With climate change, overcrowding in cities, deforestation, and the ease of long-distance travel, contact between humans and pathogens will continue to bring new and devastating diseases into our civilization. In light of this, neither ecologists nor physicians can afford to ignore one another. A combined effort between these communities will become more crucial as we move into the twenty-first century.

References

Acha, P.N., and Szyfres, B. (1980). *Zoonoses and Communicable Diseases Common to Man and Animals.* Pan American Health Organization/World Health Organization, Washington, D.C.

Anderson, R.M. (1991). Populations and infectious diseases: Ecology or epidemiology? *J. Anim. Ecol.* 60:1–50.

Anderson, R.M., and May, R.M. (1979). Population biology of infectious diseases: Part I. *Nature* 280: 361–367.

Anderson, R.M., and May, R.M. (1991). *Infectious Diseases of Humans: Dynamics and Control.* Oxford University Press, New York.

Anderson, R.M., Donnelly, C.A., Ferguson, N.M., Woolhouse, M.E.J., Watt, C.J., Udy, H.J., Mawhinney, S., Dunstan, S.P., Southwood, T.R.E., Wilesmith, J.W., Ryan, J.B.M., Hoinville, L.J., Hillerton, J.E., Austin, A.R., and Wells, G.A.H. (1996). Transmission dynamics and epidemiology of BSE in British cattle. *Nature* 382(6594):779–788.

Bannerjee, J. (1988). Kyasanur Forest Disease. In *The Arboviruses: Epidemiology and Ecology,* Vol. III (T.P. Monath, ed.), pp. 93–116. CRC Press, Boca Raton, Florida.

Barbour, A.J., and Fish, D. (1993). The biological and social phenomenon of Lyme disease. *Science* 260:1610–1616.

Bennett, F.J., Kagan, I.G., Barnicot, N.A., and Woodburn, J.C. (1970). Helminth and protozoal parasites of the Hazda of Tanzania. *Trans. R. Soc. Trop. Med. Hyg.* 64:857–880.

Black, F.L. (1966). Measles endemicity in insular populations: Critical community size and its evolutionary implication. *J. Theor. Biol.* 11:207–211.

Brown, A.J.L. (1990). Evolutionary relationships of the human immunodeficiency viruses. *TREE* 5:177–181.

Carter, S.D., Hughes, D.E., Bell, S.C., Baker, J.R., and Cornwell, H.J.C. (1990). Immune responses of the common seal (*Phoca vitulina*) to canine distemper antigens during an outbreak of phocid distemper viral infection. *J. Zool.* 222:391–398.

CDC. (1977). *Viral hemorrhagic fever: Sudan and Zaire.* Morbidity and Mortality Weekly Report 26(26):209–210.

CDC. (5/15/95). *Ebola Virus Hemorrhagic Fever: General Information.* CDC Press Release. Document #920001.

Cohen, J.E. (1995). Population growth and Earth's human carrying capacity. *Science* 269: 341–346.

Croll, N.A. (1983). Human behavior, parasites, and infectious diseases. In

Human Ecology and Infectious Diseases. (N.A. Croll and J.H. Cross, eds.). Academic Press, New York.

Daily, G.C., and Ehrlich, P.R. (1992). Population, sustainability, and Earth's carrying capacity. *BioScience* 42:761–771.

Dobson, A.P., and Carper, R. (1992). Global warming and potential changes in host–parasite and disease–vector relationships. In *Global Warming and Biological Diversity* (R.L. Peters and T.E. Lovejoy, eds.). Yale University Press, New Haven.

Dobson, A.P., and Carper, R. (1993). Health and climate change: Biodiversity. *Lancet* 342:1096–1099.

Ehrlich, P. R., Ehrlich, A.H., and Daily, G.C. (1993). Food security, population, and environment. *Population and Development Review* 19(1):1–32.

Fenner, F. (1982). Transmission cycles and broad patterns of observed epidemiological behavior in human and other animal populations. In *Population Biology of Infectious Diseases* (R.M. Anderson and R.M. May, eds.). Springer- Verlag, New York.

Fiennes, R. N. T. W. (1978). *Zoonoses and the Origins and Ecology of Human Disease.* Academic Press, London.

Garrett, L. (1994). *The Coming Plague: Newly Emerging Diseases in a World Out of Balance.* Farrar, Straus & Giroux, New York.

Gillett, J.D. (1974). In *The Effects of Meteorological Factors upon Parasites* (A.E.R. Taylor and R. Muller, eds.), Symposia of the British Parasitological Society 12, pp. 79–95. Blackwell Scientific, Oxford.

Haddow, A. J., Smithburn, K. C., Mahaffy, A. F., and Bugher, J. C. (1947). Monkeys in relation to yellow fever in Bwamba County, Uganda. *Transactions of the Royal Society of Tropical Medicine and Hygiene* 40:677–700.

Heide-Jørgensen, M.P., Härkönen, T., Dietz, R., and Thompson, P.M. (1992). Retrospective of the 1988 European Seal Epizootic. *Diseases of Aquatic Organisms* 13:37–62.

Heymann, D.L., Weisfeld, J.S., Webb, P.A., Johnson, K.M., Cairns, T., and Berquist, H. (1980). Ebola Hemorrhagic Fever: Tandala, Zaire, 1977–1978. *Journal of Infectious Diseases* 142:372–376.

Ho, D.D. et al. (1995). Rapid turnover of plasma virions and CD4 lymphocytes in HIV-1 infection. *Nature* 373:123–126.

Houghton, R.A., and Woodwell, G.M. (1989). Global climate change. *Scientific American* 260(4):36–44.

Krause, R.M. (1992). The origin of plagues: Old and new. *Science* 257:1073–1078.

Marshall, E. (1993). Hantavirus outbreak yields to PCR. *Science* 262:832–833.

McNeill, W.H. (1976). *Plagues and People.* Anchor Press/Doubleday, New York.

Mech, L.D., and Goyal, S.M. (1993). Canine parvovirus effect on wolf population change and pup survival. *J. Wildlife Dis.* 29:330–333.

Morbidity and Mortality Weekly Report. (1992). 40:505.

Morse, S., and Schluederberg, A. (1990). Emerging viruses: The evolution of viruses and viral diseases. *Journal of Infectious Diseases* 162:1–7.

Murphy, F. A. (1994). New, emerging, and reemerging infectious diseases. *Adv. Vir. Res.* 43:1–52.

Murray, K. (1995). A *Morbillivirus* that caused fatal disease in horses and humans. *Science* 268:94–97.

Myers, N. (1992). Synergisms: Joint effects of climate change and other forms of habitat destruction. In *Global Warming and Biological Diversity* (R.L. Peters and T.E. Lovejoy, eds.). Yale University Press, New Haven.

Nathanson, N., McGann, K.A., Wilesmith, J., Desrosiers, R.C., and Brookmeyer, R. (1993). The evolution of virus diseases: Their emergence, epidemicity, and control. *Virus Research* 29:3–20.

Nichol, S.T., Spiropoulou, C.F., Morzunov, S., et al. (1993). Genetic identification of a Hantavirus associated with an outbreak of acute respiratory illness. *Science* 262:914–918

Nowak, M. A. (1991). Antigenic diversity thresholds and the development of AIDS. *Science* 254:963–969.

Peters, R.L., and Lovejoy, T.E. (eds.). (1992). *Global Warming and Biological Diversity.* Yale University Press, New Haven.

Peterson, R.O., and Page, R.E. (1988). The rise and fall of Isle Royale wolves, 1975–1986. *J. Mamm.* 69:89–99.

Preston, R. (1994). *The Hot Zone.* Random House, Inc., New York.

Quetal, C. (1990). *History of Syphilis.* Polity Press, Great Britain.

Roelke-Parker, M.E., Munson, L., Packer, C., Kock, R., Cleaveland, S., Carpenter, M., O'Brien, S.J., Pospischil, A., Hofmann-Lehmann, R., Lutz, H., Mwamengele, G.L.M., Ngasa, M.N., Machange, G.A., Summers, B.A., and Appel, M.J.G. (1996). A canine distemper virus epidemic in Serengeti lions (*Panthera leo*). *Nature* 379:441–445.

Sankurathi, C.S., and Holmes, J.C. (1976). Effects of thermal effluents on the population dynamics of *Physa gyrina* Say (Mollusca: Gastropoda) at Lake Wabamum, Alberta. *Can. J. Zool.* 54:582.

Schantz, P.M. (1983). Human behavior and parasitic zoonoses in North America. In *Human Ecology and Infectious Diseases* (N.A. Croll and J.H. Cross, eds.). Academic Press, New York.

Simon, N. (1962). *Between the Sunlight and the Thunder: The Wildlife of Kenya.* Collins, London.

Smith, J.S., Reid-Sanden, F.L., Rounillat, L.F., Trimarch, C., Clark, K., Baer, G.M., and Winkler, W.G. (1986). Demonstration of antigenic variation among rabies virus isolates by using monoclonal antibodies to nucleocapsid proteins. *J. Clin. Microbiol.* 24:573–580.

Stommel, H., and Stommel, E. (1979). The year without a summer. *Scientific American* (June):176–186.

Stone, R. (1993). The mouse-pinon nut connection. *Science* 262:833.

Torrence, M.E., Jenkins, S.R., and Glickman, L.T. (1992). Epidemiology of raccoon rabies in Virginia, 1984 to 1989. *J. Wildlife Dis.* 28:369–376.

Walsh, J. F., Molyneux, D. H., and Birley, M. H. (1994). Deforestation: Effects on vector-borne disease. *Parasitology* 106:S55–S75.

Wei, X. et al. (1995). Viral dynamics in human inmunodeficiency virus type 1 infection. *Nature* 373:117–122.

Wenzel, R.P. (1994). A new hantavirus infection in North America. *New England Journal of Medicine* 330:1004–1005.

WHO. (1990). *Tropical Disease in 1990.* World Health Organization, Geneva.

WHO. (5/22/95). Press Release.

CHAPTER 5

Biodiversity Loss in and around Agroecosystems

JOHN VANDERMEER

In 1954, a clever advertisement appeared in the Spanish language edition of *Life Magazine*. A boll weevil was attached with an insect pin to a map of Central America, and a short explanation noted "The case of *Anthomonis grandis*" (the boll weevil), cause of all the problems in cotton production. It went on to promise an end to this plague through the use of the insecticide Aldrin. It's message was clear. The problem was the boll weevil, neatly pinned to a map of Central America, and the solution was Aldrin, made by the Shell company.

A dozen years later a billboard outside of Guatemala City noted a very similar problem with cotton production. An insect pest and its insecticide solution. But there were two crucial differences here: the pest was no longer the boll weevil, it was the *bellotero,* the cotton boll worm, and the solution was Sevin rather than Aldrin, Union Carbide replacing Shell as the savior of the day. "Put an end to the bollworm with Sevin," was the message.

Some 15 years later, I took a photograph of a billboard in Managua, Nicaragua. The rhetoric was much more catchy—"An explosive death to pests," it said—but the underlying message was the same. A new chemical to kill a new group of pests, by now too numerous to mention explicitly by name.

This basic pattern has now been repeated in many places throughout the world. A pesticide kills natural enemies and provokes resistance in the pest it was intended to kill. The absence of natural enemies allows formerly benign insects to increase their population levels to become pests and permits an especially large explosion of those pests that acquire resistance, since the pesticide eliminates their natural controlling factors, the predators and parasites that normally regulate their populations. This pattern is known as the pesticide treadmill. The case of cotton in Central America is perhaps the best known example (Sweezy and Daxl 1983).

This is the transformation of an agroecosystem, its modernization to hear the technocrats tell it. And along with that transformation has come a reduction in biodiversity. A host of predatory and parasitic arthropods were removed from this system—we don't know how many. With hindsight we see clearly that this diverse assemblage of arthropods was functional in a very profound sense, and its loss created an ever worsening problem, to the point that the cotton industry of Guatemala, El Salvador, and Nicaragua virtually disappeared. Those three countries are well-known for their economic devastation, for their continuing political violence, and for the human health problems associated therewith. Such problems are due to a variety of complicated forces, to be sure. But among those forces was this dramatic ecological/economic force, which at its base is the loss of biodiversity through agroecosystem transformation.

The state of human health indicators is directly and indirectly related to the changes in biodiversity associated with agricultural transformation. Traditional farmers produce a variety of crops, a practice that is thought to contribute positively to general nutritional status (Dewey 1990), while the tendency toward monocultural production may reduce the diversity of diet, at least for rural populations. Farmers and farmworkers are poisoned by the increasing quantities of pesticides that are necessitated by the almost inevitable operation of the pesticide treadmill (Perfecto 1992), a direct consequence of the loss of biodiversity. Future benefits of biodiversity, now well publicized as new genes for crop improvement and new molecules with pharmacological potential, exist as much in agroecosystems as in other systems (perhaps more so, as suggested below), and the negative health consequences of losing this biodiversity are identical to the consequences of losing the biodiversity of more charismatic ecosystems such as rain forests. Finally, and most important, the same forces that led to our particular trajectory of modernism have also led to impoverishment of many rural populations and the spread of urban shanty towns, the human health consequences of which are well-known (Garrett 1994; see also Janzen, Chapter 14).

There are three primary issues involved in the interrelation of biodiversity and agriculture. First, there is far more biodiversity in agroecosystems than is usually acknowledged by those concerned with biodiversity loss.

Second, agriculture is probably the single most important force contributing to the general global decline in biodiversity. Third, the health of the human population is both directly and indirectly affected by agroecosystem biodiversity—by the biodiversity contained therein as well as by the biodiversity losses provoked by agriculture's transformation.

Biodiversity in Agroecosystems

The biodiversity that is planned by a farmer ranges from the famous backyard gardens of Java, with more than 40 species of productive plants specifically tended by the family (Marten 1990), to the infamous pesticide drenched cotton fields of Nicaragua, where the only biodiversity remaining is cotton and its many devastating pests which no longer are killed by any pesticides.

While there are many intermediate stages, this range from the highly diverse Javenese garden to the Nicaraguan cotton field represents a change in intensification of production, frequently simply referred to as "modernization" of agriculture. I believe that the term "modernization" is ill chosen since I hope we can some day look to a modern, or, perhaps, a postmodern, agriculture in which the rules of production will be different. But given contemporary political and economic realities, we can clearly see a gradient of production activities from a traditional backyard garden in Java to a traditional coffee farm in El Salvador, to a slash and burn tradition in Mexico, to an organic vegetable farm in Costa Rica, to a pesticide-drenched cotton farm in Nicaragua. This is what is frequently referred to as "an intensification gradient," from the least intense backyard garden to the most intense cotton farm (Swift et al. 1996). While the term "intensification" may not be the best, that is what is normally used in historical treatments of agricultural development, and I continue its usage here.

With respect to biodiversity, a simple count of the types of plants purposefully tended in each of these systems is almost surely a monotonically decreasing function (Figure 5.1). One can argue about the exact shape of the curve, which, of course, depends on how you scale the intensification gradient, but the general pattern is a monotonically decreasing curve of biodiversity as agriculture is intensified. Of course, this need not be the case, it simply has been, on average, over the past 500 years of agricultural development.

There is now an enormous literature on the practical significance of this trend in biodiversity loss. As traditional varieties of crops are replaced with modern varieties we are losing much genetic material, some of which

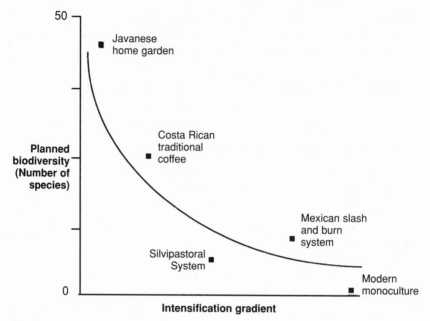

FIGURE 5.1

Approximate position of various agroecosystems on the intensification gradient.

could be important in future plant improvement programs (Merrick 1990). As monocultures replace polycultures we are losing production tradition, much of which could be important in future production planning (Altieri 1990). In short, the tendency to reduce the biodiversity of agricultural production poses a variety of problems, all of which have been explicitly articulated in the past, and all of which are under intense scrutiny by researchers across the globe (Kloppenburg and Kleinman 1987).

However, this focus is perhaps a bit myopic. The crops planted by the farmer represent only a fraction of the biodiversity that is really in the agroecosystem. The previous example is a case in point. The original cotton agroecosystems of the 1950s in Central America were viewed by farmers as having two species, cotton and boll weevils. But we now know that they had a large biodiversity including at least 30 species of arthropods that eventually became pests in the system and at least another 30 species of predators and parasites that evidently held those potential pests under control before their death at the hands of the pesticide enthusiasts the 1950s. This is what is refered to as the "associated biodiversity" (Swift et al. 1996).

As the farmer plants crops, he or she is introducing the "planned" biodiversity into the agroecosystem. But inevitably other organisms enter the system—the "associated" biodiversity. Both planned and associated may be involved in ecosystem function, but the effect of associated biodiversity, which is the *indirect* effect of the planned biodiversity, may be largely invisible, at least initially, to the farmer. And if the cotton agroecosystem is typical, this associated biodiversity may have important functional consequences, as suggested in Figure 5.2.

If we now wish to explore what pattern of associated biodiversity exists vis-à-vis the intensification gradient, we encounter a dramatic absence of relevant data. It has not been very popular to study associated biodiversity *per se* in agroecosystems, and with a handful of exceptions (Pimentel et al. 1992; Perfecto et al. 1996) we know very little. We can, of course, speculate based on some simple biological intuition. The pattern is likely to take on one of four distinct forms (Figure 5.3). Form I or III is what many proponents of organic agriculture suggest is true. As soon as the process of intensification begins, there is a dramatic loss of associated biodiversity, perhaps leveling off at some intermediate intensification level, as in curve III, perhaps not, as in curve I. The other extreme is that as the traditional system is intensified, not much happens to the biodiversity until some threshold level of intensification is attained, at which point a dramatic decline occurs, perhaps occurring at some intermediate level of intensification as

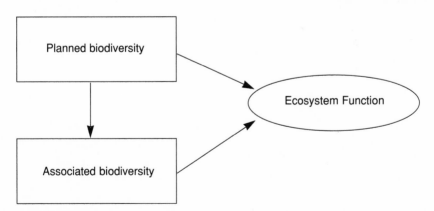

FIGURE 5.2
Theoretical relationship between planned and associated biodiversity and their relationship to ecosystem function. Planned biodiversity directly affects ecosystem function, but also indirectly affects it through the effect of the associated biodiversity.

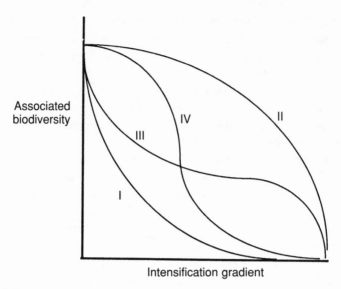

FIGURE 5.3
Possible patterns of change in associated biodiversity with respect to the intensification gradient.

in curve IV, or at some higher level, as in curve II. As to which of these
curves represents reality, we have no idea. The virtual absense of interest on
the part of conservationists in the question of associated biodiversity in
agroecosystems has resulted in a virtual absence of research activity aimed
at this most common of all ecosystems. On the other hand, the few studies that *have* been done are highly suggestive.

For example, Perfecto et al. (1996) examined general arthropod biodiversity in the coffee agroecosystem, comparing the traditional coffee production system with more modern, intensified ones. Traditional coffee production in Central America has been characterized by a high level of
planned biodiversity. A traditional farm typically has a large mixture of
shade trees and fruit trees, coupled with plantains and other understory
crops, in addition to the coffee bushes. When a shade tree is harvested for
lumber, the light gap so created is sometimes used as a site for growing
basic grains while the new shade tree is growing to fill the canopy again,
not unlike gap phase dynamics in a natural forest. These traditional production systems thus have various features that are reminiscent of a natural
forest.

At the other extreme, all the Central American countries have been promoting a highly intensive form of production, based on modern varieties

that respond favorably to light. This form of production has very low planned biodiversity, is heavily dependent on chemical control of weeds, requires high application rates of nitrate, and calls for a great deal of labor for maintenance. Between these two extremes a variety of production forms can be found.

In coffee farms ranging from the traditional to the modern, arthropod biodiversity was sampled, using a variety of techniques ranging from insecticidal fogging, to pitfall traps. The results have been somewhat surprising. From insecticidal fogging of shade trees, the number of beetle species ranges from a high of 128 species in a single tree in the traditional system, to zero species, in the modern system—obviously, since the shade trees have been removed from the system, there can be no arthropods in them. Similarly, the number of wasps ranges from a high of 103 species in a single shade tree, to zero in the modern system. There are over 25 species of ground foraging ants from the traditional system, while the modern system counts a total of 8 species, with just two dominating over 95% of the ground area in any given plantation (Perfecto et al. 1996; Perfecto and Snelling 1995; Perfecto and Vandermeer 1994). The pattern seems clear. There is a tremendous loss of associated biodiversity as this system is intensified.

The function of all of this biodiversity remains a mystery. There may be potential pests lurking among the hundreds of species of beetles, but there also may be potential natural enemies among the hymenoptera. Currently there is a devastating beetle pest that is making its way through Central America. It is a small beetle that burrows into the coffee bean and can easily cause well over 50% productivity loss in an infected plantation. Recent early work in northern Nicaragua has suggested that organic coffee farms are experiencing far less of a problem with this pest than the modernized ones. Furthermore, it appears that the ground foraging ant community may be responsible for this effect (J. Monterrey, personal communication). The beetle pest must reside in the litter of the plantation during the part of the year when fruits are not developing. Apparently the efficiency of ant predation is reduced in the modern plantation, presumably because of the dramatic change in the biodiversity.

The associated biodiversity of arthropods in the coffee agroecosystem is only an isolated example. Biodiversity of microbes in the soil, of associated plant matter, and of vertebrates also deserve detailed study, in ecosystems ranging from backyard gardens to migratory slash and burn agriculture to various levels of intensification. The two basic questions—what is the nature of the associated biodiversity and what is the function of that biodiversity—deserve to be asked of all taxa in all of these ecosystems.

Throughout the world, we are witnessing a growing realization that agricultural production must be put on a more sustainable footing than has developed since World War II. If the biodiversity of more traditional production systems, both planned and associated, is lost, what might be the effect on our ability to modernize in an ecologically sustainable way? Are we eliminating the potential natural enemies of the future? Are we losing the genetic material that may form the basis of pest-resistant crops in the future? Are we losing the traditional knowledge that evolved through hundreds or thousands of years of cultural evolution? These are all questions associated with biodiversity loss in transforming agroecosystems, and they are all crucial to the future of sustainable agricultural development in the tropics. Unfortunately they do not attract a great deal of attention, either in the public imagination or in the research agendas of ecologists concerned with biodiversity conservation.

Agriculture As a Force in Biodiversity Loss

Charismatic megafauna such as elephants and jaguars attract a great deal of attention. If I were to choose between a job as a fundraiser to save the elephants, or a job as a fundraiser to save the ants that eat the beetle that eat the coffee beans, I would certainly be economically better off with the elephants. This is an issue that has seen a great deal of discussion and analysis. The loss of charismatic megafauna—elephants, lions, spotted owls, snow lepoards—invokes in all but the most cynical of us deep feelings of sadness. I too have gone on the African game drives, and I too lament the almost certain extinction of these remarkable creatures.

Similarly, charismatic ecosystems, such as rain forests and coral reefs, also attract a great deal of attention. Fundraisers who seek to save the rain forest have much more success than fundraisers who seek to save the traditional soil management system of the Almolonga valley of Guatemala (Wilkin 1987). These are the charismatic ecosystems, the prime example of which is the tropical rain forest. Only a brute is not moved to tears when contemplating the loss of this cradle of biodiversity, the closest thing we have in the natural world that feeds our deep cultural belief in Eden (Slater 1995). Yet the lay public does not fully understand what every tropical ecologist knows quite well. The loss of this charismatic ecosystem is largely a function of agricultural transformation, frequently the same sorts of agricultural transformations that are involved in the intensification gradient discussed earlier (Vandermeer and Perfecto 1995).

Recent studies in the Atlantic coast rain forests of Nicaragua are instructive. Several permanent plots have been monitored since the landfall of Hurricane Joan in 1989 (Vandermeer et al. 1990; Vandermeer et al. 1995). The main conclusion of that work has been that first, the hurricane did a great deal of physical damage with 100% of the trees experiencing dramatic physical damage. But second, and most importantly, the recuperation of the forest after the hurricane has been dramatic. The forest today is still quite clearly a severely damaged forest, but the structural features of the old growth forest have been essentially restored, just six years after the hurricane, and the forest is well on its way to establishing a structure that is very similar to that which existed prior to this devastating event.

In modern ecological theory such disturbances are regarded as a normal part of ecological dynamics. Indeed many ecologists hold that such disturbances are a necessary part of biodiversity maintenance (Connell 1978; Huston 1979; Vandermeer et al. 1996). But for our purposes here I merely wish to point out that the devastation of a hurricane is in some ways similar to the devastation of a logging operation, with important obvious exceptions. In short, a selectively logged area, or even a clear-cut area, could recuperate much as the forest damaged by the hurricane has recuperated.

But this is not really the crucial issue given the realities of the contemporary world. For example, shortly after establishing three permanent plots at one site in the hurricane-damaged rain forest in Nicaragua, a logging road was constructed nearby. That forest plot today is a cassava field. A group of peasant farmers used the logging road to gain access to the area, cut the forest that was in the process of recuperation, burned the slash, and planted cassava. This is another form of disturbance. Obviously this form of disturbance has a far more significant direct effect on the forest and its ability to grow back than did the hurricane, and much more than a simple lumbering operation. However, even this form of disturbance, if viewed over a long-term perspective, does not result in permanent damage. An abandoned slash and burn system, five or six years after abandonment, begins looking remarkably like a forest. While any tropical ecologist can see that it is not really a tropical rain forest, it nevertheless is an early successional stage on its way to becoming one. And, provided the appropriate seed sources are available, within a short 90 or 100 years it is likely to become what most people would refer to as an old growth tropical rain forest.

There is yet a more extreme form of disturbance, a modern technified export agriculture operation—for example, the modern banana plantation. To establish a banana plantation, a truly spectacular transformation of nature is required. Massive movement of earth is required for leveling the land and constructing drainage ditches. Cement footings are positioned to hold metal monorails that stream throughout the plantation. The forest is not simply cut and burned, the landscape itself is dramatically altered. And today a modern banana plantation has the soil subsurface crisscrossed with plastic drainage tubes. Production includes the extensive use of herbicides when the plantation is initiated and a continuous intensive use of pesticides for pest control, especially of fungi and nematodes. Phosphorous is mined excessively from an already phosphorous-poor soil. In short, this is a disturbance event whose impact on biodiversity is not only immediate, but may take on a more permanent nature. It is not at all clear that one will see a forest in 100 years after abandonment, nor can we say for sure that eventually the same forest will return. In short, this modern agriculture does extreme damage to the biodiversity of the system, and probably extends the time of recuperation very significantly. Many other forms of so-called modern agriculture, from intensive cattle production to chocolate, to pineapple, have a similar structure.

Thus we can see again the effect of management intensification on planned and associated biodiversity. Logging and leaving a forest does little ultimate damage to biodiversity since most tropical forests are well-adapted to similar natural disturbance regimes. Peasant farming certainly does more direct and indirect damage, but likely far less than a modern chemically intensive system.

However, there is more to the story than this simple intensification gradient. There are important and inevitable higher order interactions among various points along that gradient. The logging road that was built near the designated three plots provided access to the area for peasant farmers. The banana expansion currently underway in eastern Costa Rica attracts workers from all over Costa Rica and Nicaragua, creating more peasant farmers every time the world price of bananas declines. For example, in the Sarapiquí region of Costa Rica there is a recently established community of ex-banana workers. There are currently 350 families living in this community, trying to farm on very poor soils. New families arrive to the community every day, mainly displaced banana workers, many of whom have come from Nicaragua, escaping an economic condition created by 15 years of some very cynical international diplomacy. And there agricultural activities have now expanded up to the fence line of two of the important biological reserves in the region (Vandermeer and Perfecto 1995). Logging roads are being expanded in eastern Nicaragua and an estimated 90% unemployment rate in the main city of Bluefields pushes more and more workers into the countryside looking for a piece of forest to cut down to establish a homestead. Workers from Nicaragua continue to arrive in the northern counties of Costa Rica, where they are hired by the banana companies for a while, then released to fare for themselves in an ecosystem that is difficult, at best, to do agriculture. It is not an encouraging scene.

The Connection to Human Health

As outlined above, agriculture and biodiversity are interrelated in two general ways—(1) agroecosystems contain a great deal of biodiversity, both that planned by the farmer and the associated biodiversity that arrives independently of the farmer's direct action, and (2) the particular socioeconomic dynamics of the contemporary world causes biodiversity loss in agro- and other ecosystems through a systematic transformation of agroecosystems. The health of the human population is related to both of these issues.

Along with the transformation of agroecosystems to more high input technified forms has been an inevitable tendency to produce crops in monocultures. Such a change has suggested to some that there would be a direct change in human nutrition, given that a farm family would have less direct access to a diversity of food types, and would be especially vulnerable to economic instability, being able to purchase at most basic staples in times of economic crisis. Actually demonstrating that such an effect exists in the real world has not been as easy as speculating about its existence. Dietary diversity may actually increase as agroecosystem diversity decreases, due to a greater purchasing power associated with higher participation in cash cropping. This was the pattern found in a Peruvian study (DeWalt 1983), although the increased dietary diversity did not correlate with better nutrition. Frequently a decrease in dietary nutrient quality is not a direct consequence of lowered biodiversity, but rather an indirect consequence of socioeconomic changes that accompany the change in the agroecosystem. For example, Dewey (1980) found that the modern cash crop monoculture which replaced the diverse traditional systems in southern Mexico brought with it the ability to purchase cheap substitutes for traditional foods. The diet may have been just as diverse, but the nutritional quality of substitute items was poorer.

A careful look at published studies (Dewey 1990) leaves much doubt about any direct connection between dietary quality and agroecosystem biodiversity. Had all things been equal, the simplistic notion that producing fewer types of food items would have led to consumption of fewer types of dietary items may have been true. But the economic, sociological, and political complexities of the real world suggest that a great many indirect links are inevitably involved in such issues, and generalizations are simply impossible.

On the other hand, there is growing acceptance of the idea that economic forces, which ultimately translate into human health factors, are involved in agroecosystem biodiversity. Subsistence and small scale farmers are not overly concerned with maximizing yields of crops, but rather pursue a strategy of economic risk reduction (Vandermeer 1989; Vandermeer and Schultz 1989). If the market for coffee suddenly declines, the traditional farmer may avoid economic disaster by bringing a harvest of plantains or fruits to market.

The other component of biodiversity, associated biodiversity, probably is more directly related to human health issues. Simply as another ecosystem, the agroecosystem has a great deal of biodiversity associated with it, as noted earlier. All arguments about biodiversity function and potential are, in principle, just as applicable to the agroecosystem as any other ecosystem. There is no inherent reason that biodiversity prospecting for pharmacological products should be confined to pristine ecosystems, and it is difficult to imagine a serious argument that we are more likely to find a chemical active against HIV or tumors in a rain forest beetle as opposed to one of the hundreds of beetle species that occupy the neighboring peasant farm. Indeed the fact that a greater percentage of the earth's terrestrial surface is in agroecosystems than any other type of ecosystem (Pimentel et al. 1992) might suggest that biodiversity prospecting here could even have more potential than the current plans to sample the very restricted set of physical conditions represented in the world's pristine preserves.

However, the force affecting human health most directly is probably the loss of ecosystem function implied by the loss of biodiversity. As described earlier, the pesticide treadmill is a case of biodiversity loss promoting a change in agroecosystem management that has resulted in untold numbers of pesticide poisonings (Hilje et al. 1987; Thrupp 1988). The simplification of the agroecosystem that results from pesticide applications almost inevitably causes the need for yet more pesticides, either more frequent applications or more toxic agents, and sometimes both (Wright 1990; Perfecto 1992). The human health consequences could hardly be more dire.

In addition to the direct effects of pesticide poisonings have been the well-publicized pesticide residues that may endanger the health of consumers worldwide (Weir and Schapiro, 1981). Even more troubling are the indirect effects on disease vectors, so eloquently traced in the case of cotton farming and Malaria in Central America (Chapin and Wasserstrom 1983; ICAITI 1977). Resistance to DDT by malaria mosquitoes was apparently provoked by the massive pesticide spraying program in cotton fields.

Finally, and probably more important than the above examples, the same forces that are causing biodiversity loss in agro- and other ecosystems are responsible for much of the negative changes in human health (Haila and Levins 1992; see also Dobson et al., Chapter 4), both contemporaneously and in the foreseeable future. The imposition of export agriculture which forces former landowners to become temporary proletarians, releasing them during economic hard times to find productive lands again, following lumbering roads to scrape out a small farm from a rain forest, is the typical pattern in today's world. Economic impoverishment is probably the major cause of negative human health conditions. It is also the major cause of biodiversity loss.

Conclusions

Unfortunately, scenes of biodiversity loss and economic impoverishment are not unusual. Indeed they seem to be the rule rather than the exception, in my experience. On the other hand, one can cite a few examples that appear more encouraging. One such example that is, to my mind, one of the most hopeful in all of Central America right now is the advances that have been made since 1980 in Integrated Pest Management in Nicaragua. Mainly inspired by the work of Kristen Nelson (1995), the current IPM program sponsored by CATIE has made remarkable advances in reeducating scientists and technicians regarding the need to work closely with, and on an equal footing, with small farmers. Modeling her work after the writings of Robert Chambers, Nelson was able to put together an experimental team of scientists, technicians, and farmers to approach the problem of pest management in a remarkably unique way. Her results were nothing short of spectacular and the program in Nicaragua now is rapidly expanding. In its area of influence, this unique approach to research and development is creating a new atmosphere in which the sustainability of high input agriculture is under serious question, and in which visions of an alternative future are growing, not just among the educated elite, but among farmers who see the new potential pathways to development. If it continues, its effect could become even more generalized, with indirect effects on the biodiversity of Nicaragua's most common ecosystem. The jury is still out, but here is a hopeful, if small, example.

A second hopeful example is the transformation currently underway in Cuba. Faced with a virtual elimination of all of its high technology inputs from the Soviet Union and the Eastern Block since 1989, Cuba has been forced to transform its agriculture wholesale. All pest control is currently via biological control or IPM, soil amendments are almost entirely organic, weed management is based on managing plant competition, *Mycorrhizae* substitute for imported phosphorous, and vermiculture makes mounds of worm compost from farm and city waste (Perfecto 1994; Rosset and Benjamin 1994; Vandermeer et al. 1993). Polycultural production has become the rule in vegetable and basic grain production, and experiments are underway with intercropping in sugar cane fields. The transformation is still in its infancy, but represents a microcosm (if 11 million people can be called a microcosm) worthy of considerable study. At least so far, Cuba seems on the road to recovering some of the biodiversity it had lost due to the high input nature of the agriculture it had developed in the previous 30 years.

What, under current global conditions, might we hope for in biodiversity conservation, considering the entire terrestrial landscape? First, the preservation of whatever pristine ecosystems and charismatic fauna that remains on this earth is an obvious goal, on par, I believe, with the preservation of the world's art treasures. Much has been written about it. An endless series of nature shows appear weekly on PBS about it. Who can be against it, in principle? But second, and of critical importance in today's rapidly changing world, is what is happening *outside* of the world's biological preserves—in agroecosystems and other managed systems. As managed ecosystems are intensified, biodiversity is lost. At what rate? We do not know. At what cost to the productive value of those ecosystems? We do not know. At what cost to human health? We do not know. At what cost to long-term sustainability? We do not know. And eventually, at what cost to the security of the very national parks and biological preserves that we take for granted today? We do not know. We know very little about biodiversity in agro- and other managed ecosystems.

Former astronomer and current ecologist Robert May once noted that we know more about stellar diversity in the sky than about biodiversity on earth. I wish to add to that lament that we probably know more about biodiversity in the ever decreasing fraction of the earth's terrestrial surface that we refer to as pristine than we know about the biodiversity and its function in the managed ecosystems that provide our basic sustenance and cover a vast majority of the earth's terrestrial surface. We court disaster with such ignorance.

References

Altieri, M. A. (1990). Why study traditional agriculture? pp. 551–564 in Carroll, C. R., J. H. Vandermeer, and P. Rosset (eds.). *Agroecology,* McGraw-Hill, New York.

Chapin, G. and R. Wasserstrom. (1983). Pesticide use and malaria resurgence in Central America and India. *Soc. Sci. Med.* 17:273–290.

Connell, J. H. (1978). Diversity in tropical rain forests and coral reefs. *Science* 199:1302–1310.

DeWalt, K. M. (1983). Income and dietary adequacy in an agricultural community. *Soc. Sci. Med.* 17:1877–1886.

Dewey, K. G. (1980). The impact of agricultural development on child nutrition in Tabasco, Mexico. *Med. Anthro.* 4:21–54.

Dewey, K. (1990). Nutrition and agricultural change. pp. 459–480 in Carroll, C. R., J. H. Vandermeer, and P. Rosset (eds.). *Agroecology,* McGraw-Hill, New York.

Garrett, L. (1994). *The Coming Plague: Newly Emerging Diseases in a World Out of Balance.* Farrar, Straus & Giroux, New York.

Haila, Y. and R. Levins. (1992). *Humanity and Nature: Ecology, Science and Society.* Pluto Press, London.

Hilje, L., L. Castillo, and L. A. Thrupp. (1987). El uso de los plaguicidas en Costa Rica. Editorial, Universidad Estatal a Distancia, San José, Costa Rica.

Huston, M. (1979). A general hypothesis of species diversity. *Am. Nat.* 113:81–101.

ICAITI (Instituto Centroamericano de Investigación y Tecnología Industrial). (1977). An environmental and economic study of the consequences of pesticide use in Central American cotton production. ICAITI Proj. No. 1412: 1–295.

Kloppenburg, J. and D. L. Kleinman. (1987). Seeds of struggle: The geopolitics of genetic resources. *Technology Review* 90:47–53.

Marten, G. G. (1990). Small-scale agriculture in southeast Asia. pp. 183–200 in Altieri, M. A. and S. B. Heccht (eds.). *Agroecology and Small Farm Development.* CRC Press, Boca Raton, FL.

Merrick. L. C. (1990). Crop genetic diversity and its conservation in traditional agroecosystems. pp. 3–11, in Altieri, M. A. and S. B. Heccht (eds.). *Agroecology and Small Farm Development.* CRC Press, Boca Raton, FL.

Nelson, K. (1995). Farmer-first technological innovation in tomato production in the Sebaco Valley of Nicaragua. *Ag. and Human Values* 12(4).

Perfecto, I. (1992). Pesticide exposure of farm workers and the international connection. pp. 177–203 in Bryant, B. and P. Mohai (eds.). *Race and the Incidence of Environmental Hazards.* Westview Press, Boulder, CO.

Perfecto, I. (1994). The transformation of Cuban agriculture after the Cold War. *Am. J. of Alt. Agric.* 9:98–108.

Perfecto, I. and R. Snelling. (1995). Biodiversity and the transformation of a tropical agroecosystem: Ants in coffee plantations. *Ecol. Applications* 5:1084–1097.

Perfecto, I. and J. H. Vandermeer. (1994). The ant fauna of a transforming agroecosystem in Central America. *Trends in Agricultural Science* 2:7–13.

Perfecto, I., J. H. Vandermeer, P. Hansen, and V. Cartin. (1996). Arthropod biodiversity loss and the transformation of a tropical agroecosystem. *Biodiversity and Conservation* (in press).

Pimentel, D., V. Stachow, D. A. Takacs, H. W. Burbaker, A. R. Dumas, J. H. Meaney, J. A. S. O'Neal, D. E. Onsi, and D. B. Corzinus. (1992). Conserving biological diversity in agricultural/forestry systems. *Bioscience* 42:354–362.

Rosset, P. and M. Benjamin. (1994). *The Greening of the Revolution: Cuba's Experiment with Organic Agriculture.* Ocean Press, Melbourne.

Slater, C. (1995). Amazonia as edenic narrative. pp. 114–131 in Cronon, W. (ed.). *Uncommon Ground: Toward Reinventing Nature.* Norton & Co., New York.

Swezey, S. L. and R. G. Daxl. (1983). Breaking the circle of poison: The integrated pest management revolution in Nicaragua. Institute for Food and Development Policy, San Francisco.

Swift, M., J. H. Vandermeer, R. Ramakrisnan, C. Ong, J. Anderson, and B. Hawkins. (1996). Biodiversity and agroecosystem function. In Mooney, H. A. (ed.) *Biodiversity and Ecosystem Function.* SCOPE. John Wiley, Chichester, U.K.

Thrupp, L. A. (1988). Pesticides and policies: Approaches to pest control dilemmas in Nicaragua and Costa Rica. *Latin American Perspectives* 15:37–70.

Vandermeer, J. H. (1989). *The Ecology of Intercropping.* Cambridge Univ. Press. Cambridge, U.K.

Vandermeer, J. H. and I. Perfecto. (1995). Breakfast of biodiversity: The true causes of rain forest loss. Institute for Food and Development Policy, Oakland, California.

Vandermeer, J. H. and B. Schultz. (1989). Variability, stability and risk in intercropping: Some theoretical explorations. In Gliessman, S. (ed.). *Approaches in the Ecological Study of Agriculture.* Ecological Studies Series, Springer-Verlag, New York.

Vandermeer, J. H., N. Zamora, K. Yih, and D. Boucher. (1990). Regeneración inicial en una selva tropical en la costa caribeña de Nicaragua después del huracan Juana. *Revista Biología Tropical (Costa Rica)* 38:347–359.

Vandermeer, J. H., J. Carney, P. Gesper, I. Perfecto, and P. Rosset. (1993). Cuba and the dilemma of modern agriculture. *Agriculture and Human Values,* 10:3–8.

Vandermeer, J. H., M. A. Mallona, D. Boucher, K. Yih, and I. Perfecto. (1995). Three years of ingrowth following catastrophic hurricane damage on the Caribbean Coast of Nicaragua: Evidence in support of the direct regeneration hypothesis. *J. of Tropical Ecology* 11:465–471.

Vandermeer, J. H., D. Boucher, I. Perfecto, and I. Granzow. (1996). Theory of periodic disturbance and the preservation of species diversity: Evidence from the rain forest of Nicaragua subsequent to Hurricane Joan. *Biotropica* (in press).

Weir, D. and M. Schapiro. (1981). Circle of poison: Pesticides and people in a hungry world. Institute for Food and Development Policy, Oakland, California.

Wilkin, G. (1987). *Good Farmers.* University of California Press, Berkeley.

Wright, A. (1990). *The Death of Ramón González: The Modern Agricultural Dilemma.* Univ. of Texas Press, Austin.

PART II

Drug Discovery from Biological Diversity

Loss of current and future sources of medicines is one of the most important repercussions of biodiversity loss. Current and projected species extinctions may represent the loss of cures for diseases that kill thousands and thousands of people each year. In the United States alone over 40,000 people die of AIDS every year. Again in the United States alone 538,000 people died of cancer in 1994, while 1,208,000 new cases were reported in the same year. The situation is far more grave in most other parts of the world. Even diseases we once thought of as cured may still threaten our well-being. Tuberculosis and numerous other "curable" diseases have reemerged in both the United States and abroad, and resistance of many of these diseases to previously effective therapies is climbing dramatically. Hence the search for new treatments is as important as ever.

Although natural products have formed the basis of medicinal therapies for millennia, investment in this area by the major pharmaceutical companies waned in the seventies and eighties and resurged only in the last decade. The resurgence of natural products is due in part to technological advances that have made the research process more efficient, and in part due to increasing need for novel therapies to treat increasingly resistant pathogens and newly emerging diseases. Very recent technological advances in chemistry have led some to question the long-term role of natural products once again. In fact, most researchers agree that natural products chemistry and the developing science of combinatorial chemistry are complementary techniques. Natural products will, in all likelihood, continue to be the ultimate source of the novel structures from which new drugs are designed.

In this section Francesca Grifo and colleagues (Chapter 6) present a new study which documents the contribution of natural products to the most prescribed drugs in the United States. This is followed in Chapter 7 by an in-depth discussion by Cathy Laughlin and Alex Fairfield of the historical role of natural products in the development of antiinfective agents. Anthony Artuso (Chapter 8) completes the section with an analysis of the economic values of the chemical diversity that the natural world continues to offer and some suggestions as to how we may utilize those values to stimulate the sustainable use and conservation of biodiversity.

CHAPTER 6

The Origins of Prescription Drugs

FRANCESCA GRIFO, DAVID NEWMAN, ALEXANDRA S. FAIRFIELD,
BHASWATI BHATTACHARYA, AND JOHN T. GRUPENHOFF

Introduction

Extinction of biological diversity risks the loss of the raw materials for existing and new weapons in the fight to alleviate human suffering and prevent death. This is true for both the products of pharmaceutical companies, and of traditional medicine, still an important source of medical care for much of the world's population. As a result, one of the primary components of any discussion of the value of biodiversity is both its past and future value as a source of therapeutics. The question of past, current, and future values of biodiversity to the pharmaceutical industry has been addressed many times (Artuso, Chapter 8; Duke 1993; Farnsworth 1977; Farnsworth et al. 1985; Principe 1996; and Simpson 1996). Although the use of active compounds from natural sources has varied according to the economics and trends of the marketplace and methodologies and discoveries of industry, one central point has remained. Whether synthesized based on natural compounds or from natural precursors, compounds taken from biodiversity are critical to our health. In this chapter we present a simple, yet quantifiable and repeatable analysis of the origins of the top 150 prescribed drugs in the United States in 1993 in order to demonstrate this point.

An overview of the drug discovery process is presented to provide the background necessary to understand the terms and concepts later in the chapter. It also demonstrates that although the process as it occurs in pharmaceutical firms is highly sophisticated and technology dependent, like all biotechnology, it remains equally dependent on raw materials. In particular, it requires the input of novel molecules most reliably found in nature.

The Drug Discovery Process

The process of developing a new pharmaceutical from a natural product in the United States, from initial discovery to final approval by the Food and Drug Administration, is long, complex, and expensive. Figure 6.1 is a generalized depiction of the drug discovery and development process.

Although this figure illustrates a sequential progression, drug discovery is not necessarily a linear process: selection criteria and testing options can create a complex "decision tree" that is unique to each screening program. In addition, there are numerous pitfalls in the process, such as missing a "hit" because of limitations of the screening process, or realizing the active principle is already a well-known drug from a different source. Also, some drugs may simply be too expensive to produce for the projected share of the market and thus are never commercialized. These and other challenges will be described more fully below.

The first step is to identify both the selection process and the natural sources to be screened, i.e., taxonomic, ethnobiological, or random samples of plants, microbial fermentation broths, insects, etc. The success of identifying potential new drugs in higher plants and other natural sources next depends on the preparation of extracts. Chemicals differ in their solubility in various types of solutions. Although the choice of solvents can vary widely, typically extracts are prepared by mashing or grinding a natural product in organic and aqueous solvents (separately) such as methanol and water. Screening programs differ widely in the treatment of soluble and insoluble material from extractions, but usually several phases are examined for activity.

Determining whether an extract has "bioactivity," which, loosely defined, means the extract has a direct effect on living cells or to components of living cells, is entirely dependent on the construction of the screening system. General screens look for crude indicators of activity, such as the effect of the extract on brine shrimp viability or crown gall growth on potato discs. Much more sophisticated screens can and have been developed, depending on the goals of a particular drug discovery program. For example, in searching for a specific anti-infective agent, microbial cultures of the target organism can be exposed to an extract to see whether the extract specifically inhibits growth of the microbial pathogen. Biochemical targets can also be used as mechanistic assays or targeted screens, whereby the activity of specific enzymes or proteins whether from a cancer cell, a pathogen, or normal tissue is tested in the presence of the extract. The issues of chemical novelty and selectivity are important to finding drugs that operate by unique mechanisms of action and are not toxic.

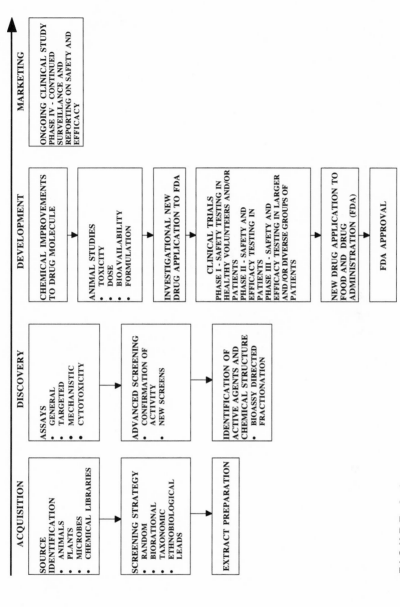

FIGURE 6.1
Drug discovery and development.

Whichever approach is used, the screening systems typically have the capability to handle large numbers of extracts to maximize the chances for success (finding a "hit"). High-throughput screening, particularly at pharmaceutical companies, has become extremely sophisticated and is in part responsible for the renewed interest in natural products on the part of industry. In order to be economical, large numbers of extracts are required, and natural products are an essential success. (Caporale 1995). The use of robotics and computer-controlled assays has become extremely popular despite the multimillion dollar investment needed for hardware and software. Fully automated screening systems, such as that run by the National Cancer Institute, can process thousands of extracts each month (Cragg and Boyd 1996).

Various criteria must be established within a screening program to decide whether to proceed with an extract or discard it at certain stages of investigation. Selection criteria are used at multiple stages of the drug discovery process and can include such measurements as the level of observed activity (specific activity), comparison of activity to a known drug or positive control, and the spectrum of activity (e.g., how many types of organisms or targets does the extract inhibit). Criteria for selectivity include testing an extract for toxicity, yet toxic drugs need not necessarily be dropped from the drug discovery process, as it may be possible to chemically modify the compound to reduce toxicity.

Having passed the initial selection criteria in the screening program, an extract is then chemically dissected to try to isolate the active component in what is otherwise a complex mixture of compounds. The goal of this process is to identify fractions that show improved activity with fewer and fewer chemical components. A good fractionation scheme will also check for cytotoxicity at multiple stages. In most cases, the successful chemist will eventually purify the active agent as a single chemical entity separated from inactive components.

To the chemical prospector, determining the chemical structure is perhaps the most intellectually satisfying stage of the drug discovery process. From a practical standpoint of pharmaceutical development, it is also the most critical stage of determining the true potential of a chemical in a commercial production setting. This step proves whether earlier criteria used to avoid the pitfall of isolating an already well-known chemical were successful. An examination of the chemical structure will indicate whether this compound can be chemically synthesized in the laboratory or must continually be harvested from the cultivated or wild source for commercial scale manufacture. The structure may also determine the potential of this chemical in the production of semisynthetics. These are the pharmaceuticals that use a natural product, usually easily purified, as starting material for a drug that is then made into its final form using various chemical reactions to add or delete specific molecules.

The combination of chemistry with computer modeling is the basis for examining structure–activity relationships that guide the chemist in making chemical improvements to the drug molecule. Knowledge of the structural or biochemical target of the drug gives the chemist an additional advantage, as the drug may be designed to provide a better "fit" with the microbial target that improves drug activity and lowers its toxicity relative to any similar host targets (Des Jarlais et al. 1987).

Animal studies are used to start the long process of assuring safety and efficacy of the new drug in humans. Animal studies commonly determine (1) what levels of the drug are well-tolerated and what levels are considered toxic, (2) dose-ranging for efficacy in an animal model of the targeted human disease, and (3) the bioavailability of the compound, which is the measure of how well it is absorbed by the body and affects how it can be administered to patients.

In the United States, all of the experimental data regarding the production and testing of a new drug are compiled into an Investigational New Drug (IND) Application for review by the FDA. Approval of this application is required for advancing into formal clinical trials. Chemical compounds and other therapies that are designated as INDs then go into protocol development for clinical trials, often in healthy volunteers first and then in volunteers who have the targeted medical problem. The three phases of clinical trials depicted in Figure 6.1 are (1) safety testing in a small group of healthy volunteers, (2) safety and efficacy testing using a small group of patients, and (3) safety and efficacy testing using larger groups of patients which may also vary in age, gender, disease severity, etc. The fourth phase includes continued surveillance and reporting on safety and efficacy once the FDA has approved a new drug application and a drug has been marketed. The time line for these phases can be extremely varied, and typically take years.

Analysis of Prescription Drug Origins

This section analyzes the top 150 proprietary drugs from the National Prescription Audit published by IMS America. The audit is a compilation of virtually every prescription filled in the United States for the period from January through September of 1993. Data are collected from chain and independent drug stores, pharmacies in foodstores, mass merchandisers, long-term care facilities, and health maintenance organizations. The list is based on the *number of times a prescription for the drug was filled,* not the amount prescribed, and not the cost. A matrix of these drugs, their generic name, disease category, and medical area they are most commonly prescribed for, source organism, origin and its relationship to a natural product can be found in the chapter appendix. Drugs are listed from most prescribed to least prescribed by the specific brand name used by the manufacturer and are not grouped in the matrix by their generic names.

Each drug on this list was then classified as to its origins. In doing this, we investigated the discovery process leading to each drug and classified them in the following three categories: *Natural Product* if the drug compound consisted of essentially the unaltered natural product; *Semisynthetic* if the structural lead preceding the drug's creation came from nature or if the compound is a combination of a lead from nature with a synthetic structure; and *Synthetic* if the drug was entirely synthesized without specific reference to a compound found in nature. It is worth noting, however, that even drugs in this last category owe some part of their origin to nature, from which we have learned most of what we know about organic chemistry.

The 150 brand names from the IMS survey represent 99 discrete compounds or compound mixtures. Fifty-one are generic drugs that appear more than once under a specific manufacturer's name. Tables 6.1 and 6.2 demonstrate the similarity of results regardless of whether the analysis is based on brand or generic names. We found that 57% of the top 150 brand names prescribed in this time period contained at least one major active compound now or once derived or patterned after compounds derived from biological diversity. In other words, we retraced the historical process leading to each named drug to determine the role(s) played by one or several natural products. This is different from the number of brand names with active major compounds that are found in nature, which would be much higher.

Table 6.3 lists the taxa from which these brand name drugs were de-

TABLE 6.1

Origins of Top 99 Prescribed Compounds

Origin	Total Number of Compounds	Natural Product	Semi-synthetic	Synthetic	Percentage
Animal	20	3	17	—	20.5%
Plant	19	6	13	—	19.5%
Fungus	10	2	8	—	10%
Bacteria	4	3	1	—	3%
Marine	1	1	0	—	1%
Synthetic	45	—	—	45	46%
TOTAL	99	15	39	45	100%

TABLE 6.2
Origins of Top 150 Prescription Drugs

Origin	Total Number of Compounds	Natural Product	Semi-synthetic	Synthetic	Percentage
Animal	27	6	21	—	23%
Plant	34	9	25	—	18%
Fungus	17	4	13	—	11%
Bacteria	6	5	1	—	4%
Marine	2	2	0	—	1%
Synthetic	64	—	—	64	43%
TOTAL	150	26	60	64	100%

rived. Approximately equal numbers were derived from animals and plants, with approximately half that many coming from fungi, and only a very few percent from bacteria, and one from sea water.

Table 6.4 summarizes the ethnobotanical uses of the plants from which the 150 prescribed brand names were derived. The base compound in almost all had or has a demonstrated use in traditional medicine related to the primary therapeutic use for which a physician might prescribe the drug.

Table 6.5 groups the brand names according to the primary ailment(s) they are prescribed to treat. Although the groupings in Table 6.5 certainly do not provide definitive conclusions due to their limited sample size and other factors, there are three therapeutic areas for which there are over 20 brand names in the matrix. Of the cardiovascular drugs, about half were synthesized and half were derived from nature. Of the infectious diseases drugs, about three-quarters were from nature and one-quarter were synthesized, and of those prescribed for general medicine, including analgesics (pain medication), three-quarters were from nature and one-quarter were synthesized.

Comparison with Results from Similar Studies

More than anything else, the statement that "57% of the top 150 brand names prescribed in this time period contained at least one compound now or once derived or patterned after compounds derived from biological diversity" will be used, cited, and remembered by readers of this chapter. Consequently, it is worth mentioning how important it is to be clear about how such numbers are calculated. Different numbers emerge depending on the data set used. Data sets may be defined as prescription or over-the-counter drugs, plant-derived compounds or those derived from

TABLE 6.3
Derivitive Organism

Scientific Name	Kingdom	Common Name	Ranks of Derived Drugs	Total # Derived Drugs
Equis caballus	Animal	horse	1	1
Ovis aires		domestic sheep	5, 140	2
Bothrops jararaca		Brazilian fer-de-lance	9, 22, 34, 55	4
Homo sapiens		people	31, 138, 142	3
Various mammals			3, 15, 19, 21, 39, 40, 48, 68, 77, 88, 89, 90, 100, 101, 104, 112, 133	17
ANIMAL TOTAL				27
Penicillium notatum	Fungi	bread mold	2, 8, 11, 30, 33, 52, 87, 91, 150	9
Cephalosporium acremonium			10, 53, 57, 58, 70, 78, 94	7
Monascus ruber			97	1
Aspergillus terreus			97	1
FUNGI TOTAL				17
Ephedra sinica	Plants	joint fir	13, 20, 23, 46, 59, 69, 102, 103, 127, 139, 149	11
Melilotus officinalis		yellow sweet clover	25	1
Ammi visnaga		bisnaga	118	1
Digitalis lanata		Grecian foxglove	6	1
Papaver somniferum		opium poppy	28, 32, 41, 64, 65, 96, 86, 98, 113, 119, 125, 126, 128, 130, 136	15
Camellia sinensis		tea	66, 134	2
Atropa belladonna		deadly nightshade	74	1
Paullinia cupana or *Coffea arabica*		guarana or coffee	84	1
Spirea/Salix alba		bridal wreath/ white willow	125	1
		vegetable oils	85	1
PLANT TOTAL				34
Streptomyces erythreus	Eubacteria		61, 81, 122, 123	4
Streptomyces clavuligerus			11	1
Pseudomonas fluorescens			105	1
BACTERIA TOTAL				6

TABLE 6.4
Ethnobotanical Uses of Derivitive Plants

Scientific Name	Chemical Compound	Common Name	Ranks of Derived Drugs	Total # Derived Drugs	Use of Traditional Medicine	Correlation Between Trad. Use and Pharm. Use
Ephedra sinica	ephedrine	joint fir	13, 20, 23, 46, 59, 69, 102, 103, 127, 139, 149	11	acute bronchitis	YES
Melilotus officinalis	coumarin	yellow sweet clover	25	1	?	?
Ammi visnaga	khellin	bisnaga	118	1	asthma	YES
Digitalis lanata	digoxin	Grecian foxglove	6	1	similar use made of similar compound from closely related species	INDIRECT
Papaver somniferum	codeine	opium poppy	32, 41, 64, 86, 113, 125, 126, 128, 130, 136	15	analgesic, sedative	YES
Camellia sinensis	theophylline	tea	66, 134	2	diuretic, stimulant	YES
Atropa belladonna	atropine	deadly nightshade	74	1	dilate pupil of the eye	YES
Paullinia cupana or *Coffea arabica*	theobromine	guarana or coffee	84	1	diuretic	YES
Coffea arabica	caffeine	coffee	125	1	stimulant	YES
Spirea/Salix alba	salicylic acid	bridal wreath/white willow	125	1	analgesic	YES
		vegetable oils	85	1	?	?

other organisms, whether major or minor ingredients are scored, how the relationship of the compound to the natural product is defined, and whether the rankings are calculated based on weight, cost, volume, or as in the case of our analysis, number of prescriptions.

TABLE 6.5
Informal Grouping of Top 150 Pharmaceuticals According to Estimated Primary Use Therapeutic Areas

Area of Medicine	Ranks of Drugs Primarily Used in This Area	% Natural and Semisynthetic	% Synthetic
Allergy/Pulmonary and Respiratory	13, 18, 20, 59, 66, 73, 74, 77, 89, 102, 104, 118, 129, 149	79	21
Antiinflammatory	12, 38, 44, 47, 49, 51, 72, 116, 143, 146	0	100
Cardiovascular	4, 6, 9, 16, 22, 23, 28, 34, 37, 42, 46, 55, 56, 60, 62, 65, 71, 79, 84, 93, 95, 96, 97, 103, 108, 110, 115, 119, 127, 137, 144	44	56
Dermatology	85, 112, 133	100	0
Endocrine Disorders	5, 26, 31, 35, 36, 138, 140, 142	25	75
Gastroenterology/ Gastrointestinal	3, 19, 40, 54, 68, 109	67	23
General Medicine Analgesics	17, 29, 32, 41, 45, 50, 63, 64, 86, 98, 113, 120, 126, 128, 130, 132, 136, 139, 141, 147	75	25
Gynecological	1, 15, 21, 39, 48, 88, 90, 100, 101	100	0
Hematology	25	100	0
Infectious Disease	2, 8, 10, 11, 24, 30, 33, 52, 53, 57, 58, 61, 70, 76, 78, 80, 81, 83, 87, 91, 94, 105, 111, 117, 122, 123	76	24
Nephrology	114	0	100
Neuro/Psychiatry	7, 14, 27, 43, 67, 75, 82, 92, 99, 106, 121, 124, 125, 131, 135, 145	6	94
Oncology	107	0	100
Ophthalmology	69	100	0

Almost any data set based on number of prescriptions will rank the compounds used in estrogen replacement therapy and oral contraceptives fairly highly based on an increasing aging female population seeking relief from menopause problems and a younger female population searching for convenient birth control. In addition to the demography, however, the manner in which these drugs are prescribed also contributes to their positions in the matrix. For example, some are normally prescribed over very long time periods, but often with small changes in dosage, ingredients, etc. Both of these factors are responsible for the high number of prescriptions written and hence the predominance of these compounds in our analysis.

In this analysis we used only prescription drugs, we included compounds derived from all natural sources, scored only major active ingredients, and included all compounds which were originally identified from nature, whether they still come from those sources or not, and whether the active compound is the one found in nature, or a compound derived from it. Thus the 57% figure is based on different data than the other most often cited figure of 25% from Farnsworth (1977) and Farnsworth et al. (1985). Although both are based on prescription audits, Farnsworth's analysis is based on only the drugs with plant origins, while we include all natural sources. A reanalysis of Farnsworth and analysis of the products of America's largest pharmaceuticals firms listed in the *Physician's Desk Reference* (1991), *Canadian OTC Drugs,* and *Sittig's Encyclopedia* (1988) by Duke (1993) reaffirms the 25% number, but concludes that only 10% of these compounds contain major compounds still derived from plants.

The derivation process can be quite complex, as the following example of steroids illustrates. Although the early chemistry leading to the determination of the structure of estrone and other related compounds was largely done on material from animal sources (the urine of pregnant mares, etc.), the yam *Dioscorea tokoro* led to the current markets for steroidal intermediates used in the synthesis of these compounds because of their cheapness and availability.

In other instances, in spite of the difference in calculations, it is surprising how consistent the emerging numbers really are. For example, a study recently completed by Cragg et al. (submitted 1996) at the National Cancer Institute included analysis of all new chemical entities (NCEs) for the years 1983–1994 that were approved by the FDA or its equivalent worldwide. Of the 521 NCEs in this time period, using the same definitions that we employed in our analysis 56% were synthetic, 11% were natural products, and 34% were semisynthetic.

Also worth mentioning are all the important biodiversity-derived drugs that for various reasons are not prescribed to the degree of those in the matrix, but nonetheless represent compounds that save thousands of lives.

These include vincristine, vinblastine, taxol, the family of antimalarials derived from quinine and artemisinin, and the countless other drugs used for rarer, but no less devastating diseases, both in the United States and abroad. Although all are responsible for saving many lives, none of these are prescribed with sufficient frequency in the United States to make the IMS audit.

The importance of information from ethnobiology, systematics, and chemical ecology to the drug discovery process is significant. Table 6.4 demonstrates this for ethnobotanical knowledge, but equally significant are the contributions that can be made by systematics and chemical ecology. Much of the world's biodiversity, however, is not traditionally used by cultures for medicinal purposes. Random screening of countless invertebrates, marine organisms, and other organisms can be made more efficient using both systematics and chemical ecology (Eisner 1994; Lewis and Elvin-Lewis 1995). This is yet another argument not only for the conservation of biological diversity *in situ* where these processes may be studied but for concern for the massive losses of cultures and their knowledge that is so often linked to the loss of forests, and other habitats they call home.

Significance of the Results for Physicians

While all of the information below will be of interest to physicians, Table 6.5 will be especially effective in demonstrating the diversity of medical specialties which depend on pharmaceuticals derived from natural products. Medical specialists will be able to see that many of the pharmaceuticals they prescribe on a regular basis to treat their patients have a direct connection to the natural world. It should be noted that the pharmaceuticals derived from the natural world indicated in this sample are utilized by some medical specialists more than others. Examples include those practitioners who specialize in respiratory, cardiovascular, gynecological, and infectious diseases.

Conclusions

Most of us in the developed world assume that when we are ill, there will almost always be something our physician can prescribe that will have us feeling better in short order. Very few of us give much thought to prescription drugs beyond that. It is becoming increasingly important, however, to consider prescription drugs as a resource that is simultaneously increasing in its significance to our health and diminishing in abundance and power. The resurgence of common ailments, once considered all but cured, their resistance to currently available drugs, and the emergence of new ailments are all increasing the demand for new drugs.

Habitat conversion and the resulting fragmentation, introduction of non-native species and other biological disasters due to patterns of consumption and population growth, and indifference to the consequences of our actions all contribute to the loss of biodiversity from which we might gather the ingredients for new drugs. Whether we seek natural products to use directly as prescribed or as over-the-counter pharmaceuticals, as phytomedicines, or as herbal remedies, compounds to use in combinatorial chemistry or biotechnology, or as the precursors for the manufacture of pharmaceuticals, or simply to learn more about medicine, we are losing our most valuable source, our biological diversity.

This analysis does not look at the use of actual plant materials in the manufacture of pharmaceuticals because this is almost irrelevant to the importance of biodiversity to our continued access to these raw materials. In carefully monitored instances, certain medicinals, particularly those harvested in small quantities for local use, may be harvested sustainably (see Peters, Chapter 15). Also, the reclamation of marginal lands, or restoration of already ecologically dead lands, for use in cultivation of plants needed for the drug manufacturing process might in some instances be possible. It is, however, more generally the case in the developing world, where land is more fragile, and techniques for restoration more rare, that if a market is created for a crop, then forest or other biologically intact lands will be transformed for this use. Consequently, synthesis by a pharmaceutical company is not necessarily a bad thing for biodiversity conservation. When resources return to local communities for the conservation of their lands from royalties from drugs discovered with their input, or from biodiversity collected on their lands, such an incentive is created (Grifo 1996; see also Rosenthal, Chapter 13).

The last conclusion we would like to draw is that mother nature, whether you conceive her as the process of evolution or with religious or pagan conviction, is a far better, more ingenious chemist (and many other things as well) than we will ever be. So until we can know which bits of nature hold which information, we are playing roulette each time a species goes extinct. Are the species we lose potentially the cure for AIDS, the remedy that will finally really rid you of athlete's foot, "just" a species that made you smile, or maybe a species you will never miss? With extinction rates at 1000 to 10,000 times the background "natural" rate, we can be sure that both species whose significance we can grasp and species whose importance we cannot fathom are leaving the face of our planet forever.

APPENDIX TO CHAPTER 6
Origin of Pharmaceuticals Matrix

Rank	Proprietary Drug Name (Manufacturer)	Generic Name	Disease Category	Area of Medicine
1	Premarin (Wyeth-Ayerst)	conjugated estrogens, USP	estrogen replacement therapy	gynecology
2	Amoxi (SKB Pharm) See also #8, 30, 33, 91	amoxicillin	antibacterial penicillin	infectious disease
3	Zantac (Glaxo Pharm)	ranitidine	histamine (H2) antagonist	gastroenterology
4	Procardia XL (Pratt Pharm)	nifedipine	Ca channel blocker/smooth muscle relaxant	cardiovascular
5	Synthroid (Boots Pharm) See also #140	levothyroxine	thyroid hormone	endocrine
6	Lanoxin (Burroughs Wellcome)	digoxin	cardiotonic	cardiovascular
7	Xanax (Upjohn)	alprazolam	antianxiety/sedative	psychiatry
8	Trimox (Apothecon)	amoxicillin	(See #2)	
9	Vasotec (Merck & Co)	enalapril maleate	antihypertensive (ACE inhibitor)	cardiovascular
10	Ceclor (Lilly)	cefaclor	antibacterial/cephalosporin	infectious disease
11	Augmentin (SKB Pharm)	amoxicillin + clavulanic acid	antibacterial	infectious disease
12	Naprosyn (Syntex) See also #51	naproxen	nonsteroidal antiinflammatory/analgesic	inflammation
13	Proventil (Aerosol) See also # 20, 102, 149	albuterol (salbutamol)	bronchodilator/ß-agonist	respiratory/pulmonary
14	Prozac (Dista)	fluoxetine HCl	antidepressant	neurology/psychiatry

Source Organism	Product Origin	Natural Product Relationship	References
animal	taken from urine of pregnant horses	natural	p373 Cutting's
fungus	synthetic deriv of penicillin, from the mold *Penicillium notatum*	semisynthetic	p42,95 Reiner
animal	systematic screening of mammalian histamine-like compounds against different preparations demonstrated effect	semisynthetic	p904 G&G1996
synthetic	synthetic dihydropyridines	synthetic	p774 G&G1990 p249–250 Cutting's
animal	sheep thyroid extracts used by Murray (1891) as replacement therapy. Thyroxine identified in 1915	natural	p365 Cutting's
plant	*Digitalis lanata* (foxglove)	natural	p167 Cutting's
synthetic	triazolo-benzodiazepine structure analog of diazepam	synthetic	p433 G&G1985
animal	prodrug deriv of captopril, found in pit venom of *Bothrops jararaca*	semisynthetic	p41 Burger p760 G&G1990
fungus	synth deriv of 7-ACA, derived from ceph C from *Cephalosporium acremonium*	semisynthetic	p114 Reiner
fungus + bacterium	synthetic deriv of penicillin, from the mold *Penicillium notatum* isolated from *Streptomyces clavuligerus*	semisynthetic plus natural	p42,95 Reiner
synthetic	phenylpropionic acid derivative	synthetic	700 G&G1985
plant	analog of sympathomimetics, with modified catecholamine nucleus, based on ephedrine from *Ephedra sinica*	semisynthetic	p37 McFadden
synthetic	synthetic phenyltolylpropylamine	synthetic	p406 G&G1990

Appendix Continues

APPENDIX TO CHAPTER 6 (CONTINUED)

Rank	Proprietary Drug Name (Manufacturer)	Generic Name	Disease Category	Area of Medicine
15	Provera (Upjohn)	medroxyproges- terone acetate	estrogen therapy and contraception	gynecology
16	Cardizem CD (Marion Merrell Dow) See also #108, 115	diltiazem	Ca channel blocker	cardiovascular
17	Mevacor (Merck)	lovastatin	cholesterol reducer	general medicine
18	Seldane (Marion Merrell Dow)	terfenadine	antihistamine (H1)	allergy/pulmonary
19	Tagamet (SKB Pharm)	cimetidine	histamine H2 antagonist	gastroenterology
20	Ventolin (Allen and Hanburys)	albuterol	(See #13)	
21	Ortho-Novum 7/7/7 28 (Ortho)	ethinyl estradiol/ norethindrone	oral contraceptive	gynecology
22	Capoten (Squibb)	captopril	antihypertensive (ACE inhibitor)	cardiovascular
23	Lopressor (Geigy)	metoprolol tartrate	cardiac ß-blocker	cardiovascular
24	Cipro (Miles Pharm)	ciprofloxacin	antibacterial quinolone	infectious disease
25	Coumadin Na (DuPont Pharm)	warfarin Na	anticoagulant	hematology
26	Micronase (Upjohn) See also #35	glyburide (glibenclamide)	oral hypoglycemic	endocrine
27	Dilantin (Parke-Davis)	phenytoin	anticonvulsant	neurological
28	Calan SR (Searle) See also #96, 119	verapamil	Ca channel blocker smooth muscle vasodilator	cardiovascular
29	Propoxyphene- Napsylate w/ APAP(Mylan) See also #50, 63	propoxyphene napsylate	opiate agonist	analgesic

Source Organism	Product Origin	Natural Product Relationship	References
animal	synthesized from mammalian steroidal intermediates	semisynthetic	p381 Cutting's
synthetic	synthetic benzothiazepine	synthetic	p774 G&G1990
fungus	isolated in 1970s from cultures of *Monascus ruber* and *Aspergillus terreus*	natural	p881 G&G1990
synthetic	screening for antihistaminic activity	synthetic	p618 G&G1985
animal	systematic screening of mammalian histamine-like compounds against different preparations demonstrated effect	semisynthetic	p904 G&G1996
animal	synthesized from mammalian steroidal intermediates	semisynthetic	p244 Taylor p154 Ross & Brain
animal	prodrug found in venom of pit viper *Bothrops jararaca*	semisynthetic	p41 Burger
plant	analog of adrenergics, with modified catecholamine nucleus, based on ephedrine from *Ephedra sinica*	semisynthetic	p37 McFadden
synthetic	quinolone synthetic	synthetic	p1666 PDR1995
plant	deriv of dicoumarin, from *Melilotus officinalis*	semisynthetic	p27 CIBA185 p149 Ross & Brain
synthetic	synthetic sulfonylurea	synthetic	p1484,1485 G&G1990
synthetic	analog of phenobarbital	synthetic	p139 Burger
plant	remote derivative of papaverine from *Papaver somniferumc*	semisynthetic	p247 Cutting's p489 G&G1990
synthetic	synthetic methadone derivative	synthetic	p517 G&G1985

Appendix Continues

APPENDIX TO CHAPTER 6 (*CONTINUED*)

Rank	Proprietary Drug Name (Manufacturer)	Generic Name	Disease Category	Area of Medicine
30	Polymox (Apothecon)	amoxicillin	(See #2)	
31	Humulin N (Lilly) See also #138, 142	NPH human insulin	antidiabetic	endocrine
32	Acetaminophen w/ Codeine (Purepac) See also #41, 64	acetaminophen plus codeine phosphate	nonsteroidal anti-inflammatory: analgesic, antipyretic, narcotic antitussive	general medicine
33	Amoxicillin trihydrate (Bicroft Labs)	amoxicillin	(See #2)	
34	Zestril (Stuart) See also #55	lisinopril	antihypertensive (ACE inhibitor)	cardiovascular
35	Diabeta (Hoechst-Roussel)	glyburide	(See #26)	
36	Glucotrol (Pratt Pharm)	glipizide	oral hypoglycemic	endocrine
37	Dyazide (SKB Pharm) See also #56, 93, 148	triamterene/ HCTZ	diuretic/ antihypertensive	general medicine/ cardiovascular
38	Ibuprofen (Boots) See also #143, 146	ibuprofen	nonsteroidal antiinflammatory	inflammation
39	Triphasil 28 (Wyeth-Ayerst) See also #100	levonorgestrel/ ethinyl estradiol	oral contraceptive	gynecology
40	Pepcid (Merck)	famotidine	histamine H2 antagonist	gastroenterology
41	Codeine + Acetaminophen (Lemmon)	codeine & acetaminophen	(See also #32)	
42	Lasix (Hoechst-Roussel) See also #60, 95	furosemide	diuretic	general medicine/ cardiovascular
43	Zoloft (Roerig)	sertraline	antidepressant	neuro/psychiatry
44	Relafen (SKB Pharm)	nabumetone	nonsteroidal antiinflammatory	inflammation

Source Organism	Product Origin	Natural Product Relationship	References
animal	recDNA human insulin	natural (biological)	p1487 G&G 1996
synthetic/ plant	synthetic analgesic plus opiate alkaloid from *Papaver somniferum*	synthetic plus natural product	p495 G&G1985 p541 Cutting's
animal	modified struc of enalapril, deriv from captopril, prodrug of which found in pit venom of *Bothrops jararaca*	semisynthetic	p41 Burger p760 G&G1990
synthetic	synthetic sulfonylurea	synthetic	p1484,1485 G&G1990
synthetic	synth analog of folic acid plus a structure derived from sulfanilamides	synthetic	p727 G&G1990/ p82 Beyer
synthetic	phenylpropionic acid derivative	synthetic	p700 G&G1985
animal	synthesized from mammalian steroidal intermediates	semisynthetic	p1432 G&G1996
animal	systematic screening of mammalian histamine-like compounds against different preparations demonstrated effect	semisynthetic	p904 G&G1996
synthetic	synthetic deriv of anthranilic acid	synthetic	p722 G&G1990
synthetic	synthetic	synthetic	p2000 PDR1994
synthetic	naphthyl substituted methyl ethyl ketone	synthetic	p641 G&G1996

Appendix Continues

APPENDIX TO CHAPTER 6 (*CONTINUED*)

Rank	Proprietary Drug Name (Manufacturer)	Generic Name	Disease Category	Area of Medicine
45	K-Dur (Key Pharm) See also #120	potassium chloride	oral electrolytes	general medicine/ emergency medicine
46	Tenormin (Zeneca Pharm) See also #127	atenolol	cardiac ß-blocker	cardiovascular
47	Voltaren (Geigy)	diclofenac	nonsteroidal antiinflammatory	inflammation
48	Estraderm (Ciba)	estradiol patch	estrogens	gynecology
49	Toradol (Syntex)	ketorolac	nonsteroidal antiinflammatory	inflammation
50	Darvocet-N 100 (Lilly)	propoxyphene & APAP	(See #29)	
51	Anaprox DS (Syntex)	naproxen	(See #12)	
52	Veetids (Apothecon) See also #87, 150	penicillin V	antibacterial	infectious disease
53	Ceftin (Allen&Hanburys)	cefuroxime axetil	antibacterial/ cephalosporin	infectious disease
54	Prilosec (Merck)	omeprazole	duodenal ulcer, GERD	gastrointestinal
55	Prinivil (Merck)	lisinopril	(See #34)	
56	Triamterene/HCTZ (Rugby Labs)	triamterene & hydrochlor-thiazide	(See also #37)	
57	Duricef (Princeton Pharm)	cefadroxil	antibacterial/ cephalosporin	infectious disease
58	Cephalexin (Biocraft Labs) See also #70	cephalexin	antibacterial/ cephalosporin	infectious disease
59	Seldane-D (Marion Merrell Dow)	terfenadine/ pseudoephedrine	antihistamine/ decongestant	allergy/pulmonary
60	Furosemide (Mylan)	furosemide	(See #42)	
61	Biaxin (Abbott Pharm)	clarithromycin	antibacterial	infectious disease

Source Organism	Product Origin	Natural Product Relationship	References
marine	component of sea water	natural	
plant	analog of adrenergics, with modified catecholamine nucleus, based on ephedrine from *Ephedra sinica*	semisynthetic	p37 McFadden
synthetic	phenylacetic acid derivative specifically for inflammation	synthetic	p636 G&G1996
animal	synthesized from mammalian steroidal intermediates	semisynthetic	p1419 G&G1996
synthetic	heteroaryl acetic acid	synthetic	p636 G&G1996
fungus	first obtained from the mold *Penicillium notatum*	natural	p6,32 Reiner
fungus	synth deriv of 7-ACA, derived from ceph C from *Cephalosporium acremonium*	semisynthetic	p94 Reiner
synthetic	synthetic substituted benzimidazole	synthetic	p902 G&G1990
fungus	synth deriv of 7-ACA, derived from ceph C from *Cephalosporium acremonium*	semisynthetic	p94 Reiner
fungus	synth deriv of 7-ACA, derived from ceph C from *Cephalosporium acremonium*	semisynthetic	p209 Ross & Brain p44, 94 Reiner
synthetic and plant	screening for antihistaminic activity/ plus stereoisomer of ephedrine from *Ephedra sinica*	synthetic and semisynthetic	p575,583 G&G1990/ p105,107 Taylor p347 Holmstedt
bacterium	semisynth deriv of erythromycin, from bacterium *Streptomyces erythreus*	semisynthetic	p1135 G&G1996

Appendix Continues

APPENDIX TO CHAPTER 6 (*CONTINUED*)

Rank	Proprietary Drug Name (Manufacturer)	Generic Name	Disease Category	Area of Medicine
62	Nitrostat (Parke-Davis) See also #110	nitroglycerin (glyceryl trinitrate)	antianginal	cardiovascular
63	Propacet 100 (Lemmon)	propoxyphene & APAP	(See #29)	
64	Tylenol w/ Codeine (McNeil)	codeine & APAP	(See #32)	
65	Hytrin (Abbott Pharm)	terazosin	antihypertensive	cardiovascular
66	Theo-Dur (Key Pharm) See also #134	theophylline	bronchodilator	respiratory/allergy
67	Klonopin (Roche)	clonazepam	anticonvulsant	neurology
68	Axid (Lilly)	nizatidine	histamine (H2) antagonist	gastroenterology
69	Timoptic (Merck)	timolol	cardiac ß-blocker glaucoma	ophthalmology
70	Cephalexin (Apothecon)	cephalexin	(See #58)	
71	Lozol (Rhone-Poulenc Rorer)	indapamide	diuretic/ antihypertensive	cardiovascular
72	Lodine (Wyeth-Ayerst)	etodolac	nonsteroidal antiinflammatory	inflammation
73	Hismanal (Janssen)	astemizole	antihistamine	allergy/pulmonary
74	Atrovent (Boeh-ringerIngelheim)	ipratropium bromide	bronchodilator	respiratory/ pulmonary
75	Amitriptyline HCl (Mylan)	amitriptyline	antidepressant	neurology/ psychiatry
76	Lotrisone (Key Pharm)	clotrimazole	antifungal, external	infectious disease
77	Beconase AQ (Allen & Hanburys) See also #89	beclomethasone dipropionate	corticosteroid/ antiasthma	pulmonary
78	Suprax (Lederle)	cefixime	antibacterial/ cephalosporin	infectious disease

Source Organism	Product Origin	Natural Product Relationship	References
synthetic	organic nitrate synthesized 1846	synthetic	p764 G&G1990 p241-242 Cutting's
synthetic	piperazinyl quinazoline deriv of prazosin, a papaverine relative, from *Papaver sominiferum*	synthetic	p222 G&G1990 p239 Cutting's
plant	found in tea, *Camellia sinensis*	natural	p101 Ross & Brain
synthetic	synthetic benzodiazepine with a nitro substitutent	synthetic	p372 G&G1996
animal	systematic screening of mammalian histamine-like compounds against different preparations demonstrated effect	semisynthetic	p904 G&G1996
plant	analog of adrenergics, with modified catecholamine nucleus, based on ephedrine from *Ephedra sinica*	semisynthetic	p37 McFadden p230 G&G1990
synthetic	benzothiadiazide	synthetic	p82 Beyer
synthetic	variation on Indomethacin	synthetic	p635 G&G1996
synthetic	screening for antihistaminic activity	synthetic	p590 G&G1996
plant	semisynthetic structure based on atropine from *Atropa belladonna*	semisynthetic	p154 G&G1996
synthetic	dibenzazepine analog originally synthesized in 1940s	synthetic	p433 G&G1996 p631 Cutting's
synthetic	synthetic imidazole	synthetic	p1169 G&G1990
animal	semisynthetic analog of cortisol from mammals	semisynthetic	p154 Ross & Brain p1449 G&G1990
fungus	semisynth deriv of 7-ACA, from *Cephalosporium acremonium*	semisynthetic	p1284 PDR1995

Appendix Continues

APPENDIX TO CHAPTER 6 (*CONTINUED*)

Rank	Proprietary Drug Name (Manufacturer)	Generic Name	Disease Category	Area of Medicine
79	Lopid (Parke-Davis) See also #137	gemfibrozil	antihyperlipidemic	cardiovascular
80	Trimethoprim/ Sulfameth-oxazole (Biocraft Labs) See also #117	trimethoprim sulfameth-oxazole	antibacterial	infectious disease
81	Ery-Tab (Abbott Pharm) See also #122, 123	erythromycin	antibacterial	infectious disease
82	Buspar (Mead Johnson Pharm)	buspirone	antianxiety	psychiatry
83	Zovirax Caps (Burroughs Wellcome)	acyclovir	antiviral	infectious disease
84	Trental (Hoechst-Roussel)	pentoxifylline	peripheral vascular disease	cardiovascular
85	Retin-A (Ortho Derm)	tretinoin (retinoic acid)	antiacne	dermatology
86	Hydrocodone w/ APAP (Watson Labs) See also #98, 126, 128, 130, 136, 147	dihydrocodeinone bitartrate + acetaminophen	narcotic plus analgesic	general medicine
87	Penicillin VK (Mylan)	penicillin V	(See #52)	
88	Ortho-Nov 1/35 28 (Ortho)	norethindrone/ ethinyl estradiol	oral contraceptive	gynecology
89	Vancenase AQ (Schering)	beclomethasone	(See #77)	
90	Lo/Ovral-28 (Wyeth-Ayerst)	norgestrel/ ethinyl estradiol	oral contraceptive	gynecology
91	Amoxicillin Trihydrate (Warner Chilcott)	amoxicillin	(See #2)	
92	Tegretol (Basel Pharm)	carbamazepine	anticonvulsant	neurology
93	Triamterene/ HCTZ (Geneva Pharm)	triamterene & hydrochlor-thiazide	(See #37)	
94	Cefzil (Bristol Labs)	cefprozil	antibacterial	infectious disease

Source Organism	Product Origin	Natural Product Relationship	References
synthetic	fibric acid deriv of clofibrate	synthetic	p886,887 G&G1990
synthetic	an antimalarial coupled to a sulfonamide	synthetic	p64 Cutting's
bacterium	isolated from bacterium *Streptomyces erythreus*	natural	p31 Reiner
synthetic	azapirone class related to halopiridol	synthetic	p425 G&G1996
synthetic	synthetic acyclic analog of adenosine; patterned after the marine natural product Ara-C	synthetic	p18 Reiner p125–126 Came
plant	derivative of theobromine from *Paullinia cupana* or *Coffea arabica*	natural	p676 G&G1996
plant	originally found in vegetable oils	natural	p95 Leung
plant	synthetically derived from codeine from *Papaver somniferum* plus synthetic antipyretic	semisynthetic	p153,159 Leake p312,657 G&G1990
animal	synthesised from mammalian steroidal precursors	semisynthetic	p1433 G&G1996
animal	synthesized from mammalian steroidal precursors	semisynthetic	p1433 G&G1996
synthetic	iminostilbene derivative	synthetic	p592 Cutting's p473 G&G1996
fungus	semisynthetic cephalosporin, derived from7ACA taken from *Cephalosporium acremonium*	semisynthetic	p717 PDR1995

Appendix Continues

APPENDIX TO CHAPTER 6 (*CONTINUED*)

Rank	Proprietary Drug Name (Manufacturer)	Generic Name	Disease Category	Area of Medicine
95	Furosemide (Geneva Pharm)	furosamide	(See #42)	
96	Verapamil SR (Goldline)	verapamil	(See #28)	
97	Pravachol (Squibb)	pravastatin	antihyperlipidemic	cardiovascular
98	Vicodin (Knoll)	hydrocodone & APAP	(See #86)	
99	Nortriptyline HCl (Schein Pharm)	nortriptyline	antidepressant	neurology/ psychiatry
100	Tri-Levlen 28 (Berlex Labs)	levonorgestrel/ ethinyl estradiol	(See #39)	
101	Estrace (Mead Johnson Labs)	17ß–estradiol	estrogen therapy	gynecology
102	Albuterol (Lemmon)	albuterol	(See #13)	
103	Corgard (Bristol Labs)	nadolol	cardiac ß-blocker	cardiovascular
104	Azmacort (Rhone-Poulenc Rorer)	triamcinolone acetonide	bronchodilator	respiratory/ pulmonary
105	Bactroban (SKB Pharm)	mupirocin	antiinfective, external	infectious disease
106	Valium (Roche) See also #145	diazepam	antianxiety	psychiatry
107	Nolvadex (Zeneca)	tamoxifen	antineoplastic— (breast cancer)	oncology
108	Cardizem (Marion Merrell Dow)	diltiazem	(See #16)	
109	Carafate (Marion Merrell Dow)	sucralfate	antiulcer therapy	gastroenterology
110	Nitro-Dur (Key Pharm)	nitroglycerin	(See #62)	
111	Terazol 7 (Ortho)	terconazole	vaginal antifungal	infectious disease/ gynecology
112	Deltasone (Upjohn) See also #133	prednisone	glucocorticosteroid	dermatology/ metabolic disease
113	Roxicet (Roxane)	oxycodone/ acetaminophen	narcotic plus analgesic	pain/ general medicine

Source Organism	Product Origin	Natural Product Relationship	References
fungus	struc deriv of lovastatin, isolated in 1970s from cultures of *Monascus ruber* and *Aspergillus terreus*	semisynthetic	p881 G&G1990
synthetic	dibenzazepine analog originally synthesized in 1940s	synthetic	p631 Cutting's
animal	synthesized from mammalian steroidal precursors	semisynthetic	p373 Cutting's
plant	analog of adrenergics, with modified catecholamine nucleus, based on ephedrine from *Ephedra sinica*	semisynthetic	p37 McFadden
animal	semisynthetic corticosteroid analog of cortisol from mammals	semisynthetic	p1450 G&G1990
bacterium	produced by *Pseudomonas fluorescens*	natural	p1584 G&G1990
synthetic	benzodiazepine	synthetic	p222 G&G1985
synthetic	synthetic triarylethylene derivative	synthetic	p1395 C&G 1990/ p375,387–388 Cutting's
synthetic	synthetic gel	synthetic	p910 G&G1990
synthetic	synthetic triazole	synthetic	p1169 G&G1990
animal	synthesized from mammalian steroidal precursors	semisynthetic	p360 Cutting's
plant	synthetically derived from codeine from *Papaver somniferum*	semisynthetic	p557 Cutting's

Appendix Continues

A P P E N D I X   T O   C H A P T E R   6   (*CONTINUED*)

Rank	Proprietary Drug Name (Manufacturer)	Generic Name	Disease Category	Area of Medicine
114	Bumex (Roche)	bumetanide	diuretic (loop)	nephrology
115	Cardizem SR (Marion Merrell Dow)	diltiazem	(See # 16)	
116	Ansaid (Upjohn)	flurbiprofen	nonsteroidal antiinflammatory	inflammation
117	Trimethoprim/ Sulfamethoxazole (Mutual Pharm)	trimethoprim/ sulfamethoxazole	(See #80)	
118	Intal (Fisons)	cromolyn sodium	respiratory inhalant	pulmonary
119	Verelan (Lederle)	verapamil	(See #28)	
120	Micro-K 10 (Wyeth-Ayerst)	potassium chloride	(See #45)	
121	Lorazepam (Purepac) See also #131	lorazepam	antianxiety/ sedative/ hypnotic	psychiatry
122	Erythrocin Stearate (Abbott Pharm)	erythromycin stearate	(See #81)	
123	E.E.S. (Abbott Pharm)	erythromycin ethylsuccinate	(See #81)	
124	Halcion (Upjohn)	triazolam	sedative/hypnotic	psychiatry
125	Fiorinal w/Codeine (Sandoz)	butalbital caffeine aspirin codeine	sedative/hypnotic vasoconstrictor analgesic	psychiatry, headaches
126	Hydrocodone w/ APAP (Qualitest)	hydrocodone & APAP	(See #86)	
127	Atenolol (Mylan)	atenolol	(See #46)	
128	Lorcet Plus (Forest Pharm)	hydrocodone & APAP	(See #86)	
129	Phenergan (Wyeth-Ayerst)	promethazine	antihistamine (H1)	allergy and pulmonary
130	Lortab 7.5/500 (Whitby Pharm)	hydrocodone & APAP	(See #86)	

Source Organism	Product Origin	Natural Product Relationship	References
synthetic	synthetic deriv of anthranilic acid, like furosemide	synthetic	p722 G&G1990
synthetic	phenylpropionic acid derivative	synthetic	p638 G&G1996
plant	derivitive of khellin, from the fruit of *Ammi visnaga*	semisynthetic	p630 G&G1990
synthetic	benzodiazepine structure analog of diazepam	synthetic	p363 G&G1996
synthetic	triazolo-benzodiazepine structure analog of diazepam	synthetic	p363 G&G1996
synthetic plants	derivative of barbital caffeine isolated from *Coffea arabica* aspirin derived from salicylic acid, first taken from *Spiraea* plants and also from bark of white willow, *Salix alba* codeine from *Papaver somniferum*	synthetic natural semisynthetic natural	p8 Wolff p101 Ross&Brain p197–202 Taylor p10-11 Parnham &Bruinvels
synthetic	phenothiazine antihistamine	synthetic	p587 G&G1996

Appendix Continues

APPENDIX TO CHAPTER 6 (*CONTINUED*)

Rank	Proprietary Drug Name (Manufacturer)	Generic Name	Disease Category	Area of Medicine
131	Lorazepam (Mylan)	lorazepam	(See #121)	
132	Cyclobenzaprine HCl (Schein Pharm) See #141	cyclobenzaprine	musculoskeletal relaxant	pain/general medicine
133	Prednisone (Schein Pharm)	prednisone	(See #112)	
134	Slo-Bid (Rhone-Poulenc Rorer)	theophylline	(See #66)	
135	Compazine (SKB Pharm)	prochlorperazine	antiemetic/ antipsychotic	psychiatry
136	Vicodin ES (Knoll)	hydrocodone & APAP	(See #86)	
137	Gemfibrozil (Warner-Chilcott)	gemfibrozil	(See #79)	
138	Humulin70/30 (Lilly)	human insulin	(See #31)	
139	Entex LA (Procter & Gamble Med)	phenylpropanola-mine/guaifenesin	decongestant expectorant	general medicine
140	Levoxine (Daniels Pharm)	levothyroxine	(See #5)	
141	Cyclobenzaprine HCl (Mylan)	cyclobenzaprine	(See also # 132)	
142	Humulin R (Lilly)	human insulin	(See #31)	
143	Motrin Childrens (McNeil CPC)	ibuprofen	(See #38)	
144	Norvasc (Pfizer)	amlodipine	Ca channel blocker	cardiovascular
145	Diazepam (Mylan)	diazepam	(See also #106)	
146	Motrin (Upjohn)	ibuprofen	(See #38)	
147	Hydrocodone w/ APAP (Halsey)	hydrocodone & APAP	(See #86)	
148	Maxzide 25 (Lederle)	triamterene & hydroclor-thiazide	(See #37)	
149	Proventil (Oral Solid) (Schering)	albuterol	(See #13)	
150	Betapen-VK (Apothecon)	penicillin V	(See #52)	

Source Organism	Product Origin	Natural Product Relationship	References
synthetic	oxidized congener of amitriptyline (See #75)	synthetic	p489 G&G1985
synthetic	phenothiazines originally dyestuffs, screened as antihistamines/sedatives	synthetic	p403 G&G1996
plant	strucural analog of ephedrine, derived from *Ephedra sinica*	semisynthetic	p178 Ross&Brain
synthetic	synthetic dihydropyridine	synthetic	p831 G&G1996

References

Beyer, Karl H. (1978). *Discovery, Development, and Delivery of New Drugs.* New York: Spectrum Publications.

Burger, Alfred. (1983). *A Guide to the Chemical Basis of Drug Design.* New York: John Wiley and Sons.

Caporale, Lynn. (1995). Chemical Ecology: A View from the Pharmaceutical Industry. In: *Chemical Ecology: The Chemistry of Biotic Interaction.* (T. Eisner and J. Meinwald, eds.). Washington, D.C.: National Academy of Sciences Press.

CIBA Foundation Symposium 185. (1994). *Ethnobotany and the Search for New Drugs.* New York: John Wiley and Sons.

Cragg, Gordon M., and Michael Boyd. (1996). Drug Discovery and Development at the National Cancer Institute: The Role of Natural Products of Plant Origin. In: *Medicinal Resources of the Tropical Forest.* (Michael J. Balick, Elaine Elizabetsky, and Sarah Laird, eds.). New York: Columbia University Press.

Cragg, Gordon M., Newman, David J., and Snader, Kenneth M. (submitted 1996). Natural Products in Drug Discovery and Development. *Journal of Natural Products.*

Cutting's Handbook of Pharmacology. (1979). (6th ed.). New York: Appleton-Century-Crofts.

Des Jarlais, R.L., G.L. Seibel, I.D. Kuntz, P. Furth, J. Alvarez, P. Ortiz de Montellano, L.N. Babe, and C. Craik. (1987). Structure-based Design of Nonpeptide Inhibitors Specific for the Human Immunodeficiency Virus Protease. *Proceedings of the National Academy of Science* 87:6644–6648.

Duke, James. (1993). Medicinal Plants in the Pharmaceutical Industry. In: *New Crops.* (J. Janick and James E. Simon, eds.). New York: John Wiley and Sons.

Eisner, Thomas. (1994). Chemical Prospecting: A Global Imperative. *Proceedings of the American Philosophical Society* 138:385–392.

Farnsworth, Norman R. (1977). Problems and Prospects of Discovering New Drugs from Higher Plants by Pharmacological Screening. In: *New Natural Products with Pharmacological, Biological, or Therapeutic Activity* (H. Wagner and P. Wolf, eds.). pp.1–22. New York: Springer-Verlag.

Farnsworth, Norman R., Akerele, Olayiwola, Bingel, Audrey S., Soejarto, Djaja D., and Guo, Zhengang. (1985). "Medicinal Plants in Therapy." *Bulletin of the World Health Organization* 63(6):965–981.

Goodman and Gilman. (1985). *The Pharmacological Basis of Therapeutics* (7th ed.). New York: MacMillan Publishing Company.

Goodman and Gilman. (1990). *The Pharmacological Basis of Therapeutics* (8th ed.). New York: Pergamon Press.

Goodman and Gilman. (1996). *The Pharmacological Basis of Therapeutics* (9th ed.). New York: McGraw-Hill.

Grifo, F.T. (1996). The Role of Chemical Prospecting in Sustainable Development. In: *Emerging Connections Among Biodiversity, Biotechnology, and Sustainable Development in Health and Agriculture.* (Julie Feinsilver, ed.). Washington, DC: Pan American Health Organization.

Holmstedt, B. (1981). *Readings in Pharmacology.* Raven Press.

Leake, Chauncey D. (1975). *An Historical Account of Pharmacology to the 20th Century.* Springfield, IL: Charles C. Thomas.

Leung, Albert Y. (1980). *Encyclopedia of Common Natural Ingredients.* New York: John Wiley and Sons.

Lewis, W. H., and M. P. Elvin-Lewis. (1995). Medicinal Plants as Sources of New Therapeutics. *Annals of the Missouri Botanical Garden* 82:16–24.

McFadden, E. Regis. (1976). *Inhaled Aerosol Bronchodilators* Baltimore: Williams & Wilkins.

Parnham, M.J., and J. Bruinvels. (eds.). (1983). *Psycho- and Neuro-Pharmacology,* Vol. 1. Amsterdam: Elsevier.

Physician's Desk Reference. (1994). Montvale, NJ: Medical Economics Data Production Co.

Physician's Desk Reference. (1995). Montvale, NJ: Medical Economics Data Production Co.

Principe, Peter P. (1996). Monetizing the Pharmacological Benefits of Plants. In: *Medicinal Resources of the Tropical Forest.* (Michael J. Balick, Elaine Elizabetsky, and Sarah Laird, eds.). New York: Columbia University Press.

Reiner, Roland. (1982). *Antibiotics, An Introduction.* Stuttgart: Georg Thieme Verlag.

Ross, M.S.F., and K.R. Brain. (1977). *An Introduction to Phytopharmacy.* Pitman Medical.

Simpson, R. D., R. A. Sedjo, and J. W. Reid. (1996). Valuing Biodiversity for Use in Pharmaceutical Research. *Journal of Political Economy* 104:1548–1570.

Sittig, M. (1988). *Pharmaceutical Manufacturing Encyclopedia* (2nd ed.), 2 vols. Park Ridge, NJ: Noyes Publ.

Taylor, Norman. (1995). *Plant Drugs That Changed the World.* New York: Dodd, Mead & Company.

Wilson, E.O. (ed.). (1988). *Biodiversity.* Washington, D.C.: National Academy Press.

Wolff, Manfred E. (ed.). (1979). *Burger's Medicinal Chemistry, Fourth Edition, Part 1, The Basis of Medicinal Chemistry.* New York: John Wiley and Sons.

CHAPTER 7

Natural Products for the Treatment of Infectious Diseases

CATHERINE A. LAUGHLIN AND ALEXANDRA S. FAIRFIELD

Evidence of our appreciation of the healing properties of natural treatments for infectious diseases is well documented from earliest times. As discussed later, quinine, the classic malaria remedy, was introduced to the Europeans in seventeenth century South America and a "new" malaria treatment now in clinical testing was known to the Chinese 2000 years ago. Fortunately, today, the potential value of natural remedies central to the armamentaria of shamans and other traditional healers is increasingly being recognized as a source for the identification and development of the active ingredient as a drug. In addition to the longstanding interest of major pharmaceutical firms in natural products, several new companies, such as Shaman, Magainin, and Demeter, whose primary focus is the development of pharmaceuticals from natural products, have recently been incorporated.

The first step in pharmaceutical drug discovery is the identification of a compound with therapeutic potential. Classically this is accomplished by randomly screening large numbers of "off-the-shelf" agents for activity and, once an active compound is identified, trying to improve its efficacy and reduce its toxicity by chemical modification.[1] There are two major features which make natural products attractive for the process of new drug identification. The first is that random screening of natural products is likely to yield totally novel chemical structures which have not been synthesized, or even imagined, by a chemist and are therefore not on anyone's "shelf" to be tested. The second is that all multicellular residents of the natural world are subject to attack by microorganisms which cause infectious diseases. For example, infections of tomato ringspot virus, wheat mosaic virus, and *Phytophthora infestans,* the fungus which causes potato blight, can have a devastating impact on agriculture. Not surprisingly, many organisms have evolved natural mechanisms to defend against microbial invasion and these may prove to be effective therapies for similar infections of humans. Penicillin, an antibacterial agent synthesized by a fungus, is perhaps the best known example of this phenomenon. However, other examples include insect-derived cecropins, the magainins which were initially discovered in amphibians, the defensins from mammals, and the phytoalexins, which are produced by plants in response to infection (Gabay 1994).

A corollary to the advantages of novel chemistry and defensive evolution is the hope that solutions to the epidemic emergence of drug-resistant disease-causing microorganisms may be found in the natural world. Although the phenomenon of drug resistance was identified soon after the initial use of penicillin, the magnitude of the problem and its devastating impact on our ability to treat bacterial disease have only recently been fully appreciated. The report of a task force convened by the American Society for Microbiology concludes that "... there is little doubt that the problem (of antibiotic resistance) is global in scope and very serious" (Report of the ASM 1995).

The availability of therapeutic agents derived from natural products has been an important component of existing means to control infections caused by bacteria, viruses, fungi, and parasites. As new experimental compounds derived from natural sources to combat these infections are in all stages of development, they are also likely to be important for future control of infectious diseases. Chemical novelty has repeatedly allowed us to overcome infectious agents as they continue to challenge modern medicine, and without the chemical smorgasbord of natural products, our search for novelty would be much more difficult. An investment in preserving biodiversity is in fact an investment in the future of new anti-infective drugs.

In this chapter we will provide examples of current treatments for bacterial, viral, fungal, and parasitic pathogens, as well as experimental treatments currently "in the pipeline," which are derived from natural products. A synopsis of the drug discovery and development process will also be provided to illustrate the complex, multidisciplinary, and expensive process necessary to bring a natural product from forest to pharmacy. We will conclude that the need to identify better treatments for known infectious diseases, as well as the need to develop treatments for infections caused by drug-resistant and newly emerging microbial pathogens, support the continued importance of natural products for chemical novelty and therapeutic efficacy. This effort will require the close collaboration of infectious diseases experts, conservationists, and the pharmaceutical industry.

Antibacterials

Throughout history bacterial infections have plagued civilization. The scourges of tuberculosis and leprosy are documented in the Bible and the mummified remains of ancient Egyptians (Lappe 1995). The bubonic plague catapulted western Europe into the "Dark Ages." Although primitive control measures may seem to have relied primarily on prayer, quarantine, and conflagration, remedies derived from natural sources were also utilized. Despite these measures, people remained generally defenseless against virulent bacterial infections until the realization in the early 1940s of the significance of Alexander Fleming's 1929 discovery of penicillin, the first antibiotic.

The term "antibiotic" itself, defined as ". . . a soluble substance derived from a mold or bacterium that inhibits the growth of other microorganisms" (Hensyl 1990), emphasizes the natural derivation of these antibacterial agents. The recognition of penicillin's importance soon led to the discovery of other antibiotics such as streptomycin and later chloramphenicol, the tetracyclines and erythromycin in fungi from soil samples (Neu 1991). The vast majority of antibacterial drugs available today are either antibiotics themselves (whether synthesized by natural fermentation processes or chemically) or chemical derivatives of antibiotics. Antibacterial drugs can be divided into groups based on their mechanisms of action, as illustrated in Figure 7.1. Some inhibit synthesis of bacterial cell walls. These include cycloserine, bacitracin, and vancomycin all of which are derived from products of the *Streptomyces, Bacillus,* and *Nocardia* bacterial genera. Other drugs in this class, the cephalosporins, are derived from the fungus *Cephalosporium acremonium*. Another group of drugs interferes with bacterial membrane function. These include the polymixins, gramicidins, and polyenes which are all derived from strains of *Bacillus* and *Streptomyces* bacteria. Another class of drugs inhibits bacterial protein synthesis. Many of these, such as streptomycin, tetracycline, chloramphenicol, and erythromycin, are also derived from other strains of *Streptomyces.*

The optimistic expectation of the 1950s which followed the initial widespread availability of antibiotics, that the majority of bacterial infec-

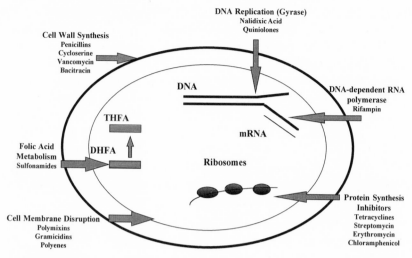

FIGURE 7.1

Examples of antimicrobial agents with different sites of action. DHFA, dihydro-folic acid; THFA, tetrahydrofolic acid. *Source:* Modified from Neu 1991.

tions could be brought under control, has largely been realized. However, there are numerous exceptions, especially for diseases which result from bacterial invasion of unusually susceptible populations such as the very young, the elderly, and the immunocompromised. For example, in the United States, the bacterium *Streptococcus pneumoniae* is a major cause of fatal pneumonia in the elderly and nonfatal pneumonia in other adults. Even though most pneumococcal strains are highly sensitive to penicillin, each year invasive pneumococcal infections cause greater than 40,000 deaths and significant disease in more than 10 times that number (Centers for Disease Control and Prevention 1994). Other bacterial diseases which remain serious global problems include cholera, pertussis, *Chlamydia* infections (both ocular and genital disease), enteric infections caused by *Salmonella, Shigella,* and highly virulent strains of *Escherichia coli* which can contaminate food products and received considerable publicity as the cause of death for several children in the United States who ate undercooked hamburgers.

Tuberculosis, once thought to be under control in the United States, has re-emerged as a serious concern in the United States and, with an estimated 8.5 million new cases each year worldwide, in 1993 was declared a public health emergency by the World Health Organization (Dolin et al.

1994; Klaudt 1995). Many factors have influenced this resurgence but the prolonged therapy required to achieve a cure relies on a combination of drugs, one of which, rifampin, is derived from *Streptomyces mediterranei*. An additional important contributor to the re-emergence of TB and other bacterial infections has been the emergence of drug-resistant organisms.

The initial success of antibiotic treatment led, perhaps inevitably, to the development of bacterial drug-resistance to such a degree that we may return to the vulnerability of the preantibiotic era. Several bacterial infections may soon be untreatable. These include methicillin-resistant *Staphylococcus aureus,* vancomycin-resistant *Enterococci* (see Figure 7.2), multidrug-resistant *Mycobacterium tuberculosis,* penicillin-resistant *Streptococcus pneumoniae,* and numerous Gram negative species. Factors leading to the emergence of resistance among nosocomial pathogens include the use of broad-spectrum antibiotics; increasing numbers of susceptible, immunocompromised patients; technologic changes (implants, catheters, i.v. lines) leading to increased exposure to resistant microorganisms; and the breakdown in hygiene, infection control, and disease control programs that lead to increased transmission of resistant bacteria.

Clearly, the control of bacterial infections is increasingly of major importance to global human health. Natural antibiotics, and more often, synthetic derivatives of these natural products, dominate today's available

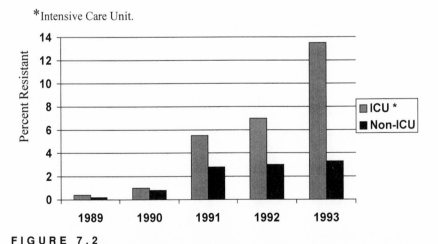

FIGURE 7.2

Increasing incidence of vancomycin-resistant enterococci in U.S. hospitals by year, national nosocomial infections surveillance system (NNIS), 1989–1993.
Source: Centers for Disease Control and Prevention, *MMWR* 42:597–599 (1994).

pharmaceutic agents. It is encouraging that new classes of antibiotics have been discovered and are currently in development, although there are no guarantees that they will prove to be clinically useful or financially feasible. These new agents include the cecropins, potent antimicrobial proteins, originally discovered in insects where they are produced in response to microbial infection (Merrifield et al. 1994). Since their initial discovery, they have also been identified in mammals and are related to the magainins found in amphibians and the defensins isolated in mammals (Ganz 1994; Jacob and Zasloff 1994). Most act by physically damaging bacteria, rather than interfering with their replication as do most currently available antibiotics. They are active against both Gram positive and Gram negative bacteria and act by the permeabilization of bacterial cell membranes. Some also kill parasites and tumor cells as well.

Plants are also a potential source for new antibacterial agents. Many plants synthesize phytoalexins, compounds with broad-spectrum antimicrobial activity, in response to infection just as humans produce a substance, interferon, in response to viral infections. Some phytoalexins have been shown to be inhibitory to methicillin-resistant and oxacillin-resistant *Staphylococcus aureus* (Iinuma et al. 1994). However, like the other natural compounds described above, it is too soon to tell if they will become useful drugs.

Antivirals

Like bacterial infections, diseases caused by viruses have had a serious impact throughout the ages and, for the most part, have not been controlled. These include influenza which causes annual epidemics and devastating pandemics every 30 years or so, the recently emerged human immunodeficiency virus (HIV), hepatitis viruses, and viruses causally associated with cancers such as papillomaviruses and Epstein-Barr virus. Other serious viral pathogens include respiratory syncytial virus which causes life-threatening infection in infants with heart or lung problems and serious respiratory infections in normal children; cytomegalovirus which congenitally infects one percent of all babies born in the United States each year, can cause birth defects, and is the major viral cause of hearing loss; and emerging viruses such as Ebola, dengue, and hantaviruses.

However, unlike bacterial infections, for which numerous treatments are now available, there are fewer than thirty drugs approved for the treatment of viral infections. Also in contrast to antibacterials which are often broad spectrum and useful against several or many different infections, antiviral drugs are usually effective against only one, or at the most two, viral infections. This is largely due to the differences in the way these pathogens replicate. For the most part, bacteria are free living organisms with the ability to generate energy, grow, and self-replicate. Accordingly, most bacteria will grow in a nutrient broth and do not require access to a plant or animal host. Viruses, however, are not independent cells. They are much smaller and simpler, have dramatically smaller genomes and lack the ability to supply their own needs for replication. Consequently, they persist and propagate by invading the cells of a host and subverting the cellular machinery normally used for the maintenance and growth of the host for the purpose of replicating the viral parasite. Since viruses thus "commandeer" the cells of the host, it is extremely difficult to identify an agent which will inhibit the virus without concomitantly harming the host.

Nonetheless, compounds with antiviral efficacy and acceptable safety have been identified and many more are under study. Several of the currently approved drugs are derived from natural sources. Vidarabine, a product of *Streptomyces antibioticus,* was the second antiviral drug approved for systemic (whole body) use. It was licensed initially as an intravenous treatment for herpes simplex encephalitis, and later also for disseminated infections caused by the varicella-zoster and herpes simplex viruses. Another natural antiviral is interferon, a protein produced by many mammals and birds to help fight viral infections. This protein has been proven useful, although not completely curative, for infections caused by papillomaviruses, two types of hepatitis viruses (B & C), and is an effective therapy for hairy cell leukemia. Although this protein initially had to be harvested from human blood, and later from virus-infected cultured cells, its gene, and those of related interferons, have now been cloned and the proteins can easily be produced in large quantities.

There are a number of natural products currently in various stages of preclinical and clinical testing as antiviral therapies. Some of these are true antivirals and others may control viral diseases by other mechanisms, such as modulating the immune response or interfering with neural transport. An example of the latter is capsaicin, a component of chile peppers, which is an active ingredient of an over-the-counter product for the control of pain resulting from shingles, a reactivation of infection with the virus which causes chickenpox. Capsaicin is being studied in animal models of herpes infection as a means to reduce recurrences of genital herpes, a result of infections with the herpes simplex virus, as well as a possible treatment for primary infection (Stanberry et al. 1994). A novel compound that inhibits the HIV enzyme integrase and therefore acts as a true antiviral, curcumin, is the source of the yellow color of tumeric and curries (Mazumder et al. 1995). This compound is now in early phase clinical testing for HIV infection. Screening of fermentation products of the fungus *Tolyplocladium niveum* at Sandoz Pharmaceuticals Corporation led to the discovery of NIM 811, a compound with anti-HIV activity. As NIM 811 acts at a different replicative stage than AZT, when the two compounds are used together in infected cell cultures they have a synergistic level of inhibition. Interestingly, although NIM 811 is an analog of another natural product, cyclosporin A, which is used to suppress the immune system of transplant recipients, NIM 811 has no immunomodulatory activity (Malkovsky et al. 1996).

As has already been described for drug-resistant bacteria, the emergence of viruses resistant to drugs has been observed. Although there are fundamental differences in the mechanisms of transmission of resistance that may reduce the prevalence of drug-resistant viruses in the general population, it is clear that antiviral resistance is of major concern for immunocompromised populations such as AIDS patients. Indeed, it has been necessary to design combination drug treatment strategies for these patients to delay the appearance of resistance which is usually a predictor of clinical failure. However, there are exceptions to the current sense that antiviral resistance is not a matter of concern for the population as a whole. A case of acyclovir-resistant herpes simplex in an immunocompetent person has recently been reported (Kost et al. 1993). Furthermore, it has been clearly documented that emergence of rimantadine-resistant influenza viruses are common only a few days after treatment initiation (Hayden et al. 1989). As we are less than twenty years into widespread use of antiviral drugs, it is too soon to evaluate the extent of the potential threat of antiviral resistance. In any event, with so few agents in the antiviral pharmacy, it is clearly essential to investigate and preserve any potential source of new, safe, and effective chemicals.

Antifungals

Fungal infections of the skin and mucosal surfaces are not uncommon in the general populace, yet they sometimes present challenges in treatment or patient management. They can also cause significant morbidity (disease) in physiologically stressed or immunodeficient people. Before the era of the AIDS epidemic, systemic fungal infections were the bane of cancer chemotherapy patients, individuals with hereditary immunodeficiencies, and organ or bone marrow transplantation recipients. Now, however, systemic infections with fungi such as *Cryptococcus, Candida,* or *Histoplasma* are also common infections in AIDS patients.

The pathogenic yeast *Candida albicans* provides an illustration of the burden of fungal infections on health. A recent review article notes that candidiasis is probably the most important human fungal infection in terms of overall prevalence and attributable morbidity and mortality (Dixon et al. 1996). *C. albicans* and other *Candida* species present a wide spectrum of clinical diseases that affects nearly 13 million people each year in the United States. To a great extent, the seriousness of a *Candida* infection depends on the virulence of the yeast strain and on the underlying health of the patient.

Both mucosal and disseminated candidiasis are on the increase. Factors which have contributed to this are underlying immunosuppression, use of broad-spectrum antibiotics, antineoplastic agents, and use of central venous catheters. Recent hospital surveys have shown *Candida* species now account for more than 10% of all nosocomial (hospital-acquired) bloodstream infections. Outbreaks in neonatal intensive care units are also increasing (Bendel and Hotstetter 1994). Three out of four otherwise healthy women will experience vulvovaginal candidiasis in their life time; women that are infected with the human immunodeficiency virus (HIV) experience more severe and frequent disease and it is often one of the earlier opportunistic infections they develop (Sobel 1985). Oral candidiasis is the most common fungal infection in HIV-infected people in general: mucocutaneous candidiasis (thrush and other manifestations) occurs in greater than 75% of AIDS patients at some time during the course of their disease (Dixon et al. 1996). Progression of localized infections to severe gastrointestinal candidiasis results in general malnutrition and dehydration of patients.

Current treatment of systemic fungal infections like candidiasis is often limited to a choice of three drug classes: amphotericin B, flucytosine, and synthetic azoles—alone or in combination (Vanden Bossche et al. 1994). There are limitations in their use, however: (1) these agents attack a limited number of targets in the fungal cell, (2) not all antifungals are effective against all types of fungi, and (3) structurally they represent only a few chemical classes which raises the question of how to circumvent resistance. In addition, toxicities of current antifungals can greatly limit their use. Lastly, certain of these drugs stop fungal growth but are not fungicidal, that is, they don't actually kill the fungus, so that long-term maintenance therapy is required for immunosuppressed people.

But of all these limitations probably the most worrisome is the problem of drug resistance. Clinicians are increasingly aware that resistance to antifungal agents is on the upswing, particularly to flucytosine, but resistance to orally active azoles, such as fluconazole, and to amphotericin B is also of considerable concern (He et al. 1994; Vanden Bossche 1994). Recent widespread use of over-the-counter antifungal preparations may also be contributing to the development and spread of antifungal-resistant *Candida*.

Due to the small number of chemical classes and fungal targets of known drugs, the prospect of developing chemical derivatives of drugs currently in use is not very promising. The likelihood exists that newer derivatives would act via similar mechanisms as the parent compounds, and therefore have limited utility against resistant and nonsusceptible pathogens.

Where are new antifungals to come from, then? Reliance on natural products for antifungal agents has had a long history. Of the most widely used antifungals, two were originally isolated as natural products from microbial fermentations: amphotericin B from *Streptomyces nodosus,* and griseofulvin from *Penicillium griseofulvum.* More recent examples of promising new antifungals isolated from microbial fermentation broths include the echinocandins (Schwartz et al. 1992; Schmatz et al. 1995) and the pradimicins (Kakushima et al. 1991).

The urgent need for new drugs against fungal pathogens has led to the consideration of a wide variety of potential natural sources beyond microbial fermentations. Thus, screening of higher plants for activity has experienced renewed interest over the past two decades. Higher plant extracts can offer the potential advantages of providing unique chemical structures with presumed novel mechanisms of action (Mitscher et al. 1987) which lowers the possibilities of cross-resistance to current drugs. Although some plant products have associated toxicities, further derivatization may lead to drugs that are better tolerated. Since many higher plants have developed defenses against fungal invaders, randomly screening their extracts for antifungal activity has met with some success.

Other investigators have reported success with plants using ethnobotanical leads. Examples of new classes of antifungal agents which have been isolated from medicinal plants include the sampangines, isolated from the West African tree *Cleistopholis patens* (Peterson et al. 1992), and pseudolaric acid, a promising lead compound found in a Chinese medicinal plant,

Pseudolarix kaempferi (Li et al. 1995). In addition, practitioners of what is often referred to as "alternative medicine" in the United States have made use of herbal preparations from traditional medicines from many ethnic/folkloric sources. Although many of these have not been tested under controlled clinical settings, these treatments are starting to attract more sophisticated investigation. One example is the use of the essential oil from *Melaleuca alternifolia,* the Australian "tea tree," as a topical treatment for mucocutaneous yeast infections. Eight components of this oil were evaluated for antimicrobial activity and were found to have significant activity against several bacterial species and *Candida albicans* (Carson and Riley 1995), thus strengthening its potential for development as a new topical antifungal.

Each of these exciting new leads has far to go on the road to approval as a safe and effective drug. And whether chemists use random screening or ethnobotanical guidance to search for new antifungals, the basic requirements of novelty, resources, and long-term investment remain the same to bring a drug to market. Of these, assurance of chemical novelty through preservation of biodiversity is perhaps the most important to overcome the advancing front of drug resistance.

Antiparasitics

Evidence of parasitic infections has been found in the fossilized remains and feces of prehistoric humans. Thus we have long been hosts to these pathogens and, in some cases, have no doubt co-evolved with them. Parasites come in a variety of forms from microscopic protozoa that can quickly cause life-threatening encephalitis to 20-foot tapeworms that can persist for years in a host with relatively mild symptoms. Across the fascinating spectrum of disease these organisms cause is the sobering reality that (1) their global health burden is staggering and (2) vaccines for parasitic diseases are at best experimental, and that we are heavily dependent on chemotherapy for control of these infections.

The basic biology and metabolism of many parasitic pathogens approximates that of human cells and therefore our armamentarium for treatment is not nearly as extensive as that for bacterial infections. Many of the older antiparasitic agents were toxic, difficult to administer, and posed a health risk themselves, such as arsenicals, antimonials, and general purgative agents. One can argue that true chemotherapy, defined as "treatment with a known chemical entity," drew its beginnings from antiparasitic remedies with the isolation of the active principles of cinchona bark for malaria (quinine) and ipecac for amoebic dysentery (emetine) in the 1820s. Other classic, plant-based remedies that were in wide use up to the twentieth century included aspidium from male fern extract (*Dryopteris filix-mas*) for liver flukes (*Fasciola hepaticum*) and for hookworm (*Necator americanus*), and pelletierine from pomegranate bark (*Punica granatum*) and ascaridole from American wormseed (*Chemoposium ambrosioides var. anthelminticum*) for various round worm infections. The twentieth century miracles of chemistry, pharmacology, and toxicology have since aided in the discovery of both synthetic and derived agents that are several degrees safer and more effective. An excellent review can be found in Campbell (1985). Current research has focused on discovery of unique biochemical differences between parasite and host that can be exploited as a target for chemotherapy with minimum toxicity: a review of this exhaustive body of literature is beyond the scope of this chapter.

Perhaps the best known of the age-old parasitic scourges is malaria, remarkable for its global breadth and impact on various civilizations, and a classic example of how chemical novelty of natural products and careful drug design can work together to develop effective drugs. Malaria is transmitted by mosquitoes in both tropical and temperate zones, and thus has long been associated with swamps and other bodies of fresh water that were thought to give rise to "*mal aire*" and subsequent sickness. Historically, it is perhaps most famous for having caused more troop casualties than did actual combat for the armies of Alexander the Great, Napoleon Bonaparte, and the U.S. forces stationed in the Pacific during both World War II and the Vietnam War. The discovery and widespread use of DDT and chloroquine in the 1940s led prematurely to the belief that the disease could be controlled or even eradicated. However, malaria has re-emerged as a pandemic in the past 30 years. The comeback of this parasitic disease was primarily due to the emergence of drug-resistant malarial parasites, the development of insecticide-resistant mosquitoes, and loss of surveillance programs.

Currently, an estimated 2.8 billion people live in malaria endemic areas where they are at risk of contracting the disease; of these, an estimated 1.9 billion live in abject poverty. An additional 20 to 30 million people from developed, industrialized countries are put at risk each year when they travel or work overseas. Estimates of the global disease burden claim that 200 million clinical cases occur each year, which result in 1 to 2 million deaths, primarily in children (Foster 1994). Millions of dollars have been spent on the search for a malaria vaccine, but prospects are poor for an effective, readily available, and inexpensive vaccine in the foreseeable future.

Not surprisingly, the multimillennial relationship of malaria and its human host gave rise to numerous herbal remedies within ancient cultures. Perhaps the best known of these is quinine, which is derived from the bark of various trees in the genus *Cinchona*. The "discovery" of quinine by the Europeans merits retelling here, if only to illustrate how indebted the world can be to the ethnobotanical knowledge of a remote culture. According to legend, in the seventeenth century the Countess of Cinchon, wife of the Viceroy of what is now Peru, became ill with malaria. Upon the advice of local healers, Spanish officials successfully treated her with a preparation of bark from a specific tree that the Aguaruna Jivaro Indians had used quite possibly for generations for this malady (Lewis and Elvin-Lewis 1995). Although the existence of the Countess has been disputed, the origins of this cure from indigenous ethnobotanical information is not. Export of cinchona bark to Europe was a lucrative business well into the twentieth century, when during World War II the Japanese cut off the bulk of the world supply of cinchona bark. This, in turn, gave great impetus to synthesizing and testing new quinine analogs by the allied forces (Campbell 1985).

To this day, quinine and its chemical analogs make up the bulk of our antimalarial armamentarium, but this may change for reasons given below. In 1994, the global market for antimalarial drugs was estimated at US $100 to 120 million, the bulk of which was spent on chloroquine, a synthetic 4-aminoquinoline derived from the structure of quinine. Chloroquine is probably the second or third most widely used drug in the world after aspirin (Foster 1994). It can be manufactured very cheaply and is therefore affordable to most of the developing world, as well as to the millions of travellers and troops that have relied on it since its discovery. Tragically, parasite resistance to chloroquine and to most of quinine's derivative drugs, including one of the newest, mefloquine, has emerged in recent decades. Numerous reports have documented the spread of drug resistance over the past 35 years from Southeast Asia to Africa and Latin America (Longworth 1995) which in turn has accelerated the spread of disease.

In light of the resurgence of malaria and decreasing efficacy of antimalarial drugs, scientists, industry, governments, and international organizations are seeking new treatments for the disease. Phillipson (1994) and colleagues have examined the potential of several dozen plants used throughout the world as antimalarials in traditional medicine and have isolated numerous active compounds in preliminary testing both in culture and in animal models. Most attention has been focused on a promising family of drugs derived from qing hao su (artemisinin), a substance which has been in use for perhaps 2000 years in Chinese traditional medicine and is prepared from the leaves of *Artemisia annua* (qing hao), or sweet wormwood (White 1994).

The use of leaves of *Artemisia annua* for fever in ancient Chinese medical texts was rediscovered in the 1970s by Chinese and Western researchers, and a primary active component, artemisinin, was purified (Qinghaosu Antimalarial Coordinating Research Group 1979). Hundreds of artemisinin analogs have been made in the past 15 years in efforts to enhance the activity and lower the toxicity of the drug. (Meshnick 1994; Posner and Oh 1992). Over a million people in Asia, Africa, and Brazil have participated in clinical trials as of 1994. Results from these trials indicate that artemisinin, or its derivatives artesunate and artemether, are the most rapidly acting of all antimalarial drugs. The World Health Organization has invested heavily in the development of yet another derivative, arteether. Despite the excellent performance of this class of drugs, cost and potential neurotoxicity will be critical factors in determining the extent to which they will be used (White 1994; Brewer et al. 1994).

We have been fortunate to have had the traditional knowledge of the Amazonian and Chinese cultures to aid us in the discovery of these essential drugs. As the malarial parasite continues to be exposed to even the most newly developed drugs, however, the development of resistance will be inevitable. Resistance to artemisinin derivatives can be induced in a laboratory setting, and the specter of resistance "in the field" will undoubtedly again appear on the horizon. With no available vaccine, and the rapidly diminishing effectiveness of known drugs, a health crisis is clearly looming—and the search for new leads for antimalarials will continue.

Conclusions

A plausible response to our urgent need to identify new drugs to treat infections caused by pathogenic microbes and, increasingly, their drug-resistant cousins is to seek natural solutions to these problems. The search for new chemical entities which can either be used as a drug or serve as a basis for modification to make a drug should include all forms of life: plant and animal, macro- and microscopic. The value of these natural products as sources for novel chemical structures is supported by the observation that, for the most part, bioactive compounds isolated from plants have vastly different chemical structures than those isolated from microorganisms (Recio et al. 1989).

Several pharmaceutical houses are known to have maintained natural products screening programs for decades to search for antibiotics against a variety of fungal, bacterial, and viral pathogens. These have traditionally looked to microbial fermentation broths, but increasingly are exploring other natural sources for novel antimicrobial agents. The plant world has also provoked the interest of both large and small pharmaceutical firms. Very few of an estimated 250,000 species of higher plants and 30 million total plant species have been evaluated for antimicrobial or any biologic activity (Harvey 1993). About 20,000 of these plant species are known to be used in traditional medicine and warrant further study (Phillipson 1994). More recently amphibians, mammals, and insects have also been the source of experimental antimicrobial agents. A review of our species' long history of combatting infectious and other diseases makes clear that there has rarely been a reliable means of predicting the source of the next important therapeutic discovery.

The drug discovery and development process from natural products has been described in detail elsewhere in this volume (see Grifo et al., Chapter 6) and in recent reviews (Waterman 1993; Cragg et al. 1994). The advent of "rational" drug design in medicinal chemistry and the development of sophisticated software for molecular modeling of inhibitors can maximize the potential of lead compounds from natural products that are themselves impractical for further pharmaceutical development (Nasr et al. 1992; Des Jarlais et al. 1987). Due to the enormity of investment and technical resources necessary for drug discovery and development, there is a strong rationale for professionals in industry, ethnobotany, and conservation to work together to maximize the chances of finding new drug leads from natural products. Combining these skills has lead to the formation of at least one pharmaceutical company (King and Tempasta 1994), and historically ethnobotanical approaches have often proven to be productive (Soejarto 1993; Duke 1995). Recent studies indicate that efforts to conserve natural resources for future study might best succeed if economic reciprocity is established with communities in source countries. This meritorious concept is examined in detail elsewhere in this volume (Cox, Chapter 9; Moran, Chapter 11; Rosenthal, Chapter 13).

New or re-emerging infectious diseases have made serious impacts on public health: lyme disease, AIDS, Legionnaire's disease, cryptosporidiosis, and drug-resistant forms of tuberculosis, pneumococcal pneumonia, and gonorrhea represent just a short list. The emergence of new infectious diseases and the rapidly increasing incidence of drug-resistant disease, as well as the persistence of infections for which therapies have not been discovered, threaten to return the status of the world's public health to a condition similar to the pre-antibiotic era. The search for chemically novel, safe and effective therapies is urgent and it is consequently essential to preserve and look to the natural world for the antimicrobial therapies of tomorrow.

Note

1.   A second strategy of drug discovery is often referred to as "rational design." This utilizes knowledge of the three-dimensional atomic structure of a microbial target enzyme or organism to design a chemical entity which inactivates the target. This approach has the significant advantage of specificity with the concomitant expectations of efficacy and minimal toxicity. Although this approach has been successfully used to design a number of new drugs, such as several inhibitors of the HIV protease recently approved by the FDA, and is likely to become increasingly important, it is often limited by the lack of information on the structure of the target.

References

Bendel, C.M., and M.K. Hostetter. (1994). Systemic Candidiasis and Other Fungal Infections in the Newborn. *Seminars in Pediatric Infectious Diseases* 5:35–41.

Brewer, T.G., J.O. Peggins, S.J. Grate, J.M. Peteras, G.S. Levine, P.J. Weina, J. Swearengen, M.H. Heiffer, and B.G. Schuster. (1994). Neurotoxicity in Animals Due to Arteether and Artemether. *Transactions of the Royal Society of Tropical Medicine and Hygiene* 88:S1-33–S1-36.

Campbell, W.C. (1985). Historical Introduction. In *Chemotherapy of Parasitic Diseases,* edited by W.C. Campbell and R. S. Rew, 3–21. New York: Plenum Press.

Carson, C.F. and T.V. Riley. (1995). Antimicrobial Activity of the Major Components of the Essential Oil of *Melaleuca alternifolia. Journal of Applied Bacteriology* 78:264–269.

Centers for Disease Control and Prevention. (1994). Summary of Notifiable Diseases, United States, 1993. *Morbidity and Mortality Weekly Report:*10/21/94.

Cragg, G.M., M.R. Boyd, J.H. Cardellina, II, D.J. Newman, K.M. Snader, and T.G. McCloud. (1994). Ethnobotany and Drug Discovery: The Experience of the U.S. National Cancer Institute. In *Ethnobotany and the Search for New Drugs*, edited by G.T. Prance, D.J. Chadwick, and J. Marsh, 178–196. Chichester: John Wiley & Sons Ltd.

Des Jarlais, R.L., G.L.Seibel, I.D. Kuntz, P. Furth, J. Alvarez, P. Ortiz de Montellano, L.N. Babe, and C. Craik. (1987). Structure-Based Design of Nonpeptide Inhibitors Specific for the Human Immunodeficiency Virus Protease. *Proceedings of the National Academy of Science* 87:6644–6648.

Despommier, D. D., R. W. Gwadz, and P.J. Hotez. (1995). Malaria. In *Parasitic Diseases*, 3rd edition, 174–189. New York: Springer-Verlag, Inc.

Dixon, D.M., M.M. McNeil, M.L. Cohen, and J.R. La Montagne. (1996). Emerging Fungal Infections and the Public Health. *Public Health Reports* III (3, May/June):226–235.

Dolin, P.J., M.C. Raviglione, and A. Koch. (1994). Global Tuberculosis Incidence and Mortality during 1990–2000. *Bulletin of the World Health Organization* 72(2):213–220.

Duke, J. (1995). Personal communications. U.S. Department of Agriculture.

Foster, S. (1994). Economic Prospects for a New Antimalarial Drug. *Transactions of the Royal Society of Tropical Medicine and Hygiene* 88: S1-55–S1-56.

Gabay, Joelle. (1994). Ubiquitous Natural Antibiotics. *Science* 264: 373–374.

Ganz, T. (1994). Biosynthesis of Defensins and Other Antimicrobial Peptides. *Ciba Foundation Symposia* 186:62–76.

Harvey, A. (1993). An Introduction to Drugs from Natural Products. In *Drugs from Natural Products: Pharmaceuticals and Agrochemicals*, edited by A. Harvey, 1–6. New York: Ellis Horwood.

Hayden, F.G., R.B. Belshe, R.D. Clover, A.J. Hay, M.G. Oakes, and W. Soo. (1989). Emergence and Apparent Transmission of Rimantadine-Resistant Influenza A Virus in Families. *New England Journal of Medicine* 321:1696–1702.

He, X., R.N.Tiballi, L.T. Zairns, S. F. Bradley, J.A. Sangeorzan, and C.A. Kauffman. (1994). Azole Resistant Oropharyngeal *Candida albicans* Isolates from Patients Infected with HIV. *Antimicrobial Agents and Chemotherapy* 38:2495–2497.

Hensyl, W.R. (1990), editor. *Stedman's Medical Dictionary.* Baltimore: Williams & Wilkins.

Iinuma, M., H. Tsuchiya, M. Sato, J. Yokoyama, M. Ohyama, Y. Ohkawa, T. Tawaka, S. Fujiwara, and T. Fujii. (1994). Flavanones with Potent Antibacterial Activity against Methicillin-Resistant *Staphylococcus aureus. Journal of Pharm. Pharmacology* 46:892–895.

Jacob, L. and M. Zasloff. (1994). Potential Therapeutic Applications of Maga-inins and Other Antimicrobial Agents of Animal Origin. *Ciba Foundation Symposia* 186:197–216.

Kakushima, M., S. Msuyoshi, M. Hirano, M. Shinoda, Aohta, H. Kamei, and T. Oki. (1991). *In Vitro and In Vivo* Antifungal Activities of BMY-28864, a Water-Soluble Pradimicin Derivative. *Antimicrobial Agents and Chemother-apy* 35:2185–2189.

King, S.R. and M.S. Tempesta. (1994). From Shaman to Human Clinical Tri-als: The Role of Industry in Ethnobotany, Conservation, and Community Reciprocity. In *Ethnobotany and the Search for New Drugs,* edited by G.T. Prance, D.J. Chadwick, and J. Marsh, 197–213. Chichester: John Wiley & Sons.

Klaudt, K., editor. (1995). WHO Report on the Tuberculosis Epidemic, 1995. Tuberculosis Programme, World Health Organization, Geneva, Switzer-land.

Kost, R.G., E.L. Hill, M. Tigges, and S.E. Straus. (1993). Brief Report: Recur-rent Acyclovir Resistant Genital Herpes in an Immunocompetent Host. *New England Journal of Medicine* 329:1777–1781.

Lappe, M. (1995). *Breakout: The Evolving Threat of Drug-Resistant Disease.* San Francisco: Sierra Club Books.

Lewis, W.H. and M. P. Elvin-Lewis. (1995). Medicinal Plants as Sources of New Therapeutics. *Annals of the Missouri Botanical Garden* 82:16–24.

Li, E., A.M. Clark, and C.D. Hufford. (1995). Fungal Evaluation of Pseudolaric Acid B, a Major Constitutent of *Pseudolarix kaempferi. Journal of Natural Products* 58: 57–67.

Longworth, D.L. (1995). Drug-Resistant Malaria in Children and Travelers. *Pe-diatric Clinics of North America* 42: 649–664.

Malkovsky, M. et al. (1996). Immunodeficiency Virus-Inhibitory Effects of Cyclosporine and its Nonimmunosuppressive Analogues. *International An-tiviral News* 4:7–10.

Mazumder, A., K. Raghavan, J. Weinstein, K.W. Kohn, and Y. Pommier. (1995). Inhibition of Human Immunodeficiency Virus Type 1 Integrase by Cur-cumin. *Biochemical Pharmacology* 49:1165-1170.

Merrifield, R.B., E.L. Merrifield, P. Jurradi, D. Andrew, and H.G. Boman. (1994). Design and Synthesis of Antimicrobial Peptides. *Ciba Foundation Symposia* 186:5–26.

Meshnick, S. R. (1994). The mode of action of antimalarial endoperoxides. *Transactions of the Royal Society of Tropical Medicine and Hygiene* 88:S1-31–S1-32.

Mitscher, L.A., S. Drake, S.R. Gollapudi, and S.K. Okwute. (1987). A Modern Look at Folkloric Use of Anti-Infective Agents. *Journal of Natural Products* 50:1025–1040.

Nasr, M., J. Cradock, and M.I. Johnston. (1992). Rational Design of Anti-HIV Drugs. *Drug News & Perspectives* 6:338–344.

Neu, H.C. (1991). Antimicrobial Chemotherapy. In *Medical Microbiology*, edited by S. Baron, 179–202. New York: Churchill Livingstone.

Peterson, J.R., J. Zjawiony, S. Liu, C.D.Huford, A.M. Clark, and R.D. Rogers. (1992). Copyrine Alkaloids: Synthesis, Spectroscopic Characterization and Antimycotic/Antimycobacterial Activity of A- and B-Ring Functionalized Sampangines. *Journal of Medicinal Chemistry* 35:4069–4077.

Phillipson, J.D. (1994). Natural Products as Drugs. *Transactions of the Royal Society of Tropical Medicine and Hygiene* 88:S1-17–S1-19.

Posner, G. and C.H. Oh. (1992). A Regiospecifically Oxygen-18 Labelled 1,2,4 Trioxane: A Simple Chemical Model System to Probe the Mechanisms for the Antimalarial Activity of Artemisin (Qinghaosu). *Journal of the American Chemical Society* 114:8328–8329.

Qinghaosu Antimalarial Coordinating Research Group. (1979). Antimalarial Studies on Qinghaosu. *Chinese Medical Journal* 92:811–816.

Recio, M.C., J. L. Rios, and A.Villar. (1989). A Review of Some Antimicrobial Compounds Isolated from Medicinal Plants Reported in the Literature, 1978–1988. *Phytotherapy Research* 3:117–125.

Report of the American Society for Microbiology Task Force on Antibiotic Resistance. (1995). American Society for Microbiology, Washington, D.C.

Schmatz, D. M., M. A. Poles, D. McFaden-Nollstadt, F. A. Bouffard, J. F. Dropinski, and P. Liberator. (1995). New Semi-synthetic Pneumocandins with Improved Efficacies against *Pneumocystis carinii* in the Rat. *Antimicrobial Agents and Chemotherapy* 39:1320–1323.

Schwartz, R. E., D. F. Sesin, H. Joshua, K. E.Wilson, A. J. Kempf, K. A. Goklen, D. Kuehner, P. Gailliot, C. Gleason, R. White, E. Inamine, G. Gills, P. Salmon, and L. Zitano. (1992). Pneumocandins from *Zalerion arboricola*. I. Discovery and Isolation. *Journal of Antibiotics* 45:1853–1866.

Sobel, J.D. (1985). Epidemiology and Pathogenesis of Recurrent Vulvovaginal Candidiasis. *Journal of Obstetrics and Gynecology* 152:924–935.

Soejarto, D. (1993). Logistics and Politics in Plant Drug Discovery: The Other End of the Spectrum. In *Human Medicinal Agents from Plants,* edited by A.D.

Kinghorn and M.F. Balandrin, 96–111. American Chemical Society Symposium Series 534. American Chemical Society, Washington, D.C.

Stanberry, L.R., N. Bourne, and F.J. Bravo. (1994). Comparison of Topical Capsaicin and Parenteral Acyclovir in the Treatment of Primary Genital Herpes: Effect of Treatment on the Subsequent Development of Recurrent Disease. *Antiviral Research* 23 (S1):73.

Vanden Bossche, H., P. Marichal, and F.C. Odds. (1994). Molecular mechanisms of drug resistance in fungi. *Trends in Microbiology* 2:393-400.

Waterman, P.G. (1993). Natural Products from Plants: Bioassay-Guided Separation and Structural Characterization. In *Drugs from Natural Products: Pharmaceuticals and Agrochemicals,* edited by A. Harvey, 7–16. London: Ellis Horwood.

White, N.J. (1994). Artemisinin: Current Status. *Transactions of the Royal Society of Tropical Medicine and Hygiene* 88:S1-3–S1-4.

WHO. (1989). *In vitro* Screening of Traditional Medicines for Anti-HIV Activity: Memorandum from a WHO Meeting. *Bulletin of the World Health Organization* 67:613–618.

CHAPTER 8

Capturing the Chemical Value of Biodiversity: Economic Perspectives and Policy Prescriptions

ANTHONY ARTUSO

For millennia nature has provided humanity with a source of medicines and other valuable products. However, it is the more recent concern over the rate of species extinction that has brought the development of new pharmaceuticals and other products from biological material to the attention of politicians and policymakers. Interest in the potential connection between biochemical prospecting and biodiversity protection was heightened in 1991 when Merck announced its contract with Costa Rica's Instituto Nacional de Biodiversidad (INBio). The Merck–INBio contract attracted worldwide attention because it provided an advance payment for access to a supply of biological samples as well as a share of the royalties generated from any new products derived from the venture. In addition, 10% of the advance payments and 50% of any royalties were pledged to a fund established to support Costa Rica's National Park Service. Other organizations, including the National Cancer Institute, Bristol Meyers-Squibb, Smith-Kline, Glaxo, and Pfizer have also agreed to provide or negotiate royalties and other forms of compensation for access to biological material in countries such as China, Surinam, Peru, Argentina, Mexico, Chile, Nigeria, and Cameroon.

These agreements, together with the size of the market for biologically derived pharmaceutical products, have raised hopes that natural product research by the pharmaceutical industry could provide substantial economic incentives to protect biodiversity while also providing biologically rich countries with a new opportunity for sustainable development. The study of natural product contributions to the pharmaceutical industry, which is summarized in Chapter 6, found that on the basis of the number of prescriptions filled each year, 57% of the top selling 150 pharmaceutical products in 1993 contained active ingredients that were natural products or derivatives or analogs of natural products. Since sales of prescription drug products in 1990 were approximately $147 billion (USITC 1991), biologically derived pharmaceuticals could reasonably generate in excess of $80 billion in revenues per year. While this estimate provides some indication of the historical importance of biodiversity as a source of new drugs, it does not provide sufficient information on the value of preserving species for their pharmaceutical potential.

Despite growing international interest, there have been very few systematic efforts to estimate the value of biological resources as a source of new chemical compounds. Farnsworth and Soejarto (1985) were perhaps the first to estimate the cost of plant extinctions in terms of lost opportunities for new drug development. Principe (1991) updated and expanded Farnsworth and Soejarto's analysis using the same basic methodology. Unfortunately, these studies do not take into account the costs of pharmaceutical research and development or consider how alternatives to natural product research might affect their estimates of the foregone benefits resulting from plant extinctions. As a result, both studies tend to overestimate the value of preserving a species for pharmaceutical research.

Aylward (1993) and Mendelsohn and Balick (1995) have developed estimates of the pharmaceutical value of rainforest species that attempt to incorporate the cost of new drug development as well as potential revenues. However, neither study outlined very detailed models of the pharmaceutical research and development (R&D) process. In particular, neither study takes into account the relationship between the cost of new drug development from biological material and the number of biological samples that must be tested to develop a new drug. Mendelsohn and Balick also fail to realize that many proprietary drug screening technologies are testing for activity in very similar therapeutic categories. As a result, their analysis overestimates the value of biodiversity to the pharmaceutical industry.

The valuation model developed by researchers at Resources for the Future (RFF) (Simpson et al. 1996) makes an important contribution to our understanding of this issue by demonstrating that the expected value of screening an additional biological sample declines as the number of samples increases. The declining value of screening additional samples results from the increasing probability of already having satisfied the demand for new drugs from samples that have already been screened. Using their model together with some estimates of the demand for and net revenues of new drugs, Simpson et al. attempt to show that, even in areas with the highest density of endemic plant species, the value of conserving natural habitat for purposes of pharmaceutical research is quite low. This conclusion rests on a number of questionable assumptions and on an incompletely specified economic model. For example, they assume that discovery of ten new pharmaceutical products per year (the average number approved annually by the FDA over the past ten years) would saturate the potential demand for new pharmaceutical products. In addition, they assume that discovery of a new drug in one therapeutic area does not affect the demand for other types of new drugs.

In fact, the demand for new drugs is rarely if ever satisfied by the few new drugs approved each year by the FDA. As might be expected, increasing the demand for new drugs in the RFF model would increase the expected pharmaceutical value of a plant species. Rather than assuming that drug discovery in one therapeutic area will have no effect on the future demand for new drugs in other therapeutic areas, it seems more reasonable to assume a positive relationship. For example, demand for new arthritis and anticancer drugs would be expected to increase as a result of the discovery of other drugs that extend life expectancy. Incorporating this positive feedback effect into the RFF model would significantly increase their estimate of the pharmaceutical value of the marginal plant sample. In fact, if the annual percentage increase in the demand for new drugs as a result of prior drug discoveries is greater than the discount rate, the RFF model would indicate that the value of preserving a species or natural area for pharmaceutical research is infinite.

The models described above were intended to provide estimates of the pharmaceutical value of biodiversity or components of biodiversity. These studies were not intended to provide insights into how to manage biochemical prospecting projects or the emerging market for biochemical resources in order to maximize net benefits. All of these studies model the pharmaceutical R&D process as a black box. Costs and success rates are treated as if they are indivisible rather than occurring in stages with decision points and new information available at each stage. In addition, biological samples and extracts are modeled as uniform commodities that are not distinguished by the quality of the information associated with them. In addition, these studies do not incorporate public benefits or option value into their valuation estimates.

To provide practical guidance to companies, countries, and international organizations that have an interest in the emerging biochemical prospecting market, it is necessary to develop more detailed and flexible models of natural product research in the pharmaceutical industry. A more systematic analysis of the political economy of this emerging market is also needed to guide international policy development. In this chapter, I present three analytical perspectives from which to consider the value of biodiversity for pharmaceutical research. I then outline several reasons why this potential value may not automatically lead to significant incentives for biodiversity protection or an equitable distribution of resulting benefits. In light of this economic analysis, I evaluate several protocols and policy options that could help promote an efficient and equitable biochemical prospecting market.

Pharmaceutical R&D As a Series of Lotteries

Pharmaceutical research and development (R&D) can be thought of as a series of lotteries, each of which represents a stage in the R&D process from preliminary research through clinical trials and FDA approval. The price of entering each lottery is the cost of testing, isolating, and/or modifying a compound in that R&D phase. The prize associated with each lottery is information; information about the chemical structure of an effective new drug and information about the expected value of proceeding with the next phase of the R&D process.

It is possible to graphically depict the series of lotteries involved in the pharmaceutical R&D in the form of a decision tree, as shown in Figure 8.1. Each branch on the tree represents the success or failure of a particular phase in the pharmaceutical R&D process (e.g., screening of biological material, identification of active compounds, preclinical and clinical trials). Success in any given phase is defined as test results that justify proceeding to the next phase.

Given statistical data or expert judgment about (1) the costs, duration, and success rates of each R&D phase, (2) the expected revenue of a new drug, and (3) the cost of capital, the decision tree model of the pharmaceutical R&D process can be used to determine the expected value of a compound at any stage in the R&D process.[1] The final step in the underlying calculations yields the expected value of screening a biological extract for potential activity in a particular therapeutic category. A simple example may help to illustrate these points.

Suppose a pharmaceutical company is considering launching a new program to screen 10,000 biological extracts for new drug leads in ten differ-

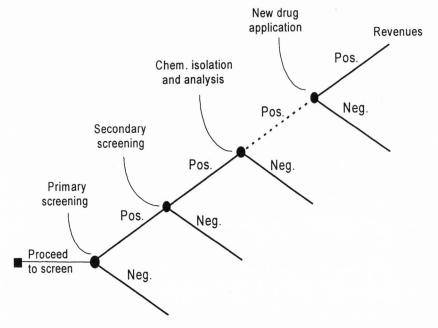

FIGURE 8.1
Decision tree of pharmaceutical R&D process.

ent therapeutic categories. The firm has developed estimates of the costs, success rates, and duration of each R&D phase which are summarized in Table 8.1. The firm has also determined that its real cost of capital (i.e., net of inflation) is 7.5%. In addition, the firm's market analysts have developed estimates of the expected annual revenues (Table 8.2) of new drugs in the therapeutic categories that are the focus of their screening program.[2]

These data together with the decision tree framework described earlier can be used to estimate the net benefits to the pharmaceutical company (or to society if social benefits are incorporated) of screening a biological extract for a particular therapeutic use. Using a real discount rate of 7.5% and the R&D costs, expected new drug revenues and other assumptions summarized in Tables 8.1 and 8.2, the pre-tax net present value to the pharmaceutical company of screening each biological extract would be approximately $1100.[3] This analysis is summarized in Figure 8.2. By multiplying the value of each compound in each screening test by the number of compounds being screened and the number of distinct therapeutic objectives being screened for, it is possible to derive a first approximation of the expected value to a private research organization of screening a certain

TABLE 8.1
R&D Phase Durations, Costs, and Success Rates—Hypothetical Example

	Effective Duration (yrs.)	Success Prob. per Trial	Mean No. of Successes	Cost per Trial
Initial Screening[a]	0.50	0.004	400.00	0.10
Secondary Screening	0.10	0.400	160.00	1
Isolation & Dereplication	0.50	0.100	16.00	20
Synthesis and Modification	1.50	0.500	8.00	250
Preclinical Trials	1.00	0.400	3.20	771
Clinical Phase I	1.35	0.750	2.40	3,137
Clinical Phase II	1.88	0.475	1.14	9,933
Clinical Phase III	2.49	0.700	0.80	18,817
NDA	3.00	0.900	0.72	1,000
CUMULATIVE	12.32	0.000007	0.72	33,930

[a]Cost per trial for this phase is $100 times the number of primary screens. Number of extracts tested, 10,000; number of indicators in primary screen, 10.

quantity of biological extracts for potential activity in relation to several therapeutic objectives.[4] In this example, the expected pre-tax net present value of testing 10,000 biological extracts in ten therapeutic categories is approximately $11 million.

Of course assumptions about R&D success rates, potential revenues of a new drug, and the cost of capital can significantly affect the estimated value of screening biological material. A reduction in the expected success rate of any phase reduces the expected value of the screening program and vice versa. However, changes in the success rates of later R&D phases have a much greater impact on the expected value of the program. If a new drug candidate fails in clinical trials, it represents the failure of all R&D expenditures necessary to identify that new drug candidate and gain approval for testing it in humans. However, if a raw biological extract fails to show significant activity in preliminary screening, very little time or money has been lost. In the current example, if the anticipated primary screening success rate is reduced by 20%, from .005 to .004, the expected value of screening each extract is reduced from $1100 to $473. But, if the conditional success rate of the first phase of clinical trials is reduced by 20%, from .4 to .32, the expected value of screening each biological extract becomes negative.

TABLE 8.2
Expected Revenues from Each New Drug (in $000's)[a]

Year	Gross Global Sales	Production and Marketing Costs	Net Revenues	Exp. After Tax Net Revenues
0	0	(42,334)	(42,334)	(27,517)
1	13,605	(21,768)	(8,163)	(5,306)
2	44,811	(51,644)	(6,833)	(4,441)
3	56,013	(49,963)	6,050	3,933
4	70,017	(44,362)	25,655	16,676
5	84,020	(52,764)	31,256	20,316
6	96,623	(60,326)	36,297	23,593
7	108,701	(67,572)	41,128	26,734
8	116,310	(72,138)	44,172	28,712
9	122,707	(75,976)	46,731	30,375
10	127,002	(78,553)	48,449	31,492
11	128,907	(77,344)	51,563	33,516
12	117,188	(70,313)	46,87	530,469
13	106,535	(63,921)	42,614	27,699
14	96,850	(58,110)	38,740	25,181
15	88,045	(52,827)	35,218	22,892
16	80,041	(48,025)	32,016	20,811
17	74,457	(44,674)	29,783	19,359
18	69,262	(41,557)	27,705	18,008
19	64,430	(38,658)	25,772	16,752
20	59,935	(35,961)	23,974	15,583
TOTAL	1,725,456	(1,148,788)	576,668	374,834

[a]Assumptions: (1) Contribution margin, 0.40; (2) global/U.S. sales, 1.90; (3) tax rate, 35%; (4) capital costs equal one-half of tenth-year sales with two-thirds of expenditures prior to product launch and the balance in years 2 through 10; (5) marketing expenditures equal 100%, 50%, and 25% of sales in years one through three, respectively.

This simple sensitivity analysis illustrates that estimates of the number of compounds that must be tested to develop one new drug provide very little information about the chemical value of biodiversity. In both of the examples outlined above, approximately 110,000 biological extracts would have to be tested in order to identify a novel compound that will eventually lead to a new drug in a given therapeutic category. The difference in the expected value of screening each extract in these two examples is due to differences in how quickly unsuccessful new drug leads are identified and dropped from further testing. Many natural product screening programs are consciously designed to have low "hit rates" in preliminary and

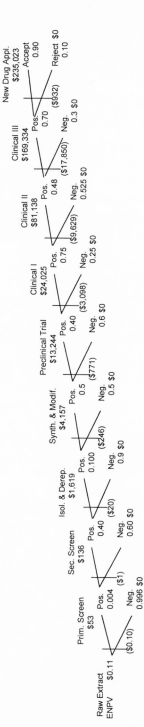

Notes
1) Figures directly below vertical lines are the costs of testing a compound in that phase, discounted to start of phase.
2) Figures below each outcome indicate success and failure rates for compounds entering that R&D phase.
3) Figures above positive results branch indicate the expected value of a compound after successful completion of that R&D phase.
4) Value of compound after regulatory approval is the pre-tax value of expected revenues from a new drug discounted to date of approval.
5) All values are discounted to the start of each R&D phase at a real rate of 7.5% based upon R&D phase durations presented in Table 1.

FIGURE 8.2

Summary of valuation analysis for each therapeutic objective in hypothetical biochemical prospecting example (all monetary values in $000's) even secondary screening so that the very substantial expenditures of subsequent R&D phases are lavished on only the most promising new leads. A well-designed screening program will seek to balance the added secondary screening and research costs resulting from too lenient a definition of a "hit" in the primary screening phase against the forfeited potential benefits of new drug leads that could go undetected if the definition of a "hit" in the primary screening phase is too stringent.

The cost of capital can also significantly affect the expected value of a proposed biological screening program simply because significant R&D expenditures must be made many years in advance of potential revenues. If the discount rate in the above example is increased from 7.5% to 8.5%, the present value of screening each compound is reduced from an expected net benefit of $1100 to less than zero. The discount rate or cost of capital must be understood not only as the return that investors in pharmaceutical companies demand for the use of their capital but also as the potential return available from other drug discovery opportunities such as those posed by gene therapy or recombinatorial chemistry.

The expected value of obtaining and screening a supply of biological extracts as computed from the model presented above should be understood as the maximum that a firm would be willing to pay to obtain exclusive rights to these extracts. The actual market price for biological extracts or samples could be significantly less given competition between suppliers. In addition, uncertainty over the costs and risks of natural product research creates incentives for firms to wait for more information before making significant investments (Pindyck 1991). However, even if the expected net present value of screening a large quantity of biological samples is negative it may still be advantageous to screen a small sample as a means of gaining additional information on screening success rates and costs. This is particularly true when there is significant uncertainty about these parameters. Given the possibility that technological improvements and scientific discoveries might increase the value of biodiversity for pharmaceutical research, it may also be economically efficient to conserve biodiversity even if the expected value of alternative uses of biologically diverse habitats is greater than the expected value to be derived from preservation (Fisher and Hanemann 1986; Artuso 1994).

Depicting pharmaceutical R&D as a series of lotteries and analyzing the decision tree that summarizes the available data on these lotteries provide a means of estimating the value of biological resources for particular screening programs. However, this model of the R&D process does not highlight the reasons why biological material can be an important input in pharmaceutical R&D programs. It is the inherent chemical diversity of biological material that is the source of its value. This can be illustrated through the use of two other analytical perspectives on the pharmaceutical R&D process.

The Option Value of Biochemical Prospecting

A call option on a stock provides the holder the right to buy a share of stock at a predefined price, known as the strike price. If the market price of the stock rises above the strike price, the holder will exercise the option to purchase the stock, thereby realizing a profit equal to the difference between the market price and the strike price. Models developed to value financial options indicate that the value of the option is positively related to the expected variation in the price of the stock over time. The relationship between the variability of the stock price and the value of the option relationship provides insights into the value of biodiversity in pharmaceutical screening programs.

Screening biological extracts or continuing to pursue R&D on a new drug lead gives the pharmaceutical company or research institute the option to continue with subsequent R&D phases. In this sense screening each biological extract is similar to purchasing a call option. If screening test results exceed some critical value it will be profitable to exercise the option to continue with further R&D. Similarly, the value of screening a supply of biological extracts will be positively correlated with the expected variation in the screening test results. The reason for this is quite simple. The potential loss involved in screening a biological extract or compound is limited to the cost of performing the screening test, but the potential gain is limited only by the potential revenues or social benefits that would result from discovery of a new drug. Consequently, a high variance in expected test results is a positive asset.

The more diverse the supply of biochemical compounds being screened, the greater the expected variance in screening test results and the greater the expected benefits of the screening program. The analogy with financial options indicates that the wide variety of chemicals produced by nature is the source of biodiversity's value to the pharmaceutical industry.

Exploring Molecular Space

The commonly observed similarity between natural product research in the pharmaceutical industry and prospecting for precious metals or drilling exploratory oil wells can also yield valuable insights about the source of biodiversity's pharmaceutical value. The goal of drug discovery is to identify chemical compounds, which I shall now refer to simply as molecules, that can inhibit a disease or activate a healing response. It is possible to think of drug discovery as the exploration of a multidimensional space where each dimension represents the possible range of a relevant molecular characteristic such as size, shape, valence, or atomic composition. Molecules with the desired therapeutic properties will occupy a very small portion of this space. In some cases, sufficient biomedical information is available to know where in this "molecular space" to look for potential new drugs. Unfortunately, for many diseases and ailments, this type of information, which is essential for a rational drug design effort, is simply not available. It is in these situations that access to a diverse array of biologically derived molecules can be particularly valuable.

Consider a molecular space where each dimension defines a molecular characteristic that is of relevance to a proposed new drug discovery effort. The screening test that will be used in this drug discovery effort can detect bioactivity of molecules that lie within a certain "distance" from the point in molecular space occupied by the assay. Libraries of synthetically derived compounds often contain hundreds of thousands of molecules, but most of these are simply slight variations of one another (Harris 1995). In the context of the spatial analogy that I have been developing, all of the molecules may be tightly grouped in one small region of the molecular space of interest. The ideal new drug may be represented by a combination of characteristics that is quite distant from the region in molecular space where the library of synthetic compounds is grouped. It is entirely possible that none of the compounds in the library will react strongly with the bioassay and little new information will be obtained about the structure of an effective new drug.

Now consider a biological screening program involving a collection of biological extracts drawn from widely divergent taxonomic groups, ecological niches, and geographic regions. Each of these extracts may contain hundreds or thousands of different molecules. Presumably, the total molecular diversity embodied in the entire collection of extracts is quite high. If the molecules are distributed randomly throughout the molecular space of interest (as shown in Figure 8.3), then testing the entire collection of extracts is likely to identify a few that react strongly with the screen. After a process of chemical isolation and characterization is completed, important new information has been obtained about the structure of molecules that are likely to be effective in the therapeutic area that is being screened for. Given this information, combinatorial chemistry, tissue culture, and

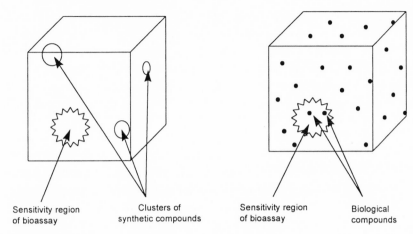

FIGURE 8.3
Drug discovery as the exploration of molecular space.

other drug discovery techniques can be used to develop new molecules that are slight variations on the original natural product. The objective at that stage is to increase the therapeutic benefits and/or reduce the toxicity of the original natural product.

This spatial analogy highlights the importance of obtaining a highly diverse supply of molecules to screen and the potential synergy between natural product research and other forms of drug discovery. By isolating and analyzing the structure of molecules that react with a bioassay, it is possible to determine what area of molecular space to search in more detail (i.e., what types of analog molecules should be developed through chemical synthesis or bioengineering and then evaluated in subsequent tests). The more diverse the collection of biological extracts or molecules being tested and the more sensitive the bioassay, the greater the region of molecular space that can be effectively explored and the more likely a new drug lead will be discovered.

Promoting an Efficient and Equitable Biochemical Prospecting Market

The signing of the Biodiversity Convention at the Rio Conference in 1992 marks the first step away from an open access approach to what had been considered the common heritage of humanity and toward a market system based on property rights. The focus of the Convention is on the establishment of sovereign rights over biological resources, developing mechanisms for their sustainable use, and ensuring an equitable distribution of the resulting benefits. While there is general agreement that countries, indigenous groups, and local communities should be compensated in exchange for providing access to biological resources and associated knowledge, there is very little consensus over what constitutes an equitable level of compensation.

The model of the R&D process presented above provides one basis for estimating the value of biological resources for pharmaceutical research. While this model can be useful for project planning and contractual negotiations, it does not incorporate the full market interaction between suppliers and users of biological material, nor does it take into account external costs and benefits. If market forces are to be the primary means by which the level of compensation for biological extracts and compounds is determined, it is important to develop a broader understanding of this emerging market and to develop policy responses for the factors that could inhibit efficient and equitable market outcomes.

Although the Biodiversity Convention recognizes sovereign rights over biological resources, these property rights are not yet well defined or easily protected. Transaction costs are also quite high, and risk spreading mechanisms are not well developed. In addition, the irreversibility of extinction truncates the opportunities for supply adjustment that exist in most other markets. Perhaps most importantly, information about biodiversity and its potential as a source of novel chemical compounds is a public good. Once the information has been developed it can, with little or no extra cost, be used by any country or firm (Arrow 1984; Aylward and Barbier 1992; Vogel 1995). While patents, breeders rights, and other forms of intellectual property rights attempt to provide some control over newly discovered biochemical information, it remains true that the international social benefits of biodiversity conservation and biochemical prospecting exceed the benefits that can be appropriated by individual firms or countries. These conditions combined with increased competition between sources of biological material can be expected to lead to underinvestment in biochemical prospecting and a low level of compensation for access to biological resources (Artuso 1994, 1996b).

Recently there has been substantial debate about whether the extension of existing systems of intellectual property rights will allow pharmaceutical companies and other biochemical prospecting organizations to exploit the biological resources and ethnobiological knowledge of developing countries and indigenous groups (Shiva 1994; Gollin 1993; Kloppenburg and Kleinman 1988). I believe that these criticisms are not so much misguided as misplaced. It is certainly true that acknowledgment of developing country researchers and indigenous groups as inventors or coinventors in patent filings can be important for psychological reasons. Attribution of inventorship can also provide the host country, indigenous group, or individual researcher with prestige that brings benefits in the form of future biochemical prospecting investments. Nevertheless, even before any discoveries are made, the rights to any discoveries and any revenues derived from ownership of intellectual property can be divided and assigned quite creatively through contractual arrangements, thereby making who is named as an inventor in the patent somewhat irrelevant. In addition, it does not take long for information on who really contributed to the discovery of a new product to circulate within the pharmaceutical industry and the broader circle of firms and organizations involved in biochemical prospecting. For these reasons, I will focus on issues of compensation and other contractual arrangements in discussing the ethics of biochemical prospecting and the equity of the emerging biochemical prospecting market.

I believe it would be counterproductive to attempt to promote biodiversity conservation and sustainable development by imposing a global system of minimum compensation for biological resources and indigenous knowledge. Pharmaceutical companies and other purchasers of biological material and indigenous knowledge will not pay more than the added value that these resources provide to their drug development efforts. The economic models outlined in the first section of this paper provide a framework for estimating this value. For example, given the data presented in Tables 8.1 and 8.2, the maximum royalty that a profit seeking pharmaceutical company would be willing to pay for access to biological extracts would be 3% of gross revenues. If required levels of compensation are set too high, firms will simply focus their attention on other forms of drug discovery. Artificially high levels of compensation will also open the door to unscrupulous brokers of biological material.

Rather than trying to coerce firms into providing a predefined equitable level of compensation, I believe it would be more worthwhile for the international community to focus on increasing the demand for biological resources, ensuring relatively equal bargaining power over the compensation paid for access to these resources, and increasing the capabilities of developing countries to conduct biochemical research and development.

In the United States, public support for natural product research has until recently been limited to the National Cancer Institute's (NCI) ongoing antitumor and now anti-HIV screening program. In 1994, NIH,

NSF, and USAID initiated the International Cooperative Biodiversity Group (ICBG) program, a highly innovative effort focusing on drug discovery, biodiversity conservation, and sustainable development. The continuity of NCI's screening activities over a period of several decades is an achievement that would have been difficult to realize in the private sector. It is also unlikely that an integrated demonstration program such as the ICBG initiative would have developed without public support. Nevertheless, the focus of the drug discovery aspects of these publicly supported programs is to use existing technologies to screen a large number of biological extracts. In order to stimulate private sector research and increase the incentives for biodiversity conservation, I believe public funding agencies must begin to give more attention to developing new biochemical screening techniques and biotechnological processes with widespread application to plants, insects, and microorganisms.

Screening biological samples with existing bioassays can be expected to yield a finite number of valuable new products. With new scientific knowledge and technological improvements, the stock of new drugs that can be developed from biodiversity will eventually be exhausted and there would be no further demand for biological material from the pharmaceutical industry. On the other hand, development of more advanced screening technologies, together with continued biological exploration (e.g., species inventories, advances in biochemistry), tends to increase the quantity of new drugs that can be discovered with available technologies. In the long term, the flow of material benefits from biodiversity, and the economic incentives for its conservation, will be a function of the rate at which scientific and technological advances can translate the unknown potential value of biodiversity into known reserves of biochemical resources and then create the tools with which to "extract" these reserves economically.

While the development of new and improved bioassays will be needed to maintain or increase demand for biological material, there is still an important niche for publicly supported biochemical screening programs. A case can be made for public support for the search for new drugs that would have substantial social benefits in excess of any revenues that such drugs would generate for pharmaceutical companies. This case can be made most strongly with regard to the search for new drugs to combat tropical diseases such as malaria. The relative lack of private sector research in this area is due to the inability of most of those that suffer from tropical diseases to afford new drugs if they were developed. Pharmaceutical companies simply have insufficient profit incentives to justify significant research expenditures in this area. Many of the projects supported by the

ICBG program include screening programs for malaria and other tropical diseases. But in relation to the enormity of the problem and the social benefits of developing effective new drugs for these diseases, the ICBG program is woefully inadequate. The time has come to initiate a global program to screen biological material for new drug leads for malaria and other tropical diseases.

A network of biochemical research and development centers should be developed to promote this global screening program and to provide ongoing support for the development of new screening technologies. Such a system would serve a function analogous to the international agricultural research centers managed by the Consultative Group on International Agricultural Research (CGIAR). However, in contrast to CGIAR's management philosophy, an international system of biochemical research centers should acknowledge sovereign rights to biological resources and develop compensation mechanisms for biological materials and associated intellectual property.

In addition to promoting research and development, policy initiatives should seek to reduce the transaction costs involved in arranging biochemical prospecting projects. The Clearinghouse that is required under Article 18 of the Biodiversity Convention could provide important benefits in this regard. But the role of the Clearinghouse must be expanded beyond a simple repository of scientific and technical information to also include economic, financial, and legal information. The Conference of the Parties (COP) to the Convention might also seek to promote the development of legitimate exchange services for biological material. Firms, research organizations, countries, and even indigenous groups could use these services to register their willingness to buy or sell research and development rights to various types of biological resources. This would greatly reduce transaction costs while providing all parties with valuable market information. These exchange services should be coordinated with the informational exchange efforts of the Clearinghouse, but the services need not be carried out by a public organization. Just as the Environmental Protection Agency has authorized the Chicago Board of Trade to list transferable discharge permits for sulfur and nitrogen oxides, one or more existing private exchanges could provide exchange services for biological material in accordance with guidelines established by the COP and with periodic reporting to the Clearinghouse.

A flexible set of protocols is also needed to guide the activities of pharmaceutical companies, academic institutions, and public organizations involved in biochemical prospecting programs. These protocols should include a basic code of conduct and contractual guidelines for such issues as informed consent for commercial use of biological material, ecologically appropriate collection practices, and respect for intellectual property rights. In addition, the protocols should include a basic framework for interpreting the general obligations for compensation and technology transfer outlined in the Biodiversity Convention. In this regard, the guidelines should seek to clarify the responsibility of prospecting organizations and host country governments to provide appropriate benefits for indigenous groups when their cultural knowledge is used in identifying species that lead to the development of new drugs. Finally, the guidelines should encourage all parties to take into account the allocated cost of protecting biological resources when negotiating compensation for biochemical prospecting projects. The protocols should not take the form of regulatory requirements. Instead they should focus on providing guidelines for designing biochemical prospecting endeavors that fulfill the requirements of the Convention.

Host countries must begin to take responsibility for regulating transfers of biological material by their own citizens and domestic organizations. This will require promulgating and enforcing regulations for collection, commercial research, export, and compensation related to biological material. These national guidelines would reinforce and complement international protocols, and international organizations should provide technical assistance in developing and implementing them. International support should also be provided to developing countries to conduct the analyses necessary to determine the direct and opportunity costs of biodiversity conservation.

International organizations under the auspices of the Biodiversity Convention should also develop a series of informational workshops and more extended training programs for developing and developed country managers with responsibility for managing biochemical prospecting activities. These workshops and training programs should cover topics such as the techniques of collecting and preparing biological samples and extracts; explanation of the phases, costs, and risks involved in development of new pharmaceuticals and other products from biological material; demonstration of technologies used for screening biological materials; assistance in valuing and negotiating various forms of compensation including royalties, profit shares, technical assistance, and technology transfer; and discussion of systems of intellectual property rights and their implications for natural product research.

Another service that could be provided under the auspices of the Biodiversity Convention would involve certification. Once a flexible set of protocols has been established, one means of promoting compliance is to provide certification for projects that do adhere to the guidelines included in the protocols. While I believe certification should be completely voluntary, there would be a strong incentive to obtain certification if it was believed that it would forestall expensive litigation or negative publicity. Indeed, for pharmaceutical companies that received it, certification could become a positive marketing tool.

Summary

Development of more rapid and less expensive screening technologies, the emergence of new diseases, and evolution of drug-resistant strains of old diseases have all contributed to a resurgence of interest in biological material as a source of new chemical compounds. This renewed interest on the part of private firms and public research institutes has the potential to translate biodiversity's long-term value into more immediate conservation incentives and sustainable development opportunities. However, impediments to the development of an efficient and equitable biochemical prospecting market create a need for involvement by the public sector. The focus should be on programs and policies that complement and catalyze private sector activity, while increasing the capabilities of developing countries and indigenous groups to benefit from this emerging market. In recognition of the underlying economic motivations and constraints of biochemical prospecting, I have recommended increased public funding for development of improved biochemical screening technologies, a global network of research centers engaged in screening biological material for value in combating infectious diseases that are prevalent in developing countries, expanded training and technical assistance programs, and international protocols and certification for biochemical prospecting projects.

Notes

1.   The calculation of the value of a biological extract or compound at any phase in the R&D process can be stated more formally as follows. Assume there are n phases in the R&D process and s_i is the probability of successful results in phase i, conditional on a compound having tested successfully in all prior phases. The conditional probability of failure in phase i, would therefore be $(1 - s_i)$. Let $enpv_n$ be the present value of net benefits expected to be received after successful completion of the final phase (i.e., the expected net

benefits associated with receiving regulatory approval to market a new drug). Biologically derived compounds that do not test positively in a particular screen or trial may still provide some benefits in the form of insights into the molecular structure of effective therapeutic agents. Therefore, let f_i be the expected present value of benefits, if any, associated with failure or negative results in the phase i. Finally, let pvc_i be the per compound present value of the cost of proceeding with the final phase. After successful completion of all but the last phase of the R&D process the expected net benefit of proceeding with the last phase is therefore:

$$(1) \quad enpv_{n-1} = (s_n)(enpv_n) + (1 - s_n)(f_n) - pvc_n$$

Similarly, after successful completion of the second to last phase, the expected benefit of proceeding with phase $n - 1$ is

$$(2) \quad enpv_{n-2} = (s_{n-1})(enpv_{n-1}) + (1 - s_{n-1})(f_{n-1}) - pvc_{n-1}$$

2. R&D costs and revenue data for this hypothetical example are adapted from studies by DiMasi et al. (1991), Grabowski and Vernon (1990), and Artuso (1994, 1996a).

3. Net present value is simply a measure of net benefits in excess of some baseline rate of return, which in this example is a real rate (i.e., net of inflation) of 7.5%

4. A simple linear formulation of the expected value of testing N biological extracts for M therapeutic objectives is appropriate only in the context of a private prospecting organization that can replace one screening target for which a new drug or several promising leads are discovered with another therapeutic target. If each extract is being screened for all possible therapeutic uses, then this linear formulation must be modified to account for the possibility that one or more successful new drugs will have been discovered before the total number of samples has been screened (Simpson et al. 1996). In addition, a more complete model would incorporate a declining success rate due to an increasing likelihood of repeatedly identifying the same active compounds as the number of samples screened increases (Artuso 1996a). Finally, simple linear multiplication by the number of screening tests (M) is only appropriate if the screening tests are statistically independent. This is an appropriate assumption if the screening tests are designed to determine potential efficacy of compounds in quite different therapeutic areas.

References

Arrow, K.J. (1984). *The Economics of Information.* Vol. 4. The Collected Papers of Kenneth J. Arrow. Cambridge, MA: Belknap Press.

Artuso, A. (1994). *An Economic and Policy Analysis of Biochemical Prospecting,* Ph.D. Dissertation, Cornell University. Ithaca, NY.

Artuso, A. (1996a). *Prospecting on the Biomedicinal Frontier: An Economic and Policy Analysis of the Search, Development, and Marketing of Drugs of Natural Origin.* Howarth Press. New York.

Artuso, A. (1996b). Economic Analysis of Biodiversity as a Source of Pharmaceuticals. In *Emerging Connections: Biodiversity, Biotechnology and Sustainable Development,* Pan American Health Organization and International Institute for Cooperation in Agriculture.

Aylward, B.A. (1993). *The Economic Value of Pharmaceutical Prospecting and its Role in Biodiversity Conservation.* LEEC Paper DP 93-03. London: London Environmental Economics Centre.

Aylward, B.A., and E.B. Barbier. (1992). *What is Biodiversity Worth to a Developing Country?* LEEC Paper DP 92-05. London: London Environmental Economics Centre.

Barton, J.H., and E. Christensen. (1988). Diversity Compensation Systems: Ways to Compensate Developing Nations for Providing Genetic Material. In *Seeds and Sovereignty.* J.R. Kloppenburg, Jr. (ed.), 339–55. Durham, N.C.: Duke University Press.

Cox, C.J., and M. Rubenstein. (1985). *Options Markets.* Englewood Cliffs, NJ: Prentice-Hall.

DiMasi, J.A., R.W. Hansen, H.G. Grabowski, and L. Lasagna. (1991). Cost of Innovation in the Pharmaceutical Industry. *Journal of Health Economics* 10:107–42.

Evenson, R.E., and Y. Kislev. (1976). A Stochastic Model of Applied Research. *Journal of Political Economy* 84:265–81.

Farnsworth, N.R., and D.D. Soejarto. (1985). Potential Consequences of Plant Extinction in the U. S. on the Availability of Prescription Drugs. *Economic Botany* 39(3):231–40.

Fisher, A., and W.M. Hanemann. (1986). Option Value and the Extinction of Species. In *Advances in Applied Microeconomics.* Vol. 4. Greenwich, CT: JAI Press.

Gollin, M.A. (1993). An Intellectual Property Rights Framework for Biodiversity Prospecting. In *Biodiversity Prospecting,* W.V. Reid, S.A.Laird, C.A. Meyer, R. Gamez, A. Sittenfeld, D.H.Janzen, M.A. Gollin, and C. Juma (eds.), pp. 159–198. Washington, D.C.: World Resources Institute.

Grabowski, H., and J.Vernon. (1990). A New Look at the Returns and Risks to Pharmaceutical R&D. *Management Science* 36:804–21.

Harris, A. (1995). Improving Molecular Diversity of a Compound Collection in a High Throughput Screening Program. *The Society for Biomolecular Screening Newsline* 1:3.

Kloppenburg, Jr., J.R., and D.L. Kleinman. (1988). Seeds of Controversy: National Property vs. Common Heritage. In *Seeds and Sovereignty,* J.R. Kloppenburg (ed.), pp. 173–203. Durham, N.C.: Duke University Press.

Laird, S.A. (1993). Contracts for Biodiversity Prospecting. In *Biodiversity Prospecting,* W.V. Reid, S.A. Laird, C.A. Meyer, R. Gamez, A. Sittenfeld, D.H. Janzen, M.A. Gollin, and C. Juma (eds.), pp. 99–131. Washington, D.C.: World Resources Institute.

Mendelsohn, R., and Balick, M.J. (1995). The Value of Undiscovered Pharmaceuticals in Tropical Forests. *Economic Botany* 49(2):223–228.

Pindyck, R.S. (1991). Irreversibility, Uncertainty, and Investment. *Journal of Economic Literature* 29:1110–1148.

Principe, P.P. (1991). Valuing the Biodiversity of Medicinal Plants. In *The Conservation of Medicinal Plants,* O. Akerele; V. Heywood; and H. Synge (eds.), pp. 79–124. Cambridge: Cambridge University Press.

Raiffa, H. (1968). *Decision Analysis.* Reading, MA: Addison-Wesley.

Reid, W.V., S.A. Laird, C.A. Meyer, R. Gamez, A. Sittenfeld, D.H. Janzen, M.A. Gollin, and C. Juma (eds.). (1993). *Biodiveristy Prospecting.* Washington, D.C.: World Resources Institute.

Sedjo, R.A. (1992). Property Rights, Genetic Resources, and Biotechnological Change. *Journal of Law and Economics* 35:199–213.

Shiva, V. (1994). Biodiversity and Intellectual Property Rights. In *The Case Against Free Trade.* Berkeley: North Atlantic.

Simpson, R.D., R.A. Sedjo, and J.W. Reid. (1996). Valuing Biodiversity for Use in Pharmaceutical Research. *Journal of Political Economy* 104:1548–70.

Subramanian, A. (1992). Genetic Resources. Biodiversity and Environmental Protection. *Journal of World Trade* 26:105–9.

U.S. International Trade Commission (USITC). (1991). Global Competitiveness of U.S. Advanced Technology Manufacturing Industries: Pharmaceuticals. *World Competition* 15(2):27–45.

Vogel, J.H. (1995). *Genes for Sale: Privatisation as a Conservation Policy.* Oxford: Oxford University Press.

PART III

Biodiversity and Traditional Health Systems

Nonprescription, nonwestern, and or "traditional" biodiversity-derived therapies continue to be the principal source of medicines for most people in the developing world and many others who are either unable to access modern medicine or prefer traditional therapies. The greatest tragedy of biodiversity loss may reside in the increased pain and suffering associated with loss of these therapies as well as the erosion of the cultural and scientific traditions that are tied to them.

In this section, Paul Cox (Chapter 9) eloquently describes the intimate relationships that exist between many indigenous peoples and the diverse species for which they have been users and stewards for thousands of years. Turning our attention to the Western world, Robert McCaleb (Chapter 10) describes the phenomenal increase in recent sales of herbal medicines in the United States and Europe and some benefits and dangers that the use of these remedies may pose for human and ecological health.

Along with the renewed attention to natural products as a direct and indirect source of therapies have come complex issues of compensation to source communities and countries. In Katy Moran's chapter (Chapter 11), and continuing in Part IV, compensation of local and indigenous peoples for their contributions to remedies and therapies is discussed and the use of such compensation as an incentive for conservation is considered.

CHAPTER 9

Indigenous Peoples and Conservation

PAUL ALAN COX

On the eve of a new millennium, the conservation of the world's biodiversity is of deep concern to scientists and thoughtful people throughout the world. Destruction of habitat and concomitant extinction of species are accelerating at such a prodigious rate that it now appears that our age will be remembered in distant epochs not for our increased use of digital data transmission and the flowering of information technology, nor for the development of an integrated global economy, or even for our initial deployment and subsequent halting steps toward the elimination of nuclear devices as instruments of warfare and terror. Though these issues are of overwhelming importance, through the next thousand years of human history, their significance may eventually become muted. The indelible mark we now etch on the chronicle of time will not, however, so easily disappear, for we have chosen to destroy biological species.

May et al. (1995) estimate that before the rise of human populations the average projected life span of any given biological species (defined as the period beginning at the inception of any species and ending at its eventual extinction) was around 5–10 million years. Current extinction rates have now been accelerated such that a predicted life span for a bird or mammal species is around 10,000 years, a decrease in life span of three orders of magnitude. Increased extinction rates are not limited to animal species; the IUCN Redbook (Smith et al. 1993) indicates that 584 seed plant species have been certified as having gone extinct in the period between 1600 and 1992, with an additional 22,137 plant species (approximately 9% of all flowering plant species) being regarded as threatened (a category combining the IUCN groupings of "vulnerable," "endangered," and "probably extinct"). Between 1990 and 1992, 163 plants were added to the list of extinct species, evidence of an accelerated extinction rate that if unchecked will result in over half of the 250,000 species of seed plants becoming extinct within 3,000 years (May et al. 1995).

At these rapid rates of extinction it is almost certain that our current period will be characterized as a time of mass extinction. An adversary of conservation could perhaps argue that this is but the sixth of such global mass extinctions, events through which life on our planet somehow survived. But the aftermath of those previous mass extinction events, which resulted in the vast reordering of global biological communities, should give even a jaded observer pause when considering that our current crisis is the first global extinction coincident with the presence of humankind on earth. It is possible that humankind could persist in a greatly simplified biological world, a world devoid of the plethora of natural species that now inhabit it, but the quality of human life in such a biologically monotonous world may be significantly diminished.

Many of us are familiar with human-generated extinctions that have occurred in the distant past. David Steadman has found evidence that numerous species of birds were driven to extinction in the Pacific islands by human colonists. In Mangaia Island three species of seabirds and eight species of land birds were extirpated or driven to extinction (Steadman and Kirch 1990). In Hawaii, up to 43 species of birds were driven to extinction by the Polynesians (Pimm et al. 1995). Even earlier, the Pleistocene extinction of giant beavers, gyptodons, mastodons, and other genera of the North American megafauna was likely triggered by the arrival of *Homo sapiens* (Martin 1967). As terrible as such cases are, these examples of ancient mismanagement of biodiversity can be looked upon as the consequence of ignorance. We cannot, however, escape culpability for what will appear in distant epochs as the largest mass extinction since the end of the Cretaceous period, for in our current destruction of biological species we have acted knowingly.

The same modernity that has endowed us with the technological ability to monitor with increasing precision the consequences of human activities on the global environment has also paradoxically drawn an insatiable materialism that has led to the destruction of those very biological entities that we monitor. Although nearly every sacred tradition of the world's religions exhorts us to care for and protect biological species (e.g., in Western tradition, the story of Noah's ark) there appears to be scarcely a species or a habitat that can withstand the modern imperatives of development, growth, and progress. Yet the embracement of modernity is also driving a different sort of extinction that has not been heralded as broadly as the loss of biological species—one which may have equally ominous consequences for the quality of human life—the loss of indigenous knowledge systems.

Indigenous Knowledge Systems and Human Health

Although few would dispute the rapid loss of biological species, precise quantification of the extinction rate is somewhat difficult, since current reasonable estimates of the number of biological species present on the planet range between three and ten million. However, there have been even fewer attempts to quantify the loss of indigenous knowledge systems: the knowledge accumulated over millennia by indigenous peoples. A rough sense of the magnitude of the problem can be deduced from rates of language loss. Estimates of the total number of different languages currently spoken on the planet vary between 3,000 and 9,000, with the discrepancy largely attributable to disputes between what constitutes a distinct language (as opposed to a dialect). It is clear, however, that most of the world's languages are spoken by relatively small numbers of people, with the median number of speakers of any language being between 5,000 and 6,000 (Krauss 1992). In Papua New Guinea, for example, there are more than 850 different languages. Inhabitants of contiguous villages may speak entirely different languages, with the result that some languages are spoken by a small number of individuals, or in the extreme only by a single family. Even languages that were once spoken by indigenous peoples who occupied large geographical areas have been reduced in many cases to only a few native speakers. The Gosiute people once occupied much of the Great Basin of the United States, but now fewer than 30 Gosiute speakers remain. The Gosiute language, despite over a century and a half of nearby European presence, has never been published for Gosiuze children in an accepted written orthography. Over 149 of the 187 languages (80%) spoken by native North Americans have almost completely disappeared. In Alaska, only 2 out of 20 indigenous languages are being learned by children. The Eyak language of Alaska is now spoken by only two living individuals. Similar rates of attrition occur in the northern regions of the former Soviet Union where only 3 out of 30 languages are being learned by children (Krauss 1992).

TABLE 9.1

Probable Extinction Rates for Plant Species and Indigenous Languages

	Number	Percentage
IUCN "Redbook" Data[a]		
Estimated Seed Plant Species	250,000	100%
Plant Species Certified as Extinct	747	3%
Plant Species Threatened	22,137	9%
Total Extinct or Threatened	22,884	12%
Predicted Plant Species Going Extinct in 3000 Years	125,000	50%
Language "Redbook" Data		
Estimated Current Spoken Languages	6,000	100%
Estimated Languages Extinct	600	10%
Estimated Languages Threatened	2,400	40%
Total Extinct or Threatened	3,000	50%
Predicted Languages Going Extinct in 100 Years	5,400	90%

[a]Plant species data from Smith et al. 1993; May et al. 1995. Language data from Krauss 1992.

We can compare language loss to the loss of biological species (Table 9.1) by defining a language as extinct when it is no longer spoken by any living individual. Linguist Michael Krauss (1992) uses the term "moribund"—the linguistic equivalent of endangered—when a language is no longer learned as the mother-tongue by little children. We can consider a language to be not "viable" when nearly all of the native speakers are bilingual in another language such as English, French, or Spanish. By lumping "moribund" and "unviable" languages as "threatened" it appears from the data of Krauss that of the languages spoken at the beginning of this century, over half are in serious trouble: 10% are extinct and 40% are threatened. The rate of language loss is so rapid that Krauss predicts that by the end of the next century, 90% of the world's languages will have become extinct. If the languages which are used to transmit indigenous knowledge systems are being lost at such a rapid rate, is it likely that the indigenous knowledge systems themselves are faring any better?

The rapid loss of plant species coupled with the even more rapid loss of indigenous knowledge systems poses a particular urgency to ethnobotanical investigations since the deleterious effects of language loss and plant species loss are not merely additive but negatively synergistic. How can indigenous peoples retain knowledge of plant uses if the plants go extinct?

And how can indigenous knowledge systems maintain knowledge concerning uses of plants if the indigenous languages go extinct?

This rapid rate of extinction of biological species combined with the rapid loss of indigenous knowledge systems has deep potential consequences for human health. Norman Farnsworth (1994) has identified over 119 plant-derived substances that are used globally as drugs, with over 25% of the drugs issued in the United States and Canada being derived from, or modeled after, naturally occurring molecules in plants (see also Grifo et al., Chapter 6). The majority of these substances were derived from studies with indigenous peoples (Cox and Balick 1994; Balick and Cox 1996). Indeed, the very course of modern medicine would have been very different if western science had not begun studying a number of bioactive molecules as a result of the studies of indigenous knowledge systems. Tubocurarine, for example, a muscle relaxant which facilitated open-heart surgery, was discovered by studies of South American arrow poisons (from *Chondrodendron tomentosum* [Menispermaceae]). One of the earliest effective drugs for glaucoma, physostigmine, was isolated in studies of a bean (*Physostigma venenosum* [Leguminosae]) used in witch ordeals in Nigeria; even today the antiglaucoma drug of choice, pilocarpine, is isolated from a tropical plant (*Pilocarpus jaborandi* [Rutaceae]) long used in traditional South American medicine. The first major drug to treat hypertension, reserpine, was isolated from a plant (*Rauvolfia serpentina* [Apocynaceae] used by indigenous peoples in India, while the first major drug used to treat heart arrhythmias, digitalis, was discovered from studies of a plant (*Digitalis purpurea* [Scrophulariaceae] used to treat dropsy by "old wives" in Britain.

These major advances in medicine resulted from only a small sampling of the plant kingdom: less than ½ of 1% of the 250,000 species of flowering plants on earth, have ever been exhaustively screened for all possible medicinal properties (Balick and Cox 1996). From the number of ethnobotanically derived prescription drugs on the market, it is clear that we can derive clues to which of the 250,000 flowering plants should be screened by studying the healing traditions of indigenous peoples. Yet very few of the world's cultures have been exhaustively examined by ethnobotanists for possible leads to new medicines. Most cultural studies have been performed by anthropologists and ethnologists, yet only rarely have anthropologists and ethnologists had the botanical sophistication to make herbarium vouchers of medicinal plants or to collect samples for pharmaceutical analysis. The lack of research on both biological species and cultures would be a concern in a situation of biological and cultural stasis, but in our current era of rapid extinction of both biological species and indigenous knowledge systems, it seems almost certain that potential new pharmaceuticals are disappearing each year before they can be "discovered."

Indigenous Knowledge Systems and the Discovery of New Drugs in Samoa

Although the loss of both biological species and indigenous knowledge systems is accelerating at a precipitous rate throughout the world, the opportunity loss for drug discovery is difficult to quantify, since by definition, it involves the loss of undiscovered drugs. Perhaps a qualitative sense of the problem can be gained by considering the potential of new pharmaceutical discovery in a relatively small place: the Samoan islands of the South Pacific.

Nearly all Samoans know of the therapeutic property of a few herbs, such as the use of a tea from *Psidium guayava* leaves to treat diarrhea. In many respects this general rudimentary knowledge of medicinal plants common to most of the populace would be comparable to the knowledge of over-the-counter remedies by Western consumers. There is, however, a specialist realm of knowledge of healing plants that could be compared with the specialist knowledge of pharmacists in Western countries. In Samoa, this specialist tradition is, with only a few exceptions, nearly entirely matrilineal: nearly all *taulasea,* or indigenous herbalists, learned their craft from their mothers, who in turn had apprenticed with their grandmothers, and so on. Formerly there were also spiritual healers known as *taula'aitu* and bone setters (*fofogau*) who were men (Cox 1990), but the indigenous specialists of herbalist (*taulasea*) and midwife (*fa'atosaga*) have almost entirely been women.

Although the number of bonafide *taulasea* is few, with perhaps one per several thousand Samoans, the knowledge of a *taulasea* is impressive, with a typical Samoan healer being able to identify between 100 and 200 species of flowering plants by name. The *taulasea* I have worked with use up to 140 different remedies. Taxonomic skills are important for a *taulasea* because Samoan herbal preparations are made from only freshly collected plants. Thus, upon making a diagnosis of illness in a patient, a *taulasea* must be able to determine the plants needed for treatment and either herself collect or direct others in the collection of the needed species. The *taulasea* must also know the proper preparation techniques, which range from maceration in water or coconut oil, to combustion of plant materials to produce healing vapors. The *taulasea* must also know how to administer the plant, determine dosage, how to advise the patient on dietary taboos (*sa*) associated with the plant, and how to therapeutically intervene in the course of the disease.

Even though their numbers are relatively few, Samoan herbalists play a very important role in Samoan culture. There are very few Samoans who have not been treated by a *taulasea,* and even in expatriated Samoan populations I have studied (i.e., in Auckland, Honolulu, and San Francisco), *taulasea* are present and functioning. What is particularly remarkable is that *taulasea* render their services without compensation, except for an occasional small gift of food or toiletries that characterize normal interaction in Samoan society. Practicing *taulasea* make their living as do other villagers, through subsistence agriculture and reef foraging.

In a Western sense, then, there is little economic incentive for a young woman to pursue the long (up to seven-year) apprenticeship with her mother or grandmother that is necessary to become a *taulasea.* It should be little surprise then that the *taulasea* practicing in Samoa are older women, few of whom have an apprentice. In my studies over the last 12 years in Samoa, the youngest bonafide *taulasea* I have met is 43 years old. Since there is no alternative to apprenticeship to become a *taulasea,* the knowledge system is dying out in Samoa. I have found Samoan herbalists to shun experimentation and innovation, restricting themselves to the corpus of knowledge passed to them from generations of women in their lineage. Although there is considerable overlap in the knowledge domains of different *taulasea,* nearly all possess knowledge of remedies unique to their families. There is also a nascent sense of remedy ownership (although this is more pronounced among Tahitian and Rotuman healers) that mitigates against healers learning new remedies from outside their family. Thus the combination of lack of economic incentive, lengthy apprenticeships, the necessity of being born within a healer's family, and some lack of sharing knowledge between families has led to imperilment of the knowledge system.

But perhaps the biggest disincentive has been either tacit or explicit denigration of Samoan healing systems by Western culture. As often occurred during the colonial period, Westerners dismissed any possible contributions of Samoan medicine. "The Samoans in their heathenism seldom had recourse to any internal remedy," wrote early missionary George Turner, "except an emetic, which they used after eating a poisonous fish. Sometimes juices from the bush were tried; at other times the patient drank on at water until it was rejected; and, on some occasions, mud, and even the most unmentionable filth, was mixed up and taken as an emetic" (Turner 1884). Denigration of Samoan herbal medicine by Europeans and by western-educated Samoan physicians has continued in recent times (Cox 1995).

The knowledge realm of Samoan herbal medicine, then, while not extinct, certainly would qualify as "threatened" if ever a ethnobotanical equivalent of the IUCN "Red" book were written. Just as the environmental impacts of the extinction of a biological species are sometimes evaluated, perhaps it would be useful to consider the impact of loss of the Samoan system of herbal medicine. Would loss of an arcane indigenous healing knowledge system from a remote set of islands have any negative impacts on the world?

An answer to the first question can be found with a simple *in vitro* survey of broad pharmacological activity in Samoan medicinal plants. Together with our collaborators in the laboratory of Lars Bohlin at Uppsala University in Sweden, we analyzed 74 different species of medicinal plants used by Samoan healers for possible biological activity. Over 86% of the plants exhibited high levels of pharmacological activity. The broad Hippocratic and guinea pig ileum screens we employed did not provide analyses for specific disease targets. Based on these preliminary results, though, I decided to test some of the plants used by Samoan healers against specific bioassays.

One plant tested was a coastal tree, *Erythrina variegata,*which Samoan healers use to treat inflammation. Only one of the two varieties the healers recognize, however, is considered useful in the treatment. A team led by Vinod Hegde and Mahesh Patel at the Schering Research Institute tested whether bark samples of the two varieties of *Erythrina variegata* inhibited phospholipase A_2, an enzyme that mediates in cellular inflammation. The Schering team found that only the bark of the species used by the healers was active. A new flavanone was isolated and is under development as a topical antiinflammatory by Schering and Phyton Catalytic.

In 1984 I collaborated with the Natural Products branch of the U.S. National Cancer Institute (NCI) in studying *Homalanthus nutans* (Euphorbiaceae), which several healers had told me was efficacious against hepatitis, a viral disease. (The healers carefully prepare hot water infusions that are then drunk by the patients.) An NCI team led by Michael Boyd found the extracts to exhibit potent activity in an *in vitro* tetrazolium-based assay designed to detect cytopathic effects of the virus (HIV-1) associated with acquired immunodeficiency syndrome (AIDS) (Gustafson et al 1992). Bioassay-guided fractionation resulted in the isolation of prostratin (12-deoxyphorbol 13-acetate), which at noncytotoxic concentrations prevents HIV-1 reproduction in lymphocytic and monocytoid target cells. Prostratin also fully protected human cells from lytic effects of HIV-1.

The identification of a phorbol as the active component was a concern because phorbols are known tumor-promoters. The NCI team found, however, that prostratin does not induce hyperplasia in mice, even though it stimulates protein kinase C. The National Cancer Institute concluded that prostratin represents a nonpromoting activator of protein kinase C which strongly inhibits the killing of human host cells *in vitro* by HIV. By these criteria, prostratin is unique (Gustafson et al. 1992). Currently, NCI considers prostratin to be a candidate for drug development and is accepting bids from pharmaceutical firms for development of prostratin as an anti-AIDS therapy, possibly as combination therapy with other drugs.

Another Samoan medicinal plant which shows strong potential for development as a pharmaceutical agent is the rainforest tree *Alphitonia zizyphoides.* Samoan healers use the bark of this tree to make a water infusion claimed to be effective as a general tonic. The healers are quite precise in decanting the infusion, using only the fraction containing the suds or soapy residue (*oaoa*). Studies conducted at Uppsala University show that bark extracts of *Alphitonia zizyphoides* increase the plating efficiency of lymphoid cell lines. Bark extracts also increase survival of bone marrow cells and normal T and B lymphocytes. Inclusion of the bark extract enhances cloning efficiency of a T-hybridoma cell line more than 30 times (Dunstan et al. 1994). Although a pure compound responsible for this observed bioactivity has yet to be structurally determined, the implications for possible development of a human immunostimulatory therapy are intriguing.

Although the eventual emergence of any of these three compounds as an approved drug is far from certain, the discovery of three major new leads in immunotherapy, antiinflammatory, and antiviral therapy from Samoan indigenous healing systems should cause us to reexamine George Turner's (1884) characterization of Samoan herbal medicine as "heathenism." Yet these leads resulted from only a preliminary investigation of a single culture by a single ethnobotanist. How many other useful leads are being lost daily with the demise of indigenous knowledge systems?

Clearly, people in the colonialist era had little indication that indigenous knowledge systems could potentially produce pharmaceutically valuable compounds. It now seems almost certain that the advent of the new century will be coincident with the release to the pharmaceutical market of new drugs that ultimately originated from the witch doctor's pouch. Commercialization of such indigenous intellectual property raises ethical issues of equitable treatment of indigenous people in the resultant economic benefits. In the case of prostratin, the National Cancer Institute has agreed to return the majority of patent income to the people of Western Samoa. And Brigham Young University, which supported the original ethnobotanical research, has agreed to return one-third of its patent income to the villages and families of the healers who initially identified *Homalanthus nutans* as active against diseases of viral origin. Similar agreements are in place for other lead compounds resulting from this research. A great deal of consideration has recently been given by different workers to the issue of indigenous intellectual property rights (Greaves 1994). Such considerations are important given the plethora of medicinal plants used by indigenous peoples. Compared, for example, with the number of plants used by native Americans, for whom Moerman (1986) records 2,095 different medicinal plant species, the Samoan pharmacopoeia is tiny. Development of exciting new pharmaceuticals from traditional medicine is almost a certainty if only the research can be completed before the plants and knowledge disappear.

Indigenous People and Conservation

The discovery of new pharamaceutical compounds from studies of indigenous knowledge systems is only a single example of the possible ways that indigenous peoples can contribute to Western societies. It is even more likely that such knowledge can be of direct benefit to residents of the developing countries, since over 80% of the world's population depends directly on plants for primary health care. Yet indigenous knowledge systems can even have a greater beneficial impact on the health and well-being of the world's population if we begin to study indigenous insights concerning conservation.

Richard Grove (1990) has argued that Western paradigms for conservation are rooted in the "wise use" ethic developed during the reign of Charles II to protect the crown's forest resources. It was this ethic that led to the development of wildlife preserves throughout Africa in the 19th century. These reserves were not established with indigenous people in mind, but instead were attempts to preserve game populations necessary to sustain safaris and to allow overseas tourists the opportunity to see African wildlife in a semiwild condition. Even today, this "wise-use" ethic animates much of the conservation policies of North America and Europe. In its most reduced formulation, the "wise-use" ethic is merely an economic optimization of long-term gains from natural resources.

Most indigenous people, however, do not perceive conservation imperatives in economic terms. Instead, they view the world as sacred and believe it a religious duty to protect entire ecosystems. Alison Wilson (1993) has documented the protection of sacred groves (called *kaya*) by villagers in East Kenya. These forests are used for religious ceremonies conducted by the village elders who have protected them against all perils. Wilson believes that a "significant" amount of biodiversity has been protected in these groves. Elsewhere, several innovative experiments have been conducted by establishing conservation areas under the direct control of indigenous peoples. In Belize, Michael Balick (Balick and Cox 1996) helped establish one of the world's first ethnomedicinal reserves, Terra Nova. It is under the control of an organization of local healers who use the forest as a source of medicinal plants and as an integral part of a new educational system to train aspiring herbalists. Extractive reserves which are based on communal use of economically valuable species have been established throughout the Amazon (Pinedo-Vasquez et al. 1990; Anderson and Ioris 1992). In all of these cases, a strong sense of community and religious stewardship guides conservation policies.

In the early 1990s, several indigenous controlled rainforest reserves were established in Samoa. In American Samoa, the United States Congress approved leasing on a 55-year basis nearly 11,000 acres of rainforest and associated coral reef from Samoan villagers to establish the nation's 50th national park (Cox 1988). Leases, rather than purchases, were necessary because Samoan culture prohibits sale of communal lands; the Samoans believe that it is sacrilege to sell their land. Samoan sensibilities were carefully considered in drafting legislation. The National Park bill passed by the U.S. Congress allows Samoans to continue to harvest medicinal plants, wild foods, and construction materials from the park property using traditional techniques and tools. An advisory committee composed of Samoan chiefs guides the formulation of management policies for the park. In Falealupo and Tafua villages in Western Samoa an even more direct form of indigenous control was implemented in the establishment of rainforest reserves. Foreign donors who established the Seacology Foundation (P.O. Box 4000, Springville, UT 84663) discovered that these forests were due to be logged so that the villagers could obtain funds for needed schools. Seacology donors and donors from the Swedish Nature Foundation built the schools in return for covenant by the villagers to protect the rainforests. The villagers completely own, control, administer, and manage the reserves with foreign donors playing merely a supportive role for indigenous conservation initiatives (Cox and Elmqvist 1991). However, such initiatives have proven controversial to some in the environmental community who do not believe that indigenous people possess the requisite knowledge to manage large conservation initiatives (Cox and Elmqvist 1993). We argue that the ethnobiological knowledge of the Samoan people (which for example pinpointed flying foxes of the genus *Pteropus* as the keystone species of the Samoan lowland rainforest [Cox et al. 1991]) is indicative of a profoundly sophisticated approach to conservation which maximizes preservation of rainforest and marine resources while allowing their sustainable use. Yet the conservation knowledge and commitment of indigenous peoples, which is so strongly non-Western in its philosophical roots, have scarcely begun to be investigated.

Conclusion

Indigenous knowledge systems are arguably disappearing at an even faster rate than biological species. Both indigenous knowledge systems and biodiversity are necessary to obtain optimal development of new plant-based pharmaceuticals. Indigenous knowledge systems may have even greater impact in informing Western conservation measures. Although modernity has been concomitant with both the development of modern technology and an increasingly materialistic orientation for Western societies, indigenous knowledge systems may offer Western culture a third alternative: the possibility of establishing a deep conservation ethic based not on economic considerations but instead on a deep sense of stewardship rather than ownership in our management of the natural world. The rapid loss of both biodiversity and indigenous knowledge systems gives us but a small window in time to seize this opportunity. If we delay, both the plant and animal species that inhabit this planet as well the indigenous knowledge necessary for us to understand them may disappear before we ever realize the depth of our loss.

References

Anderson, A.B. and Ioris, E.M. (1992). Valuing the Rain Forest: Economic Strategies by Small-Scale Forest Extractivists in the Amazon Estuary. *Human Ecology* 20(3): 337–369.

Balick, M. J. and Cox, P. A. (1996). *Plants, People, and Culture: The Science of Ethnobotany.* Scientific American Library. New York: W.H. Freeman.

Cox, P. A. (1988). Samoan Rainforest—Partnership in the South Pacific. *National Parks* 62 (1–2): 18–21.

Cox, P. A. (1990). Samoan ethnopharmacology. In Wagner, H. and Farnsworth, N. R. (eds.), pp. 123–124 *Economic and Medicinal Plant Research. Vol 4. Plants and Traditional Medicine.* London: Academic Press.

Cox, P. A. (1995). Shaman as Scientist: Indigenous Knowledge Systems in Pharmacological Research and Conservation Biology. In Hostettmann, K. Marston, A., Maillard, M. , and M. Hamburger (eds.), pp. 1–15. *Phytochemistry of Plants Used in Traditional Medicine.* Oxford: Clarendon.

Cox, P. A. and Balick, M. J. (1994). The Ethnobotanical Approach to Drug Discovery. *Scientific American* 270 (6): 88–94.

Cox, P. A. and Elmqvist, T. (1991). Indigenous Control: An Alternative Strategy for the Establishment of Rainforest Preserves. *Ambio* 20:317–321.

Cox, P. A. and Elmqvist, T. (1993). Ecocolonialism and Indigenous Knowledge Systems: Village-Controlled Rain Forest Preserves in Samoa. *Pacific Conservation* 1:11–25.

Cox, P.A., Sperry, L. R., Tumonien, M., and Bohlin, L. (1989). Pharmacological Activity of the Samoan Ethnopharmacopoeia. *Economic Botany* 43:489–497.

Cox, P. A., Elmqvist, T., Pierson, E. D., and Rainey, E. D. (1991). Flying Foxes as Strong Interactors in South Pacific Island Ecosystems: A Conservation Hypothesis. *Conservation Biology* 5:448–454

Dunstan, C. A., Anderson, J., Bohlin, L., Cox, P.A., and Gronvik, K.-O. (1994). A Plant Extract which Eliminates the Plating Efficiency of Lymphoid Cell Lines and Enhances the Survival of Normal Lymphoid Cells *in Vitro. Cytotechnology* 14: 27–38.

Farnsworth, N. R. (1994). Ethnopharmacology and drug development. In G. T. Prance, D. Chadwick, and J. Marsh (eds.), pp. 42–51. *Ethnobotany and the Search for New Drugs.* Ciba Foundation Symposium 185. London: John Wiley and Sons.

Greaves, T. [ed.]. (1994). *Intellectual Property Rights for Indigenous Peoples: A Sourcebook.* Oklahoma City: Society for Applied Anthropology.

Grove, R.H. (1990). Colonial Conservation, Ecological Hegemony and Popular Resistance: Towards a Global Synthesis. In MacKenzie, J.M. (ed.), pp. 15–50. *Imperialism and the Natural World.* Manchester: Manchester University Press.

Gustafson, K.R., Cardellina, J.H., McMahon, J.B., Gulakowski, R.J., Ishitoya, J., Szallasi, Z., Lewin, N.E., Blumberg, P.M., Weislow, O.S., Beutler, J.A., Buckheit, R.W., Cragg, G.M., Cox, P.A., Bader, J.P., and Boyd, M.R.

(1992). A Non-promoting Phorbol from the Samoan Medicinal Plant *Homalanthus nutans* Inhibits Cell Killing by HIV-1. *J. Medicinal Chem.* 35: 1978–1986

Krauss, M. (1992). The World's Languages in Crisis. *Language* 68(1):4–10.

Martin, P. S. (1967). Prehistoric Overkill. In P. S. Martin and H. E. Wright, Jr. (eds.), pp. 75–120. *Pleistocene Extinctions: The Search for a Cause.* New Haven: Yale University Press.

May, R. M., Lawton, J. H., and Stork, N. (1995). Assessing Extinction Rates. In J. H. Lawton and R. M. May (eds.), pp. 1–24. *Extinction Rates.* Oxford: Oxford University Press.

Moerman, D.E. (1986). *Medicinal Plants of Native America.* University of Michigan Museum of Anthropology Technical Reports, No. 19. Ann Arbor.

Pimm, S. L., M. P. Moulton, and L. J. Justice. (1995). Bird Extinctions in the Central Pacific. In J. H. Lawton and R. M. May (eds.), pp. 75–87. *Extinction Rates.* Oxford: Oxford University Press.

Pinedo-Vasquez, M., D. Zarin, P. Jipp, and J. Chota-Inuma. (1990). Use-Values of Tree Species in a Communal Forest Reserve in Northeast Peru. *Conservation Biology* 4(4):405–416.

Smith, F. D. M., R. M. May, R. Pellew, T. H. Johnson, and K. S. Walker. (1993). Estimating Extinction Rates. *Nature* 364: 494–496.

Steadman, D. W. and P. V. Kirch. (1990). Prehistoric Extinction of Birds on Mangaia, Cook Islands, Polynesia. *Proceedings of the National Academy of Sciences* 87: 9605–9609.

Turner, G. (1884). *Samoa a Hundred Years Ago and Long Before.* London: Macmillan.

Wilson, A. (1993). Sacred Forests and the Elders. In Kemf, E. (ed.), pp. 244–248. *The Law of the Mother: Protecting Indigenous Peoples in Protected Areas.* San Francisco: Sierra Club Books.

CHAPTER 10

Medicinal Plants for Healing the Planet: Biodiversity and Environmental Health Care

ROBERT S. McCALEB

The use of natural plant-derived medicines is perhaps the only sustainable form of medicine. It can foster a greater consciousness of the value of biodiversity while offering appropriate medicines for the developing and industrialized world and high-value crops for worldwide production. Botanical medicines are experiencing a meteoric rise in popularity worldwide, especially in the United States and Europe. The increased use of herbs for health can help save the environment, but this increased demand can also destroy local ecosystems and push threatened plants to the brink of extinction. The key to achieving the positive potential of the natural health care movement is environmentally and socially conscious development and sustainable production of the botanical raw materials which feed this rapidly growing business.

Rescued from the Brink

The ginkgo tree, *Ginkgo biloba,* is often described as "a living fossil." It is the world's most primitive tree, unchanged through millions of years of evolution. It is also one of the most successful "new" medicinal plants. An extract of ginkgo leaves increases the flow of blood and oxygen to the brain, the heart, eyes, and ears, and to the arms and legs. It is among the most widely used medicines in Europe, prescribed by German medical doctors to over 10 million patients annually. As an approved drug, it is used to help prevent loss of memory and cognitive function in the elderly, increase pain-free walking distance in those with impaired peripheral circulation, and for treating circulation-related diseases from senile dementias to tinnitis, vertigo, and macular degeneration.

We almost lost the ginkgo tree. The only remaining member of its genus and family, the magnificent ginkgo became extinct in the wild between 1,000 and 10,000 years ago. It was saved from extinction by temple plantings in China. It is the earliest and most dramatic example to date of how human intervention and conservation of a plant with then-unknown medicinal value is saving lives and improving the quality of life for the human species today. It is also generating substantial income for thousands of people involved in the collection, cultivation, production, and marketing of its successful natural medicine. Ginkgo is a powerful reminder that the plant we save today may save us tomorrow.

Ancient and Modern History

Plants have always been and continue to be our most important sources of new medicines. Their earliest use predates written history, and they were virtually our only source of medicines for over ten thousand years. Only in 20th century Western medicine did synthetic chemicals arise as major medicinal agents. Even in the modern pharmacy in the United States today, over 25% of medicines are extracted from higher plants, or are synthetic copies or derivatives of plant chemicals.[1] (See also Grifo et al., Chapter 6.) Throughout the rest of the world, their use far surpasses synthetics. According to the World Health Organization, over 80% of the world's population still relies on traditional medicines[2] including herbal medicine, for *primary* health care.[3] In the United States, plant medicines composed of whole plants (crude drugs) or complex extracts have almost vanished from the pharmacy as approved drugs, but are sold as dietary supplements. This is largely the result of the high cost of drug research and of gaining regulatory approval, combined with the lack of patent protection for plant medicines. Simply put, natural medicines are not economically viable candidates for drug research and development. A company which spends the required $50–100 million to gain new drug approval for a plant would not have the exclusive right to sell it.[4,5] Despite this, herbal medicinal products are becoming ever more popular in the United States, with annual growth rates in natural food stores as high as 60–80% for "medicinal" herbs in bulk, capsules, extracts, tinctures, tablets, and teas for medicinal use.[6] Some individual companies like Solaray (34–50% annual growth) and McZand (50% growth annually) are experiencing a phenomenal rate of growth.[7] Botanicals have migrated from the pharmacy of the early 20th century, to the health food store, and now back to the pharmacy, grocery, and discount stores which are the bastions of mass market sales. Botanical sales in the mass market have increased 300% since 1992.[8] Note that these robust growth statistics are possible because of the small size of the overall herb sales of around $1.5 billion in the United States. In Europe, botanicals are among the best selling drugs, and are usually preferred by the public over synthetics, according to David McAlpine, director of McAlpine, Thorpe & Warrier, Ltd., a management consultant company in England.

Factors Driving the Trend toward Medicinal Plant Use

Potential Cost Savings from Natural Remedies

The American health care system is in a state of crisis. The cost of high-tech modern Western medicine has increased so dramatically that currently over 35 million Americans cannot afford basic health insurance.[9] As a percentage of gross national product, Americans pay twice as much as Europeans and three times as much as Japanese for health care costs, yet lag behind both groups in terms of measurable health statistics such as longevity, infant mortality, cancer, and heart disease rates.[10] While the reasons for these excessive health care costs are complex, the direction and implications are clear. Every available option must be pursued to lower the cost of health care while maximizing the benefits.

Natural remedies from plants can serve as safe and effective direct replacements for proprietary drugs and cost substantially less. For example, compare the costs (below) of treating gastric ulcers with surgery or Tagamet™, and licorice or deglycyorhized licorice extract (DGL).[11] Note that licorice extract has cardiovascular side effects which could limit the long-term use of large amounts, however in support of its safety, it is approved by FDA as a safe food ingredient (up to 24% concentrated extracts may be used in licorice candy). DGL has none of these side effects and can be used for longer periods if necessary.

Surgery	$25,000
Tagamet™ (per year)	$1,000
DGL (per year)	$300
Licorice extract (per year)	$120

Another example is treatment of benign prostatic hyperplasia (BPH).

Benign enlargement of the prostate gland affects nearly half of all men over forty and 75% of those over sixty.[12] The following compares the FDA-approved prescription drug Proscar with the European over-the-counter remedy made from a standardized extract of saw palmetto (*Serenoa repens*):

TREATMENT	COST	RISKS
Surgery	$5000	All surgery carries with it the potential for serious or fatal complications. Complications specific to prostate surgery include incontinence, impotence, bleeding, and infection.
Proscar™ (per year)	$657	Impotence (3.7% risk), decreased libido, decreased volume of ejaculate, urinary obstruction (not proven to be a long-term risk); dangerous when used in patients with liver problems.
Serenoa std. extract	$255	None known. Saw palmetto berries were used as a staple food by Native Americans and have shown no toxicity in European use.[13]

Improved Research and Clinical Use

The use of sophisticated plant medicines by advanced countries like Germany, France, and Japan produces a wealth of both scientific documentation and clinical experience. Some botanicals have hundreds of peer-reviewed scientific studies supporting their use and many have now been used by tens of millions of Europeans under clinical supervision. European countries, exemplified by Germany, have successfully integrated natural remedies with conventional medicines in medical practice and in the pharmacy. Legislators, the public, and the medical communities are beginning to watch these countries whose citizens pay a fraction of American health care costs and enjoy better health as measured by the statistical evidence cited above.

Well-Researched Botanical Medicines

GARLIC has been shown to have many benefits, the most intensively studied of which are cardiovascular, antimicrobial, and anticancer activities.

Cardiac benefits have been clinically tested since at least the 1960s. Early experiments in India showed the volatile oil of garlic could lower blood cholesterol, even when consumed with 100 grams of butter.[14] Cooked garlic and onions have also been shown to be effective at lowering blood fat levels. Garlic oil filled "perles" also lowered cholesterol, while artery-sparing HDL increased.[15] In addition to lowering blood fats, garlic inhibits blood clotting, which reduces the risk of strokes or heart attacks. A study in the prestigious *American Heart Journal* confirmed that a garlic extract prevents clot formation in coronary arteries.[16]

Garlic is antimicrobial, effective against opportunistic microbes like *Herpes* virus and *Candida*,[17,18] as well as against at least 8 types of antibiotic-resistant bacteria. Garlic extract surpassed penicillin, ampicillin, doxy-cycline, streptomycin, and cephalexin against 8 of 9 strains of antibiotic-resistant *Staphylococcus, Escherichia, Proteus,* and *Pseudomonas.* A simple, water-based garlic extract showed "promising" activity against all these organisms.[19]

Anticancer effects are among the most intensively studied attributes of garlic. One component, DAS (diallyl sulfide), inhibits colon cancer and reduces radiation damage. Garlic may not only decrease cancer risk, it may also reduce damage from radiation treatments.[20] Garlic and onion oils both inhibit an enzyme thought to promote tumors.[21] Garlic extract inhibits chemically induced cancer, preventing chemical reactions leading to tumor formation and completely preventing the first state of skin cancer development.[22] Numerous other studies document the anticancer effects of garlic, but perhaps none so directly as a large human study which found that the more garlic, onions, and other *Alliums* people consume, the less stomach cancer they have. The study involved 1,695 humans, 564 of whom had stomach cancer. Those who regularly consumed garlic and onions had 40% less incidence of cancer than people who ate little or no garlic.[23]

MILK THISTLE (*Silybum marianum*) seed extract protects against liver damage, and helps the regeneration of liver cells damaged by toxins and by diseases such as hepatitis and cirrhosis.[24,25] Milk thistle has been used as a liver remedy for at least two thousand years.[26] Six months of treatment significantly improved liver function in thirty-six patients with alcohol-induced liver disease.[27] It protects against radiation damage caused by x-rays,[28] and gives "complete protection" against brain damage caused by the potent nerve toxin triethyltin sulfate.[29] Intravenous milk thistle extract is used in emergency rooms in Europe, successfully counteracting cases of liver poisoning including those caused by the deathcap mushroom (*Amanita phalloides*).

BILLBERRY EXTRACT (*Vaccinium myrtillus*) strengthens the human capillary system. It prevents and treats fragile capillaries and capillaries which leak either fluid or blood into the tissues. Capillary fragility can result in hemorrhage, stroke, heart attack, or blindness resulting from diabetic or hypertensive damage to the retina. Less serious effects include a tendency toward easy bruising, varicose veins, "spider veins," poor night-vision, and coldness, numbness, or cramping of the legs. Bilberry extracts help prevent such problems by strengthening the capillaries and other small blood vessels, and by increasing the flexibility of red blood cell membranes. Through this action, the capillaries are better able to stretch, increasing blood flow, and red blood cells are better able to deform into a shape which allows them to pass through very narrow capillaries.

Clinical trials have shown the effectiveness of bilberry against venous insufficiency of the lower limbs[30,31] in subjects from 18 to 75 years old. It can treat varicose veins in the legs,[32] producing a significant improvement in symptoms of varicose syndrome: cramps, heaviness, swelling of the calf and ankle, and numbness. In these trials, there were no significant side effects, even when the normal dose was exceeded by 50%. A standardized bilberry extract was studied in two clinical trials totalling 115 women with venous insufficiency and hemorrhoids following pregnancy.[33,34] Both studies documented an improvement in symptoms, including pain, burning, and pruritus, all of which disappeared in most cases. Again, no side effects were noted. Both diabetes and hypertension are notorious for causing pathological changes in the retina, which compromises vision. Bilberry produced a significant improvement in vision and measurable microcirculation in forty patients.[35]

ECHINACEA (*Echinacea* spp.) stimulates the general activity of the immune system. It has been the subject of over 400 scientific studies.[36] In a double-blind placebo-controlled study of 180 volunteers, the therapeutic effectiveness of *Echinacea* against colds and flu was "good to very good."[37] Another study showed that *Echinacea* extracts, administered orally, significantly enhanced phagocytosis in mice.[38] Water-soluble components of echinacea strongly activated macrophages[39] and enhanced the motility of immune system cells and increased their ability to kill bacteria. In addition, other immune system cells were stimulated to secrete tumor necrosis factor, and interleukins 1 and 6.[40] Another study showed that *Echinacea* polysaccharides could increase the numbers of immunocompetent cells in the spleen and bone marrow and also their migration into the circulatory system. The authors noted "these effects indeed resulted in excellent protection of mice against the consequences of lethal infections by *Lysteria* and *Candida*."[41]

GINKGO BILOBA EXTRACT (*Ginkgo biloba*) stimulates circulation and oxygen flow to the brain, which can improve problem-solving and memory. It has been proven to increase the brain's tolerance for oxygen deficiency,[42] and can increase blood flow in patients with cerebrovascular disease. No other known circulatory stimulant, natural or synthetic, can selectively increase blood flow to disease-damaged areas of the brain. Ginkgo is used principally against symptoms of aging. In a French study, "the results confirmed the efficacy of [ginkgo extract] in cerebral disorders due to aging."[43] In another experiment, ginkgo users consistently and significantly improved on all tests over the control group. Specific factors which improved included mobility, orientation, communication, mental alertness, recent memory, freedom from confusion, and a wide variety of other factors.[44] A "Digit Copying Test" and a computerized classification test confirmed the improvement in cognitive function tested at a London hospital.[45]

Ginkgo extracts also stimulate circulation in the limbs, reducing coldness, numbness, or cramping. In elderly individuals, ginkgo improved pain-free walking distance by 30 to 100%.[46] It corrected high cholesterol levels in 86% of cases tested, and prevented oxygen deprivation of the heart.[47] A German study documented benefits of long-term use of ginkgo in reducing cardiovascular risks, including those associated with coronary heart disease, hypertension, hypercholesterolemia, and diabetes mellitus.[48] Ginkgo also strongly inhibited the effects of bacteria associated with gum disease (and was more effective than tetracycline).[49] By maintaining blood flow to the retina, ginkgo extracts have been shown to inhibit deteriorating vision in the elderly. Adequate amounts of the extract can actually reverse the damage from lengthy oxygen deprivation of the retina. "The assessment by both doctors and patients of the general condition of the patients showed a significant improvement after the course of therapy. The results presented here show that damage to the visual field by chronic lack of blood flow is significantly reversible."[50]

European Plant-Based "Phytomedicines"

Saw palmetto (*Serenoa repens*)	Treats benign prostatic hypertrophy
Valerian (*Valeriana officinale*)	Mild sedative
Mistletoe (*Viscum album*)	Cancer chemotherapy
Ginseng (*Panax ginseng*)	Promotes endurance and increases resistance to environmental stress
Ginkgo (*Ginkgo biloba*)	Peripheral and cerebral circulatory stimulant, PAF antagonist

Bilberry extract (*Vaccinium myrtillus*)	Combats capillary fragility, eye disease
Milk thistle (*Silybum marianum*)	Prevents liver damage and is used to treat liver disease including cirrhosis and hepatitis
Echinacea (*E. angustifolia* or *E. purpurea*)	Immune stimulant
Garlic (*Allium sativa*)	Lowers blood cholesterol and reduces risk of stroke

Public Interest in Natural Alternatives

The trend toward increasing herb use is a consumer-driven trend and is apparent from the rapid growth of commerce in herbal products. Public opinion is partly shaped by media attention, and in this area too, increases are dramatic. Only a decade ago, the majority of media coverage of herbs was negative. This included articles in the *FDA Consumer,* which painted a sensational picture of hazards which lurked in the health food store in the form of toxic herbal products.[51] Similarly, alarmist articles appeared in numerous newspapers, magazines, and television reports. Within the last five years, however, much of the coverage has become more positive. The popular media are enthralled by alternative medicine. Within the last few years, it has received major coverage on all three major television networks and CNN, *U.S. News and World Report, Time, Newsweek, The New York Times, The Washington Post,* and countless other magazines, newspapers, and radio and television stations. Sales of books and magazines focusing on natural health care methods have increased dramatically, as have the number of available titles. Increasingly, doctors are being asked about natural alternatives or adjuncts to conventional treatment. A surprising study by Harvard Medical School showed that over a third of Americans are using some form of alternative therapy with out-of-pocket expenses of over $10 billion annually.[52] The study also identified a trend toward greater interest in health, fitness, and lifestyle changes that may result partly from a distrust of an excessive reliance on science and technology, coupled with concerns with side effects and other safety issues with conventional medicines.

The Pursuit of Earth-Friendly Health Care

A growing segment of the public believes that herbal medicines and other alternatives are safer, possibly more effective, more natural, and more in harmony with a lifestyle that promotes self-care, individual responsibility, freedom of choice, and "holistic" thinking. A part of this too is the belief that a return to more natural therapies is a return to the time in which our medicines, like our foods, came from the earth, and the use of these natural substances is more in harmony with our natural surroundings.

Many environmental groups concerned with conservation of the rainforests and other wild areas have focused on the potential economic and medicinal value of plants which may become extinct before scientists have an opportunity to study their potential.

There is some question about our ability to incorporate truly natural remedies into our health care system because of economic and regulatory obstacles. Today's medicinal plant exploration has been characterized as "chemical prospecting" in which plants are analyzed only for potentially valuable chemicals. These chemicals must be novel and therefore patentable, or synthesizable, or must represent interesting structures from which synthetic analogues can be developed. The unfortunate fact is that this leads to the development of drugs in which the plant is no longer important. Thus even if profits from the marketing of drugs originally discovered in the forests are shared with developing countries, the plant is no longer important in the production of the medicine, possibly depriving producing countries of a market in the plant, and increasing the cost of the medicine. If regulatory and economic obstacles can be overcome, the plant itself, whether cultivated or sustainably collected from the wild, can be returned to a place of prominence in the production of mainstream medicines. Among the companies which have embarked on major medicinal plant research programs are Merck, Bristol Myers Squibb, Pfizer, Monsanto, Smith Klein Beecham, and Eli Lilly. The interest of these companies in as yet undiscovered medicinal plants—or rather, the compounds from medicinal plants—is a testimony to the importance of preserving biodiversity as a source of future medicines.

The most impressive collections of biodiversity are found in true wilderness. Rainforests, whether tropical or temperate, wetlands, and other wilderness areas must be preserved intact to avoid disturbing the delicate balance of these awesomely productive ecosystems.

Saving the Plants, Saving the Planet

A major contribution to saving the rainforests and other wild places is a recognition of the value, including the *economic* value, of medicinal and other useful plants living there. Only a fraction of the world's plants have been thoroughly studied for their medicinal potential; fewer than 2% of the estimated 250,000 species of higher plants. Historically, botanicals have been our most fruitful arena in the search for new medicines. The average success rate for identifying useful medicines from plants is one in 125, while the success rate for developing new drugs from randomly synthesized chemicals is only one in 10,000. From the estimated 5,000 plants which have been thoroughly investigated to date, we have developed 43 major medicines. The average economic value of each of these medicines is over $200 million per year in the United States alone.[53] Wilderness has value that cannot be measured in dollars, an aesthetic value we feel and know when we behold it, however, it is the perceived monetary value of rainforest resources that threatens such wild places. The forests will not be saved if the people living there are not fed, and both forest-dwellers and the developing countries where they live must make a living with whatever resources they have. It is vital that we realize the potential *economic* value of saving indigenous plants, which far outweighs the land's value for farming, ranching, timber, and mining. The issues of using those resources responsibly will be considered later in this chapter.

The Cost of Loss

We are currently losing around one plant species per day.[54] If our past record of discovery (1:125) holds true, this means that each year, we will lose forever three major medicines. The economic value of that loss is clear. By the year 2000 we will have driven to extinction medicinal plants worth $40 billion annually in the United States alone.[55] The value of cattle or crops grown on cleared rainforest land would be only a tiny fraction of this amount. Beyond the economic cost, we cannot estimate the cost in terms of the positive potential benefits to human health of these lost medicines.

When governments begin to view their wilderness areas with this in mind, all other economic uses pale in comparison with the potential value of every plant in the wild which has not been thoroughly studied for medicinal use. However, once the medicinal value of such plants is discovered, and assuming that a marketable product can be developed from it, the threat of overcollection and habitat destruction becomes very real. It is critical that both the governments and herb collectors realize that their future income depends upon sustainably harvesting the plants and planning for an expanding market in those botanicals. Thus, the value of medicinal herbs is simultaneously an incentive to harvest and sell those plants recklessly, and an incentive to preserve and manage herbal natural resources sustainably. In some cases, this will mean bringing plants discovered in the wild into cultivation, while in other cases, careful collection from wild populations can provide a renewable source of medicinal plant raw materials. Regardless of whether these plants are cultivated or collected from the wild, true wilderness areas must be protected from any invasive human activities. In many countries throughout the world, there is currently available crop land for which farmers are in an urgent and continual search for marketable crops to grow.

Cooperation is essential between developed and developing countries in exploring, researching, and utilizing renewable resources, including medicinal plants. Native medicinal plants can provide both cost-effective remedies for local use, and can help to generate income through exports. Unfortunately, many expeditions of discovery launched by developed countries have been exploitative. Typically when a medicinal plant is discovered through research the drug company sponsoring that research develops and profits from the discovery, yielding little or no benefit to the country in which it was discovered. New models of cooperation in the drug discovery and development process employ agreements by drug companies to set aside a portion of profits from the discovery to benefit social and environmental programs within the country of discovery.[56]

Social Benefits of Increased Herb Use

One of the major advantages of increased herb use in social terms is the potential for improving health and lowering health care costs. There is a tremendous diversity possible within the realm of herbal medicines. They extend from simple dried plants used as teas or consumed in capsules, to a wide array of extracts—from the strong decoctions characteristic of traditional Chinese medicines, to higher potency extracts using mixtures of water, alcohol, and sometimes other solvents, to the hi-tech concentrated and standardized extracts which are at the heart of European phytomedicine today. This diversity allows for the development of appropriate health care in which production, manufacturing, and use of botanical remedies can be local and appropriate to the lifestyles and incomes of the people served by these various forms of botanical medicine.

Low-Cost Appropriate Health Care

Traditional medicine, including herbal medicine, is already practiced in most developing countries. The use of medicinal infusions and decoctions and homemade salves and poultices can be made more effective by the transfer of knowledge, appropriate technologies, and the introduction of medicinal plants which have proven valuable elsewhere. One of the major developments in herbal product sales in the United States in the 1980s was the shift from powdered herbs and capsules to hydroalcoholic extracts or tinctures. This type of medicinal extract can be made simply and economically, and preserves the medicinal properties of the plant, allowing for longer-term storage for shipping regionally or internationally, and extending the reach of traditional medicines into more urbanized areas, using convenient dosage forms of shelf-stable products.

In developed countries, too, low-cost medicines are important. One of the major trends in pharmacy is the shift toward self-care as seen in the shift of some prescription drugs to over-the-counter drug status (Ibuprofen, Immodium™, and topical mycotics for yeast infections are a few examples). Herbal remedies sold in the United States as dietary supplements are primed to fill a niche for low-cost, low-toxicity self-care remedies. A perfect example is ginger powder, clinically proven to be effective against motion sickness, morning sickness, and stomach upset caused by other drugs, including chemotherapy and anesthesia.[57,58] Pharmaceutical science has not produced a safer or more effective remedy against nausea than that delivered by ordinary ginger powder in capsules. Similarly, feverfew (*Tanacetium parthenium*) is a safe, inexpensive migraine preventive which is effective even in patients who do not benefit from the conventional therapy.[59] Again, growing the plant is frequently cheaper than making the conventional drugs for which it substitutes.

Prevention: The Neglected Modality

Speaking at a botanical medicine conference, the former Director of the Office of Alternative Medicine of the National Institutes of Health, Joseph Jacobs, described the American medical system as "a sick care system, not a health care system." He described American medicine as mostly devoid of the concepts of wellness and preventive medicine, in which even the term "preventive medicine" is usually used to describe early disease detection and in which the accepted scientific and medical tools for drug approval appear inappropriate for even studying the subject of true disease prevention through the promotion of health and well-being.[60]

One of the prominent features of the U.S. deficiency in preventive medicine is the lamentable absence of true *preventive medicines* in our pharmaceutical system. In the 50 years since our food and drug approval system has been in operation, not a single over-the-counter (OTC) medicine has been approved for internal use in the prevention of any major disease. In fact, the only FDA-approved OTC preventive medicines are fluoride toothpaste, sunscreens, and motion sickness pills. Research continues to document the potential utility of antioxidant, anticarcinogenic, and other health-protective properties of medicinal plants which help to reduce the risk of serious disease and could dramatically lower health care costs. European and Japanese consumers have access to dozens of government-approved natural remedies which have disease preventive effects. Among the best researched European phytomedicines are agents which can reduce the risk of all four of the leading natural causes of death in the United States— heart disease, cancer, respiratory disease, and liver disease. The regulatory obstacles which have prevented research on medicinal plants constitute an even greater barrier to the approval of preventive medicines because these effects are harder to prove.

In the absence of significant over-the-counter preventive medicines, dietary supplements, including herbal dietary supplements, are the only products available to healthy Americans to use for the promotion and maintenance of health (with the exception of nutritious foods, of course). In October 1994, the Congress of the United States declared:

> Improving the health status of United States citizens ranks at the top of the national priorities of the Federal Government; the importance of nutrition and the benefits of dietary supplements to health promotion and disease prevention have been documented increasingly in scientific studies; there is a link between the ingestion of certain nutrients or dietary supplements and the prevention of chronic diseases such as cancer, heart disease, and osteoporosis; legislative action that protects the right of access of consumers to safe dietary supplements is necessary in order to promote wellness.[61]

The above excerpt is a portion of the findings which preface the Dietary Supplement Health and Education Act from overly zealous regulation in allowing a freer flow of information to the public about the benefits of those supplements. This law has already had a dramatic impact on the marketplace, increasing sales of dietary supplements through all mass market outlets. The positive health impact of more widespread use of cardioprotective agents, anticarcinogens, antioxidants, liver protective agents, and immune stimulant agents, among others, may take years to document. However, the impact on international trade in the raw materials for these supplements is already being felt.

Medicinal Plant Farming

Increasing American utilization of medicinal plants creates a demand for botanical raw materials which could produce a tremendous boon to farmers, both domestically and abroad. The raw materials for European and Asian phytomedicines have already produced a lively international market which can help stimulate family businesses abroad and family farms in the United States. In many cases these botanicals must be produced using ancient and labor intensive methods including hand picking and "garbling," or manual removal of twigs, rocks, and other contaminants from the dried herb. Because of this, these commodities represent a continuing economic opportunity in developing countries for all but the few crops which can be mechanically harvested. According to the U.S. Agency for International Development, the greatest challenge facing most farmers in developing countries is finding markets for crops with sufficient value to sustain a family business.[62] In many parts of the world, agroeconomic development has shifted away from subsistence farming toward the search for specialty crops and cash crops which can be grown on farmland which is currently idle.

Domestic production of medicinal plants also shows great potential. The herb industry currently purchases raw materials from all over the world including the United States. Botanicals like mints, echinacea, feverfew, and others can be mechanically harvested, and thus serve as alternative cash crops to replace crops which are now produced in surplus, or which, like tobacco, are declining in demand. Finally, domestic production of herbs can produce export income as well, as seen by the plantation of two million *Ginkgo biloba* trees in South Carolina, which is used as a source material for European phytomedicines. Many other American botanicals are exported to Europe, further underscoring the potential for producing income from collection or farming of medicinal plants. The fact that American farmers can profit from selling botanicals is a good indication of their profitability, given the high cost of land and labor in the United States.

Environmental Hazards of Herb Use

Since prehistoric times, humans have gathered plants for food, medicine, and for trade. Wildcrafting, gathering plants from the wild for profit, has been practiced worldwide for millennia. A substantial (but not precisely known) amount of the herbs in commerce is still collected in this way.

Wildcrafting, though, can be either good or bad for the environment, depending on how it is done. The concept of sustainable wildcrafting is a very old one, and many plants have been gathered from the wild in a manner which has caused no readily apparent harm to those areas. In recent years, some of the plants have become scarce, threatened, or endangered in areas which previously provided an abundance of these plants. Wildcrafting is sustainable to the extent that the plants are abundant and/or propagate quickly and can be removed from the ecosystem without significantly impacting the food chain. There are certainly examples of completely sustainable wildcrafting. An example is the collection of blackberry leaves (*Rubus fruticosis*) from eastern Europe. European blackberry, or bramble, is so prolific and troublesome, it is considered a noxious weed by the USDA, and is illegal to grow in the United States. Closer to home, an herb which is gaining rapidly in popularity in the United States and Europe is St. John's wort (*Hypericum perfoliatum*), known in California as Klamath weed. Since decades of eradication programs based on chemical weedkillers have failed to substantially reduce the populations of Klamath Weed in California, it is unlikely that wildcrafting of this species could be considered harmful.

It is also possible, unfortunately, for plants to be overcollected, depleting their populations. Hardly confined to developing countries, this problem is felt first in highly populated countries where there is a developed market in herbs and hence competition for available plants. As they become rare, prices rise, hastening the depletion by bringing in more competition. Such is the case with wild Chinese ginseng, which is reportedly extinct. American ginseng, too, is at risk in the wild, but has come under state and federal protection. Part of the problem has been a belief that the wild roots of ginseng are more potent and effective than those from cultivated plants. There is no solid scientific support for this belief, but the tenacious bias is still reflected in much higher prices paid for wild versus cultivated ginseng.

Another threatened American medicinal plant is ladyslipper (*Cypripedium* spp.), a wild orchid valued for its sedative properties. The American Herbal Products Association has passed a resolution against the sale of wildcrafted ladyslipper. Other plants in trouble include Virginia snakeroot (*Aristolochia serpentaria*), goldenseal (*Hydrastis canadensis*), and possibly osha root (*Ligusticum porterii*). Osha root is especially threatened, because, unlike most other wildcrafted plants, no one yet has successfully cultivated the plant. Even professional botanists like Harvard's Shawn Sigstedt have been unable to propagate and grow it. Some producers of osha root products are aware of the problem and scrupulously refrain from full promotion of the plant, to keep it from becoming too popular.

Some species of *Echinacea* are rapidly disappearing from the wild in the American Southeast. Wild American ginseng is a threatened plant throughout its range, as is bloodroot (*Sanguinaria canadensis*). In parts of the Amazon, the majestic *pau d'arco* trees (*Tabebuia* spp.) have been cut down for their bark, and in Africa the harvesting of *Pygeum africanum,* used as a prostate disease remedy, is causing concern in its native Cameroon and in Madagascar.

Within this problem is the seed of its solution. As countries like Cameroon begin to see the economic value of the medicinal plants in their forests, they can better appreciate the foolishness of clear-cutting those forests for timber, ranching, or mining. In fact, one of the strongest hopes we have for saving the ancient forests is that their true economic value will now be recognized.

There is an erroneous belief—sometimes promoted by herb dealers—that wildcrafted herbs are pristine and somehow equivalent to organically grown herbs. In truth, there is no way to know if gathered herbs are from unsprayed or unpolluted areas unless they have been tested for residues. Many may be collected near roads and could have higher exposure to airborne pollutants than those grown in fields. In the words of botanist Steven Foster "wildcrafted means 'origin unknown'." Cultivation is often the answer to protecting wild populations of a plant, without eliminating the market in the botanical. Often, plants which are easily and economically cultivated are gathered in the wild simply because it is cheaper than farming.

Medicinal plant expert James Duke, formerly with the USDA, advises ways to protect wild plant populations while continuing to collect sustainably from them. The main rules include: don't take whole populations, and attempt to replenish them by reseeding where appropriate. This is frequently practiced by ginseng collectors, who collect only after seeds are mature late in the season. After digging the root, they can plant the berries nearby, in areas ginseng is known to favor. To protect the consumer, wildcrafters should avoid areas close to roads or other sources of pollution including industrial (not to mention nuclear) facilities.

Foster goes further, suggesting these guidelines for collecting from the wild:

- Know what plants are rare, endangered, or threatened in the area in which you are collecting.
- Don't collect on public lands without a permit.
- Ask permission to collect on private land.

- Never take more than 10% of a population.
- Don't harvest from a population of less than 100 plants.
- Know the exact genus and species you are seeking.
- Harvest root plants after seed is set for better potency.

Herb Cultivation

Many, if not most, of the herbs used in dietary supplements and herbal medicines can be cultivated. Bringing threatened or endangered herbs into cultivation is one way to save their wild populations, and in many cases it is more practical to produce herbs on a farm than to collect them from the wild. Agriculture is the most important source of income for much of the developing world. Herbs offer unique advantages as specialty crops because of their hardiness and their value. They can also provide low-cost medicines for the countries growing them, thereby producing both income and health care from the same herb field.

Lest we forget, in the United States we pay farmers not to farm, and we subsidize the cultivation of such questionable crops as tobacco and water-intensive crops in arid parts of southwestern United States. A shift from petrochemical medicine to natural remedies can help to revitalize American agriculture without subsidies and can help to support developing countries through agricultural commerce instead of foreign aid allocation. The ginkgo plantation in South Carolina provides raw materials for the European medicinal plant industry. The fruit of the dwarf palm called saw palmetto (*Serenoa repens*), collected from the wild in Florida, has increased dramatically in value in recent years, and efforts are now under way to establish plantations. Echinacea (*Echinacea purpurea* and *E. angustifolia*) are valuable cash crops from Washington state to the Ozarks.

The Pharmaceutical Plant

The agroeconomic effects of increasing our reliance on natural medicines thus extend from the American family farm to international cultivation. Other commercial sourcing includes the wildcrafted plants of the world. One of the most obvious environmental advantages of going natural involves the nature of the "pharmaceutical plant." Synthesizing drugs in chemical factories is capital intensive, energy intensive, and polluting.

When our "pharmaceutical plant" is growing in a field of herbs, however, the opposite is true. It oxygenates the atmosphere and can reduce the transportation and utilization of toxic chemicals, not to mention the aesthetic advantages.

Concerning the issue of sustainability, all of the environmental advantages of growing our medicines instead of synthesizing them rely upon our determination to produce or collect the plants involved in a sustainable way. In the wild, this means very careful control of collection so the populations do not become depleted. In agriculture, it means using sustainable agriculture methods like organic farming, crop rotation, and other horticultural practices designed to preserve the soils on which we rely for growing our medicine.

Social responsibility is important too. All too often industrialized nations have been guilty of exploiting the suppliers of their raw materials. To prevent this, we must add to our list of issues the issue of fair trade—of assuring that the suppliers of our health care products are fairly compensated and share in the success of any medicines, especially new medicines, that they are instrumental in developing or discovering.

Summary

There is a world of benefits to be derived from bringing nature back into our pharmacies and medicine cabinets through natural herbal remedies. Better health care at lower cost, preserving threatened and endangered plants, increasing the perceived value of the rainforests and other wild areas, supporting sustainable agriculture here and abroad, and reducing pollution are among them. To realize these benefits though, we must overcome substantial obstacles. Before natural medicines will be widely accepted, they must be better researched and results must be made more widely available through education. Regulatory changes are needed to lower the obstacles to selling herbs as medicines in the United States and programs must be in place to ensure that the production of herbs, whether through collection from the wild or cultivation, is sustainable and environmentally and socially conscious. As with most major changes, the public will dictate both the direction and the speed of change. By learning more about herbal medicines, and using them wisely and responsibly, the public is already creating a demand for these products. Only by understanding the complex nature of international herb trading can we assure that this shift is good for the health of the planet as well as its peoples.

References and Notes

1. Farnsworth, N.R. (1984). The role of medicinal plants in drug development. In *Natural Products and Drug Development*. Krogsgaard-Larsen, P., Christensen, S. Brøgger, and Kofod, H. (eds.). Proceedings of Alfred Benzon Symposium 20, pp. 17–30. Munksgaard, Copenhagen.

2. Traditional medicine, as a term in international use, refers to systems of medicine in place since ancient times. The term here is distinct from "conventional" or "orthodox" medicine as practiced in the United States today.

3. Farnsworth, N.R., O. Akerele, S. Audrey, D.D. Soejarto, and Z. Guo. (1985). Medicinal Plants in Therapy. *Bulletin of the World Health Organization* 63(6):965–1170.

4. Tyler, V.E. (1985). Plant Drugs in the Twenty-First Century. Presented at the Symposium on Economic Botany in the Year 2000, Society for Economic Botany, 26th annual meeting, Gainesville, FL, August 13, 1985.

5. *Chemical Marketing Reporter* estimates an even higher figure at $359 million.

6. *Natural Foods Merchandiser* market reports for 1991, 1992, 1993, and 1994. New Hope Communications, Boulder, Colorado.

7. Anonymous. (1991). U.S. Herbal Product Market Studied. *Marketletter* Nov. 18, p. 25.

8. Brevoort, P. (1994). The Economics of Botanicals: The U.S. Experience. Presentation at the NIH/OAM Conference on Botanicals: A Role in U.S. Health Care?, Washington D.C., December 16.

9. Lamm, R.D. (1990). The Brave New World of Health Care. Presented at the Center for Public Policy and Contemporary Issues, University of Denver, May 1990.

10. Ibid.

11. Rountree, Robert, M.D. (1992). Personal communication, Boulder, Colorado.

12. Denis, L.J. (1993). Quantification and Incidence of Benign Brostatic Hyperplasia. *Drugs of Today* 29: 328–333.

13. Anonymous. (1994). Saw Palmetto Extract vs. Proscar. *Amer. Jrnl. Natural Medicine* 1:1,8–9.

14. Sharma, K.K. et al. (1976). Effect of Raw and Boiled Garlic on Blood Cholesterol in Butter Fat Lipaemia. *Indian Journal of Nutrition and Dietetics* 13(1):7–10.

15. Barrie, S. et al. (1987). *Journal of Orthomolecular Medicine* 2(1):15–21.

16. DeBoer, L.W., and Folts, J.D. (1989). Garlic Extract Prevents Acute Platelet Thrombus Formation in Stenosed Canine Coronary Arteries. *Amer. Heart Jrnl.* 117(4):973–975.

17. Adetumbi, M. et al. (1986). *Allium sativum* (Garlic) Inhibits Lipid Synthesis by *Candida albicans. Antimicrobial Agents and Chemotherapy* 30(3):499–501.

18. Weber, N. et al. (1992). In Vitro Virucidal Effects of *Allium sativum* (Garlic) Extract and Compounds. *Planta Medica* 58:417–423.

19. Singh, K.V., and Shukla, N.P. (1984). *Fitoterapia* 55(5):313–315.

20. Baer, A., and Wargovich, M.J. (1989). Role of Ornithine Decarboxylase in Diallyl Sulfide Inhibition of Colonic Radiation Injury in the Mouse. *Cancer Res.* 49(18):5073–5076.

21. Belman, S. et al. (1989). Inhibition of Soybean Lipoxygenase and Mouse Skin Tumor Promotion by Onion and Garlic Components. *Jrnl. Biochem. Toxicol.* 4(3):151–160.

22. Nishino, H. et al. (1989). Antitumor-Promoting Activity of Garlic Extracts. *Oncology* 46(4):277–280.

23. Yu, W.C. et al. (1989). Allium Vegetables and Reduced Risk of Stomach Cancer. *Journal of the National Cancer Institute* 81:162–164.

24. Bode, J.C. et al. (1977). Silymarin for the Treatment of Acute Viral Hepatitis? *Med. Klin.* (Munich) 72(12):513–518.

25. Plomteux, G. et al. (1977). Hepatoprotector Action of Silymarin in Human Acute Viral Hepatitis. *IRCS Med. Sc. Libr. Compend.* 5(6):259.

26. Foster, S. (1991). "Milk Thistle," Botanical Series #305. American Botanical Council, Austin, TX.

27. Feher, H. et al. (1990). Hepatoprotective Activity of Silymarin Therapy in Patients with Chronic Alcoholic Liver Disease. *Orvosi Hetilap* 130:51.

28. Flemming, K. (1971). Therapeutic Effect of Silymarin on X-Irradiated Mice. *Arzneim-Forach* 21(9):1373–1375.

29. Varkonyi, Tibor et al. (1971). Brain Edema in the Rat Induced by Triethyltin Sulfate. 10. Effect of Silymarin on the Electron-Microscopy Picture. *Arsneim.-Forsch.* 21(1):148–149.

30. Guerrini, M. (1987). Report on Clinical Trial of Bilberry Anthocyanosides in the Treatment of Venous Insufficiency of the Lower Limbs. Istituto di Patologia Speciale Medica e Metodologia Clinica, Università di Siena.

31. Corsi, S. (1987). Report on Trial of Bilberry Anthocyanosides (Tegens—Inverni della Beffa) in the Medical Treatment of Venous Insufficiency of the Lower Limbs. Casa di Cura S. Chiara, Florence, Italy.

32. Gatta, L. (1982). Controlled Clinical Trial among Patients Designed to Assess the Therapeutic Efficacy and Safety of Tegens (160). Ospedale Filippo del Ponte, Varese, Italy.

33. Teglio, L. et al. (1987). *Quad. Clin. Ostet. Gynecol.* 42:221.

34. Baisi, F. (1987). Report on Clinical Trial of Bilberry Anthocyanosides in the Treatment of Venous Insufficiency in Pregnancy and of Postpartum Hemorrhoids. Presidio Ospedaliero di Livorno, Italy.

35. Perossini, M., Guidi, G., Chiellini, S., and Siravo, D. (1987). Clinical Study of Bilberry Anthocyanosides in the Treatment of Diabetic and Hypertensive Microangiopathy of the Retina. *Ann. Ottalm. e Clin. Ocul.* 12:1173.

36. Hobbs, C. (1992). *Echinacea, Nature's Immune Enhancer.* Healing Arts Press, Rochester, VT..

37. Braunig, B., Dorn, M., Limburg, E., and Bausendorf, K. (1992). *Echinacea purpureae* radix for Strengthening the Immune Response in Flu-Like Infections. *Zeitschrift fur Phytotherapie* 13:7–13.

38. Bauer, V., Jurcic, K., Puhlmann, J., and Wagner, H. (1988). Immunologische in-vivo und in-vitro untersuchungen mit echinacea- extrakten. (Immunologic *in Vivo* and *in Vitro* Studies on *Echinacea* Extracts.) *Arzneimittelforschung* 38(2):276–281.

39. Stimpel, M., Proksch, A, Wagner, H., and Lohmann-Matthes, M.L. (1984). Macrophage Activation and Induction of Macrophage Cytotoxicity by Purified Polysaccharide Fractions from the Plant *Echinacea purpurea. Infect Immun.* 46(3):845–849.

40. Roesler, J., Emmendorffer, A., Stienmuller, C., Luettig, B., Wagner, H., and Lohmann-Matthes, M.L. (1991). Application of Purified Polysaccharides from Cell Cultures of the Plant *Echinacea purpurea* to Test Subjects Mediates Activation of the Phagocyte System. *Int. J. Immunopharmacol.* 13(7):931–941.

41. Coeugniet, E.G., and Elek, E. (1987). Immunomodulation with *Viscum album* and *Echinacea purpurea* extracts. *Onkologie* 10(3 Suppl.):27–33.

42. Haas, H. (1981). Brain Disorders and Vasoactive Substances of Plant Origin. *Planta Medica* (Suppl.) 257–265.

43. Taillandier, J. et al. (1988). *Ginkgo biloba* Extract in the Treatment of Cerebral Disorders Due to Aging. In *Rökan (Ginkgo biloba). Recent Results in Pharmacology and Clinic,* E.W. Fünfgeld (ed.). Springer-Verlag, Berlin.

44. Weitbrecht, W.V., and Jansen, W. (1985). Doubleblind and Comparative (*Ginkgo biloba* versus Placebo) Therapeutic Study in Geriatric Patients with Primary Degenerative Dementia—A Preliminary Evaluation. In *Effects of Ginkgo Biloba Extract on Organic Cerebral Impairment,* A. Agnoli et al. (eds.). John Libbey Eurotext Ltd.

45. Rai, G.S., Shovlin, C., and Wesnes, K.A. (1991). A Double-Blind, Placebo Controlled Study of *Ginkgo biloba* Extract in Elderly Outpatients with Mild to Moderate Memory Impairment. *Curr. Med. Res. Opin.* 12(6):350–355.

46. Foster, S. (1990). Ginkgo. American Botanical Council, Austin, TX.

47. Schaffler, K., and Reeh, P. (1985). Long-Term Drug Administration Effects of *Ginkgo biloba* on the Performance of Healthy Subjects Exposed to Hypoxia. In *Effects of Ginkgo Biloba Extracts on Organic Cerebral Impairment*. Eurotext Ltd.

48. *Fortschr. Med.* (1992). 110(13):247–250.

49. *Bull. Tokyo Dent. Coll.* (1991). 32(1):1–7.

50. *Klin. Monatsbl. Augenheilkd.* (1991). 199(6):432–438.

51. Larkin, T. (1983). Herbs Are Often More Toxic Than Magical. *FDA Consumer* 17(8):4–10.

52. Eisenberg, D. et al. (1993). Unconventional Medicine in the United States: Prevalence, Costs, and Patterns of Use. *N. Engl. Jrnl. Med.* 328:246–252.

53. Farnsworth, N.R., and Soejarto, D.D. (1985). Potential Consequence of Plant Extinction in the United States on the Current and Future Availability of Prescription Drugs. *Economic Botany* 39(3):231–240.

54. Ibid.

55. Ibid.

56. Please refer to chapters by Katy Moran (Chapter 11) and Joshua Rosenthal (Chapter 13) for more detailed discussion.

57. Mowrey, D.B., and Clayso, D.E. (1982). Motion Sickness, Ginger, and Psychophysics. *Lancet* 655–657.

58. Bone, M.E., Wilkinson, D.J., Young, J.R., and Charlton, S. (1990). Ginger Root—A New Antiemetic: The Effect of Ginger Root on Postoperative Nausea and Vomiting After Major Gynaecological Surgery. *Anaesthesia* 45(8): 669–672.

59. Hylands, P.J., Hylands, D.M., and Johnson, E.S. (1987). In *Migraine: Clinical and Research Aspects,* J.N. Blau (ed.). Johns Hopkins University Press, Baltimore.

60. Jacobs, J. (1993). Regulatory Issues and Medicinal Plants. Presentation at the WHO/Morris Aboretum of the University of Pennsylvania Utilization of Medicinal Plants Symposium. Philadelphia, PA, April 19–21.

61. Public Law 103-417—October 25, 1994 (the Dietary Supplement Health and Education Act of 1994). Section 2.(1),(2),(3A),(15A).

62. Jerry Brown, Agribusiness Advisor, USAID Africa Bureau, September 1994. Agroindustrial Markets Workshop, Zambia.

CHAPTER 11

Returning Benefits from Ethnobotanical Drug Discovery to Native Communities

KATY MORAN

Introduction

To prevent the loss of the flora and fauna of our planet, at the Earth Summit in Rio de Janeiro, Brazil in 1992, the Convention on Biological Diversity of the United Nations Conference on Environment and Development (1992) opened for signature on June 5, 1992 (herein the Biodiversity Convention or the Convention). Objectives of the Convention are (1) the conservation of biodiversity, (2) the sustainable use of its components, and (3) the equitable sharing of the benefits resulting from the use of genetic resources (Article 1).

Development of plant-derived genetic resources into pharmaceuticals for human health can accomplish the three objectives of the Convention through appropriate incentives to encourage the preservation of plants and traditional knowledge of their medicinal uses. Incentives can take the form of social and economic benefits—compensation returned to governments of biodiversity-rich nations for the commercial use of their plants and to indigenous peoples for their traditional knowledge of the use of plants for medicines. The Convention acknowledges the importance of traditional knowledge in this process, internationally legitimizing current initiatives that attempt to accomplish this (Moran 1992).

The Healing Forest Conservancy (the Conservancy), a nonprofit organization, has developed a process—"communal compensation trust funds" —to achieve the Convention's goals by returning compensation for ethnobotanical drug discovery to native peoples. Governments will be compensated for the use of their biotic material in a separate process, since they often have legislated mandates to use compensation funds for national priorities. The Conservancy's approach recognizes that (1) most of the world's remaining tropical biodiversity is in areas inhabited by indigenous, local, or native cultures, who are, today, the primary *in situ* caretakers of the planet's biological diversity; and (2) native cultural knowledge of medicinal plant use is valuable to human health and drug discovery and requires compensation.

Biocultural Diversity

Biological Diversity and Human Health

Biological diversity refers to the number and variety of the genes, species, populations, communities, and ecosystems that provide the basis for life on earth. Biodiversity fulfills aesthetic and spiritual needs for much of humanity and benefits human welfare directly. According to the World Health Organization, 80% of the population of developing countries, about 4 billion people, depend on plant-based traditional medicine for their primary health care (Farnsworth et al. 1985). Tropical plant species provide the basis for many of today's medicines and hold the promise of supplying useful chemicals for tomorrow's therapeutics (Tempesta and King 1994a). Tropical species are valuable for the richness of their biological and chemical diversity, due, in part, to climatic conditions. In temperate climates, winter kills many plant predators, and temperate plants flourish in the spring before predator populations increase. But tropical species have no seasonal respite from predators, so they have evolved complex chemical protection from countless predators. The plant chemicals that have evolved to increase plant resistance against bacteria and other infectious organisms of tropical plants can also provide protection and be therapeutically useful for human health (Tempesta and King 1994b).

It is predicted that if present trends in the loss of tropical plant habitats continue, as many as 60,000 plants, nearly 1 in 4 of the planet's total, could be extinct by the middle of the next century (Farnsworth 1988). Ethnobotanist Paul Cox (1995) stated, "Earth has more than 265,000 species of flowering plants, shrubs and trees and less than one-half of one percent have been screened even cursorily for their potential therapeutic value."

Figure 11.1 shows the location of tropical biodiversity. Countries listed in the right circle and in the middle of the Venn diagram (Durning 1992) illustrate areas that biologists recognize as centers of biological diversity.

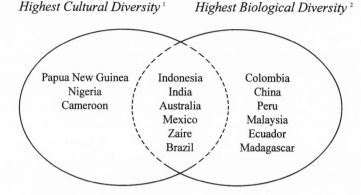

Highest Cultural Diversity [1]   *Highest Biological Diversity* [2]

Papua New Guinea	Indonesia	Colombia
Nigeria	India	China
Cameroon	Australia	Peru
	Mexico	Malaysia
	Zaire	Ecuador
	Brazil	Madagascar

Source: Worldwatch Institute
1Countries where more than 200 languages are spoken
2Countries listed by biologists as "megadiversity" countries
for their exceptional numbers of unique species

FIGURE 11.1
Cultural diversity and biological diversity, circa 1990.

These 12 countries of the tropics—Colombia, China, Peru, Malaysia, Ecuador, Madagascar, Indonesia, India, Australia, Mexico, Zaire, and Brazil—house exceptional numbers of endemic or unique plant and animal species.

Cultural Diversity and Human Health

According to most anthropologists, the best single indicator of distinct cultures is spoken language, because it explicitly encapsulates the unique view of the universe held by the group that speaks the language (Clay 1993; Berlin 1992). Of the 6,000 languages spoken today, close to 5,000 are languages of indigenous peoples. Durning's diagram shows countries, listed in the circle on the left and in the middle, that are the nine centers of cultural diversity—Papua New Guinea, Nigeria, Cameroon, Indonesia, India, Australia, Mexico, Zaire, and Brazil. Countries are so defined by the number of languages spoken within each country. As is evident, 6 of the 12 centers of biodiversity also rank at the top in cultural diversity—Indonesia, India, Australia, Mexico, Zaire, and Brazil. The relationship between biological and cultural diversity, or "biocultural diversity," is demonstrated by this Venn diagram. Tropical ecosystems where indigenous cultures live are also where much of the world's biodiversity is found (Durning 1992).

If the fate of biodiversity is tied to its conservation at the local level, then the fate of indigenous cultures is tied to it as well. This was articulated during the WHO/IUCN/WWF International Consultation on Conservation of Medicinal Plants. Reaffirming commitment to the collective goal of "Health for All by the Year 2000," participants focused international attention to the consequences of "the continuing disruption and loss of indigenous cultures, which often hold the key to finding new medicinal plants that may benefit the global community" (Akerele et al. 1991).

In discussing the biocultural use of plants for medicines, the terms indigenous, tribal, or native are used interchangeably in this paper, as well as culturally and/or linguistically distinct culture groups or communities (Colchester 1994). Although it is recognized that the traditions of peasantries, rural, and local communities have made valuable contributions to the medicinal use of plants (Elisabetsky and Castilhos 1990), the following discussion focuses on indigenous medicinal knowledge.

Some 600 million indigenous peoples in about 6,000 groups are spread around the world in more than 70 countries; they represent 10–15% of the world's population (Clay 1993). They are typically the descendants of an area's original inhabitants and are geographically isolated and distinct from dominant groups in language, politics, culture, and religion (Colchester 1994). They often describe themselves as custodians, rather than owners, of their land and resources, which generally are the base of their subsistence. Their social organization is described as tribal, referring to collective resource management and group decision making by consensus (Clay 1993).

Indigenous peoples are not, as is sometimes romantically viewed, hands-off preservationists of nature. Humans use nature for their own benefit, and indigenous cultures are no exception (Clad 1984). But typically, their local levels of biodiversity have not decreased, because indigenous resource use is commonly diversified and resource-extensive (Oldfield and Alcorn 1991). Indigenous resource use often does not deplete or concentrate pressure on a single biological resource as do resource-intensive systems such as monoculture, but generally distributes human impacts across the larger forest ecosystem (Denslow and Padoch 1988). Within their habitats, indigenous peoples see themselves as part of a bountiful nature that provides for them laboratories of raw material that they use for food, fiber, medicines, and other subsistence needs. Consequently, their traditional knowledge of the use of biological species represents libraries of information on the use of local biodiversity (Moran 1992).

This knowledge is embedded in indigenous peoples' cultural systems and intimately connected to their religious beliefs and cosmology (Reichel-Dolmatoff 1976). Often, a group's biological knowledge of their natural environment reaches a high level of abstraction, far from biological facts, but is symbolically represented in their myths, rituals, and religion. Accumulated over millennia and passed down generationally within communities, traditional knowledge of medicinal plants is as rich and diverse as biotic resources, and as threatened. In Brazil alone, poverty, acculturation, outside encroachment, and loss of habitat have resulted in the extinction of an average of one indigenous culture each year. One-third of the cultures of Brazil that existed at the turn of this century are now extinct (Clay 1993).

The developing world's loss of biocultural diversity is a portentous loss to all humanity, particularly for human health (Linden 1991). At risk is health care for 80% of the world's population today, and the tropical species that can guide future generations in drug discovery (Balick 1990; Grifo et al., Chapter 6).

Biocultural Diversity for Drug Discovery

Richard Evans Schultes is the co-author of *The Healing Forest* (Schultes and Raffauf 1990), for which the Conservancy is named. Schultes, now professor emeritus at Harvard, stated in 1988:

> The accomplishments of aboriginal people in learning plant properties must be a result of a long and intimate association with, and utter dependence on, their ambient vegetation. This native knowledge warrants careful and critical attention on the part of modern scientific methods. If phytochemists must randomly investigate the constituents of biological effects of 80,000 species of Amazon plants, the task may never be finished. Concentrating first on those species that people have lived and experimented with for millennia offers a short-cut to the discovery of new medically or industrially useful compounds.

Every continent offers rich traditions in the use of medicinal plants. In the United States, modern medicine evolved from a mix of Native American, African, Eastern, and European botanical traditions (Hogan 1979; Ford 1978). Botanist Richard Spruce (1908) documented the extensive pharmacopoeias of the Aztec, Maya, and Inca of central and middle America. The therapeutic use of plants in Africa is sophisticated and extensive, rivaling that of Asia. The plant-based medical systems of Asia include the

Ayurvedic system of India and Chinese traditional medicine and others that incorporate from 1000 to 2000 species of medicinal plants that have been in use for thousands of years (Etkin 1986).

Seventy-four percent of the chemical compounds used as drugs today are used for the same or related purposes in traditional medical systems (Farnsworth 1988). A few examples are the antimalarial, quinine, found in *Cinchona,* the tranquilizer reserpine from the East Indian snakeroot, cardiac glycosides from *Digitalis,* and the analgesics codeine and morphine. Although they are based on natural products, traditional medicines are not "found" in nature. They are products of traditional knowledge. According to Elisabetsky:

> To transform a plant into a medicine, one has to know the correct species, its location, the proper time of collection (some plants are poisonous in certain seasons), the solvent to use (cold, warm, or boiling water; alcohol, addition of salt, etc.), the way to prepare it (time and conditions to be left on the solvent), and finally, posology (route of administration, dosage) (1991).

Protection for Traditional Knowledge

Economic analysis by Anthony Artuso (1994) documents how indigenous knowledge adds value to plants screened for biological activity in the search for therapeutic chemicals. Rather than random screening, it is more efficient to use traditional knowledge as a lead to pinpoint promising plants (Cox and Balick 1994). However, the contributions that indigenous peoples have made to medicine have yet to be recognized, much less protected and paid for. Although several processes are in play to alter past practices (Moran 1992), there is not yet an international legal instrument that specifically protects indigenous intellectual property (Greaves 1994). The 1993 draft of the proposed U.N. Declaration on the Rights of Indigenous Peoples states that groups are entitled to the recognition of the full ownership, control, and protection of their cultural and intellectual property, including their sciences, technologies, medicines, and knowledge of the properties of fauna and flora, among others (Colchester 1994). Another pending document, the International Labor Organization's Convention 169, may also provide a more solid legal base for protection.

The Biodiversity Convention, which formalized the sovereignty of nations over their biodiversity, merely "encourages" equitable sharing of benefits arising from traditional knowledge, innovations, and practices. In its framework stage, the Convention does not yet establish mechanisms to accomplish this equitably within nations. It has not resolved, nor does it yet adequately acknowledge, the difference between ownership of biotic material by national governments and ownership of the vital contribution of traditional medicinal use of the material by indigenous cultures. Although equity is a critical term in the spirit of the Convention, indigenous peoples' contribution to the drug discovery process has not yet been adequately addressed (Ten Kate 1995). Signatories to the Convention are obliged to develop a national biodiversity strategy, and legislation in the Philippines represents a progressive attempt to satisfy the three goals of the Convention. The Philippines' national access legislation (Executive Order No. 247) requires researchers, whether commercial or academic, to seek prior informed consent from and to share any benefits with indigenous peoples.

During the last decade, regional coalitions and federations of indigenous peoples worldwide have joined forces in a spirit of solidarity to discuss these and other issues (Colchester 1994). Federations have committed their discussions into Charters, Declarations, and other statements and disseminated them for public use. The Charter of the Indigenous–Tribal Peoples of the Tropical Forests was promulgated in Malaysia in 1992. Article 44 of the Charter states that the traditional technologies of members can make important contributions to humanity, including developed countries: they claim control over the development and manipulation of their traditional medicinal knowledge (Colchester 1994). In 1988, the Kuna Indians of Central America produced guidelines for foreign scientists that describe Kuna terms for research undertaken in their territories. A manual details the process for gaining permission to enter an area, establishes guides for research conduct, and encourages foreign scientists to transfer knowledge and technology to the Kuna during their research (Chapin 1991). Coalitions and federations comprise groups such as the Coordinating Body for the Indigenous Peoples' Organization of the Amazon Basin (COICA) in Amazonian countries where more than 200 tribes are located; the South and Meso American Indian Information Center (SAIIC); and the World Rainforest Movement (Morris 1992).

Communal Compensation

In 1990, The Healing Forest Conservancy was founded as a nonprofit organization to work with federations to devise and implement a compensation strategy that promotes the conservation of tropical forests, particularly medicinal plants, and the conservation and welfare of tropical forest cultures, particularly knowledge of medicinal plant use (Moran 1994). No nongovernmental organization had previously existed specifically to provide a formal and consistent process to compensate countries for use of their biological diversity and culture groups for use of their ethnobotanical leads to commercial development of biotic products.

The Conservancy consults with indigenous federations, tropical country governments, professional associations of scientists, foundations, and other nonprofit organizations involved in the conservation of biocultural diversity (Soejarto and Gyllenhaal 1992). Basically, all acknowledge that indigenous plant knowledge is owned not by an individual healer, but by the community of the culture group; not only the present community, but communities of the past and of the future. Compensation, then, must benefit the whole culture, of the past, present, and future. While compensation is not intended to resolve the historic injustices visited on indigenous peoples worldwide, it can be catalytic for indigenous cultures to use while moving on to attain goals articulated in Declarations. The Conservancy focuses on developing communal compensation options, defined largely by indigenous groups, in the form of programs that have multiple objectives:

1.   To promote sustainable development by local harvesting of natural products in forests that might otherwise be cut for timber or cleared for farmland

2.   To generate local employment programs (where appropriate, to focus on women) by providing training in technical skills for species collection and identification and inventory of local genetic resources by merging traditional and nontraditional scientific methods and processes

3.   To provide resources to survey, demarcate and deed historic territories to indigenous groups

4.   To develop local markets for nontimber forest products such as medicinal plants

5.   To build and strengthen indigenous federations and institutions through education and communication between forest societies and the outside world

6.   To link the United States and international practitioners and policymakers in initiatives that foster the health and welfare of indigenous cultures and tropical forests.

The Conservancy was founded through a donation from Shaman Pharmaceuticals, Inc., a California-based natural products company focused on the discovery and development of novel pharmaceuticals from plants with a history of native use. As implied by its name, Shaman uses the science of ethnobotany, as well as isolation and natural products chemistry, medicine, and pharmacology to create a more efficient drug discovery process. Typically, this process takes from 12 to 15 years at a cost of close to $300 million. Although the young company has not yet brought a product to market, the use of ethnobotanical leads has brought potential products to clinical trials within a record time frame.

Since Shaman established operations in May 1990, the company has advanced two products from concept to clinic. Virend ™ is a topical formulation derived from a plant that is a pioneer species used by cultures in South America to treat herpes lesions. Late in 1993, Shaman completed a pilot phase I/II clinical trial in AIDS patients that demonstrated the safety of Virend. In October 1995, a phase II randomized, double-blind, placebo-controlled study involving 45 patients was completed. Thirty-eight percent of the patients receiving Virend had lesions that healed completely, versus 14% in the placebo-controlled group. With safety and preliminary indications of efficacy determined in a controlled setting, Shaman began a pivotal phase III trial of Virend in 1996. More than 30 million people have genital herpes, and each year almost 500,000 new cases are diagnosed. Currently, oral acyclovir, with 1994 sales of $1.4 billion, is the only drug that has been approved to treat herpes.

Another Shaman product in clinical trials is Provir™, an oral product for the treatment of secretory diarrhea. Worldwide, this condition affects over 16 million people, and it can be fatal when severe dehydration results. Additionally, Shaman has focused its basic research efforts on establishing an in-house diabetes research program. Efforts to identify lead compounds for the treatment of non-insulin–dependent diabetes mellitus have produced a number of extracts that are active *in vitro* and *in vivo*. The company works with native cultures in areas where diabetes is prevalent, and field teams find that many patients successfully manage their disease through plant-derived medicines.

Shaman is also addressing the interdependent issues of biodiversity conservation, sustained economic growth, and drug discovery that improves human health in tropical countries. To participate in that effort, Shaman joined a U.S. government-sponsored International Cooperative Biodiversity Group (ICBG) to work with the governments and research institutes of Nigeria and Cameroon. The ICBG team investigates plant-based therapeutics for diseases of developing countries such as malaria, schistosomiasis, and leischmaniasis (Grifo 1996).

To date, Shaman has seen an unprecedented initial indication of potential therapeutic activity in more than half the plants it has screened. The company has collected over 1,000 plants, screened 800, and identified activity in 420. Many compounds hold promise in Shaman's targeted therapeutic area, and eight are currently a priority for preclinical studies (King 1996). The company's experience substantiates that plant-derived drugs with a history of medicinal use can be developed for pharmaceuticals more quickly and cost effectively than traditional pharmaceuticals.

Agreements with cultures, groups, and countries that Shaman works with secure compensation for traditional knowledge both during the drug discovery process and after a product has been commercialized (Conte 1994). At the beginning and during research expeditions, Shaman provides specific up-front compensation that responds to immediate needs of country and indigenous collaborators. Long-term compensation will be available through the Healing Forest Conservancy after a product is commercialized (King and Carlson 1995).

Converting the Biodiversity Convention into praxis, Shaman pioneered a novel concept for long-term compensation (Artuso 1994). The company will return a percentage of company profits to all of the indigenous communities and countries with which it has worked, regardless of where the actual plant sample or traditional knowledge originated (King and Carlson 1995). Through this process, the risk of receiving no compensation for a commercialized product in individual countries or communities that did not contribute to product discovery is lessened. In a financially unpredictable industry, spreading the benefits and risks among all Shaman collaborators increases opportunities for compensation and hastens compensation returns. All Shaman collaborators benefit equally from risk-sharing provisions, acknowledging the spirit of the Biodiversity Convention.

After a product is commercialized, Shaman will channel a percentage of profits for compensation through the Conservancy. Following indigenous systems of resource use, the Conservancy will deliver to Shaman collaborators communal compensation programs through trust funds that benefit the culture group as a whole, rather than cash payments to an individual healer.

Pilot Projects

The Conservancy has successfully sought support for compensation pilot projects from foundations, nonprofit groups, and environmental organizations. Funders view compensation programs as having merit as stand-alone conservation programs. Pilot programs were developed through input from three sources: (1) indigenous federations and their Declarations; (2) international conventions signed by governments of species-rich countries; and (3) professional associations that have, or are in the process of, developing Codes of Conduct for ethnobiological research, such as the Society for Applied Anthropology, the International Society for Ethnobiology, and the Society for Economic Botany.

Pilot projects are conducted to test the feasibility of compensation processes and to gain knowledge from practical experience acquired during projects. Results are shared with culture groups and countries involved in the sustainable development of their biodiversity. They can then relate experiences from pilot projects to their specific circumstances and make an informed choice among compensation options.

Establishment of an Ethnobiomedical Plant Reserve

Demarcation of reserve boundaries offers indigenous associations the security of permitting or denying access to resources—the first step in sustainable management. Demarcation and management rights to territories are perhaps the highest priorities for most indigenous groups, and this concern has been addressed in the Cayo District of Belize. The Belize Association of Traditional Healers (BATH) sought a way to protect what many call the world's first ethnobiomedical reserve with plants integral to the BATH's primary work of providing health care to poor Belizeans and refugees from countries of Central America. A 6,000 acre tract, named Terra Nova, was donated by the government. The reserve's terrain is suitable for its purpose, which is the rescue and propagation of the medicinal plants that are vital to the survival of traditional healing throughout Central America, and a repository of plants for elderly healers who are too old to comb ever-receding forests for their medicines. The BATH also holds monthly public education meetings for local villagers where traditional healers discuss medicinal plants and traditional remedies for a variety of common ailments.

To halt current poaching and logging near Terra Nova, the Conservancy and the Rex Foundation funded the surveying and demarcating of the land after a request to do so from the BATH. Management of Terra Nova was designated to BATH following an indigenous pattern of communally managed resources. Although national governments vary greatly in granting resource rights to their indigenous populations, Terra Nova offers an enlightened example of cooperation between a government, an indigenous healers association, and outside funding sources—a tribute to the value of traditional health care. All project participants share the value of preserving traditional medicinal plants that many governments cannot otherwise afford to maintain, and providing health care to populations that may not otherwise be served.

Pilot projects in the planning stage will test ways for indigenous groups to use low cost global positioning systems (GPS) for demarcation. Now available to remote culture groups, such technologies enable communities to reference land boundary data and locate boundaries to an accuracy of between several centimeters and 25 meters. Once referenced this way, data can be entered into a global environmental database such as the Global Environmental Monitoring System (GEMS), the Global Resource Information Database (GRID), or the Earth Observing System (EOS). Computer software can also be used by groups for natural resource assessments, conservation analysis, and land-use planning (Poole 1995).

Training in Ethnobotany

In 1994, the Asia Foundation sponsored a grant to the Conservancy to participate in the Environmental Fellowship Program, a component of the United States–Asia Environmental Partnership of USAID. As part of the Fellowship, the Conservancy funded a Medicine Woman pilot project in Lucknow, Uttar Pradesh, India, for students from different regions of the country. The pilot project was held in India because it was the site of the fourth International Congress of Ethnobiology, which the students also attended. India is rich in biocultural diversity, housing more than 400 unique ethnic groups, of which 75% are tribal, and 45,000 species of plants, of which 30% are higher plants. The course was intended to (1) increase and diffuse the knowledge of ethnobiology in areas such as basic technical skills in ethnobiology, (2) provide knowledge of critical issues in applied ethnobiology through discussions on ethical issues, and (3) assist students in establishing a network for contact with other Indian tribals or with those associated with ethnobiology worldwide.

The Conservancy, in cooperation with the Asia Foundation, the Rex Foundation, and the National Botanical Research Institute of India, supported participation by twenty-eight trainees, half of whom were women. A gender ratio for the course was specified because women have specialized gender-related knowledge about plant use for contraception, aborti-

facients, nutrition, and infant and child care that many male ethnobiolo-
gists miss because of gender restricted information from female informants
in the field. What is more important, if training, education, and employment opportunities are available to women, ecological impacts can be
magnified positively. Women who have options between having a job or
having another child typically choose the former, effectively reducing population pressures on natural resources (World Resources Institute 1994).

Mornings were spent in classes discussing methods in ethnobiology with
varied international specialists in the natural and social sciences from India.
Other experts arrived in India before the congress to help teach the
course. Discussions were presented on topics such as medical anthropology,
linguistics, and ethnomedicine. Afternoons were spent in the field for technical training in collection methods such as how to make and use a plant
press and plant drying techniques. During a visit to the national herbarium, procedures such as care and storage of dried plants and simple herbarium laboratory techniques were demonstrated. Evening discussions focused on ethical issues in applied ethnobiology such as intellectual
property rights.

Evaluations by students at the end of the Medicine Woman course were
positive; they suggested future training courses of greater length to broaden
training opportunities. Virtually all female trainees were grateful that the
course specifically targeted women and commented that they probably
would have been excluded otherwise. Evaluations also showed that students recognized that information on plant and animal species, in itself, is
valuable and requested more skills to add value to the biotic resources in
their communities.

Another Medicine Woman program was held at the Limbe Botanic Garden in Cameroon, in October of 1995. The program, part of an "Ethnobiology and Field Taxonomy Training Course" for Pan-African participants, was organized by the Bioresources Conservation and Development
Programme (BCDP). The BCDP is a West African not-for-profit nongovernmental organization that links the development needs of tropical
peoples with the protection of the environment (Iwu 1995). Training focused on field methods in ethnobotany and how to perform simple tests
to detect the biological and pharmacological activity of plants locally. Currently, little of the chemical analysis of medicinal plants is completed *in
situ,* despite advantages in analyzing fresh plant material. For example,
when plant material is air dried so it can be analyzed later in a laboratory,
its chemistry changes, and some volatile compounds can be lost through
sun drying. Likewise, the separate analysis of individual plants in a con-

coction ignores the chemical reactions that occur when chemicals from different plants are mixed together (Prance 1991).

Communal Compensation Trust Funds

The Healing Forest Conservancy has used the described pilot projects to test possible programs for communal compensation that respond to local needs. When a Shaman product has been brought to market, a percentage of profits will be channeled through the Conservancy to the company's collaborating culture groups. A separate process will be employed to compensate governments for the use of their biotic material, because they often have legislated mandates to use compensation funds for national priorities such as a biological inventories. Through Compensation Trust Funds, the Conservancy will deliver biannual funding to indigenous communities for as long as Shaman shows a profit. For this process to succeed, the Conservancy has created a structure that can deliver long-term funding to finance programs in a consistent manner, while maintaining the flexibility to enable programs to respond to local needs that can be managed locally.

Communal compensation trust funds have a precedent in the five years' experience with environmental funds that were established to finance environmental programs in countries that participated in "debt-for-nature swaps" (Moran 1991; IUCN 1994). In the form of a foundation, nonprofit corporation, or common-law trust, such trusts are financial mechanisms created to receive funds. They can operate as permanent endowments, revolving or sinking trust funds, in hard currency to avoid inflation shrinkage, and either off-shore or in the country of origin.

The flexibility of trust funds is particularly useful for communal compensation and appropriate for Shaman, a company that views traditional knowledge of plant resources as the primary source for leads in drug discovery. In considering compensation, it must be recognized that most culture groups have a social and cultural identity distinct from the dominant society. They are particularly vulnerable to outsiders, because they are often geographically and linguistically isolated. Management of trust funds, however, can be independent of outsiders and empower indigenous communities to determine how to use their compensation without outside interference.

Communal compensation trust funds can be administered locally by a board of directors that is selected democratically by participating culture groups. Board members have fiscal responsibility for funds and serve for a specified term that is determined locally. Typically, boards do not implement programs or projects, but have the power to decide which projects are funded, as well as a project's appropriate financial scale and time frame. Advisory boards, comprising *ex officio* or nonvoting members, can provide expertise on financial and legal issues, including grievance procedures.

Compensation trust funds controlled locally to supply long-term funding of programs identified by Shaman collaborators as being important to them offer many advantages. They discourage corruption and promote democracy, because boards are chosen by indigenous communities in a transparent and participatory process. Monitoring and evaluation of the use of funds are built into the process, because funds will be distributed long-term on a designated schedule. Because compensation trust funds provide a stable source of funding for as long as Shaman shows a profit, culture groups have greater responsibility and capacity to identify and manage projects independently. A periodic addition of a smaller amount of funds shared by all Shaman collaborators, rather than one large donation, alleviates the problem of swamping beneficiaries with an amount of money that cannot be absorbed responsibly. It also offers the choice of regularly setting aside a stated percentage to create a permanent endowment for the culture group.

More important, communal compensation trust funds offer the added value of attracting and managing sources of finance from other NGOs or corporations that work with indigenous communities. Accounts within standing trust funds can target finances for non-Shaman communities; similarly, compensation trust funds can be created within existing environmental funds under a separate account.

Conclusions

When biodiversity is discussed in the context of human health, it too often means the health of residents of industrialized nations. Even earnest efforts tend to reflect an unconscious attitude that biodiversity must be preserved as genetic resources to enlarge the pharmacopoeia of Western medicine, which provides therapeutics primarily for Western societies. Less discussed is the vitality of biodiversity to the health of 80% of the world, the populations who have no alternatives to natural medicine, and who depend solely on medicinal plants for their primary health care. Terra Nova demonstrates that regional medical needs can be addressed by assuring the status and continuity of plant resources that provide health care through traditional medicines. Reserves such as Terra Nova offer a place for traditional healers to train and to pass on the healing forest legacy to future generations of practitioners. As tropical forests disappear worldwide, reserves offer to healers' apprentices tangible proof of the continuation of a traditional profession—an economic future assured because of the preservation of medicinal plants.

When we speak of biodiversity and human health, we must speak of those whose cultures have understood the value of biological resources for healing for millennia. Preserving biodiversity for the benefit of human health means preserving it for those in the tropics already using it, as well as for distant populations that will know it only in refined or synthetic form at some unspecified future date. Compensation in the form of training programs such as Medicine Woman offers an opportunity to merge both Western and non-Western systems. Training builds local capacity and supplies income-producing opportunities to communities that are best located to collect, identify, analyze, and protect species. Inventories that describe local use can be undertaken in the local language, which will protect the information from outsiders and leave a documentation of cultural knowledge within the community.

Collecting skills taught in programs such as Medicine Woman supplement local systems of plant use. Because collecting is labor intensive, training enhances employment opportunities with national governments that are obliged under the Biodiversity Convention to conduct country biological inventories. Training also generates employment by outsiders interested in developing bioresources commercially and increases the capability of gene-rich countries to perform these services. As important, participation in the collection process supplies communities with greater control over the use of their resources and provides a basis for more advanced training procedures.

National governments benefit from training programs by gaining a technological infrastructure for science and commerce, yielding jobs and taxes. Fees can be charged to outsiders with commercial or research interests in the sustainable development of biotic resources, allowing debt-ridden nations to forego short-term profits from logging, cattle-grazing, and monoculture development projects that destroy forests. Countries can choose whether to supply natural products in the form of extracts, rather than raw unprocessed material, to foreign investors or to establish their own medicinal plant, phytochemical, or pharmaceutical industry. Whichever, adding value locally lowers the total cost of drug discovery from natural products in a high-risk high-gain industry. There are several levels of entry into the sustainable development of biodiversity, including: collecting, inventories, bioassays, recollection, harvesting, herbarium storage and collections, and more. Training increases the capacity of biodiversity-rich countries to perform these services and eases their entry into the field of natural products if they choose to do so. Proprietary arrangements can ensure that compensation from commercialization of natural compounds or synthesized chemicals reverts fairly to the peoples and nations of original production.

Economic benefits that accrue to communities that have harbored medicinal plants and traditional knowledge of their medicinal use certainly are an incentive to conserve biocultural diversity. But another "e," representing equity, is perhaps the most critical component, and is rarely discussed. Equity means not only equitable compensation, but equal standing among participants in making decisions about what form compensation should take. Choices for Conservancy pilot projects reflect the seriousness that should be accorded to Declarations articulated by Federations. Options for compensation that respond to indigenous needs strengthen traditional cultures by increasing their capacity to become active participants in the process. Likewise, trust funds cannot be considered equitable or legitimate unless the people who contribute to the process are the same ones who control the funds. The participation of Shaman collaborators in defining compensation and controlling the communal compensation trust funds reinforces the value of biodiversity at the local level, which enhances efforts for its conservation. Another conclusion of Conservancy pilot projects is that recognition and acknowledgment of the contribution from indigenous societies in the discovery of new medicines that have historically benefited humankind are missing. Recognition validates indigenous systems within countries and internationally, just as it does for Western scientists. It also spotlights the unique identity of traditional cultures at a time when they are organizing to pursue self-determination.

There is much faith today in the ability of market forces to "save the rain forests," but market forces can be a double-edged sword when Western concepts such as markets are introduced into nonmarket economies. At the same time, however, many communities seek greater access to markets. These differences should never be an excuse to exclude indigenous groups from the sustainable use of biodiversity, for this is their, and only their, decision to make. How can we expect support for biodiversity conservation if it is perceived as "us" trying to prevent "them" from achieving what we already have?

Although the Biodiversity Convention, in its framework stage, has generated productive discussions on how to achieve its stated goals, today, the sustainable development of biotic resources in the tropics is still a work in progress, euphemistically called "an emerging science." The Conservancy has conducted three pilot projects in Africa, Asia, and Latin America which have generated observations to add to discussions on biodiversity and human health: our guiding principles must be recognition, participation, and equal standing among all concerned.

Acknowledgments

I gratefully acknowledge the critical input to this paper by numerous indigenous federations of indigenous peoples that have committed local discussions on compensation into Declarations for public use. Other valuable input came from Codes of Conduct developed by professional associations of conscientious scientists dedicated to the conservation of biological and cultural diversity. Generous support for compensation pilot projects was provided by Shaman Pharmaceuticals, Inc., the Aveda Corporation, the Rex Foundation, the Asia Foundation, the Leland Fikes Foundation, the Nelson Talbott Foundation, the Jocarno Fund, and numerous individual supporters.

References

Akerele, Olayiwola, Vernon Heywood and Hugh Synge (eds.). (1991). The Chiang Mai Declaration. In: *The Conservation of Medicinal Plants*, p. 1. Cambridge: Cambridge University Press.

Artuso, Anthony. (1994). Economic Analysis of Biodiversity as a Source of Pharmaceuticals. PAHO/IICA Conference on Biodiversity, Biotechnology and Sustainable Development. San Jose, Costa Rica. April 12–14.

Balick, Michael. (1990). Ethnobotany and the Identification of Therapeutic Agents from the Rainforest. In: *Bioactive Compounds from Plants* (D. J. Chadwick and J. Marsh, eds.), pp. 22–39. New York: John Wiley and Sons.

Berlin, Brent. (1992). *Ethnobiological Classification: Principles of Categorization of Plants and Animals in Traditional Societies.* Princeton: Princeton Press.

Chapin, Mac. (1991). How the Kuna Keep Scientists in Line. *Cultural Survival Quarterly.* Summer: 15–17.

Clad, J. (1984). Conservation and Indigenous Peoples: A Study of Convergent Interests. *Cultural Survival Quarterly* Vol. 8, No. 4:68–73.

Clay, Jason W. (1993). Looking Back to Go Forward: Predicting and Preventing Human Rights Violations. In: *State of the Peoples: A Global Human Rights*

Report on Societies in Danger. (Marc S. Miller, ed.) pp. 64–71. Boston: Beacon Press.

Colchester, Marcus. (1994). Salvaging Nature: Indigenous Peoples, Protected Areas and Biodiversity Conservation. Discussion paper for UN Research Institute for Social Development. World Rainforest Movement, unpublished.

Conte, Lisa. (1994). Testimony for Hearing on U.S. Ratification of the Convention on Biological Diversity. U.S. Senate Foreign Relations Committee. U.S. Senate, Washington, D.C.

Cox, Paul Alan. (1995). Biodiversity and Human Health Conference. Session III: Biodiversity and Traditional Health Systems. Smithsonian Institution, Washington, D.C.

Cox, Paul Alan and Michael J. Balick. (1994). The Ethnobotanical Approach to Drug Discovery. *Scientific American* June 1994: 60–65.

Denslow, Julie and Christine Padoch (eds.). (1988). *People of the Tropical Rain Forest.* Berkeley: UC Press.

Durning, Alan. (1992). *Guardians of the Forest.* Washington, D.C.: Worldwatch.

Elisabetsky, E. (1991). Folklore, Tradition, or Know-How? *Cultural Survival Quarterly* Summer: 9–13.

Elisabetsky, E. and Z. C. Castilhos. (1990). Plants Used as Analgesics by Amazonian Cabaclos as a Basis for Selecting Plants for Investigation. *International Journal of Crude Drug Research* Vol. 28: 49–60.

Etkin, Nina L. (ed.). (1986). *Plants in Indigenous Medicine and Diet: Behavioral Approaches.* Bedford Hills, N.Y.: Redgrave.

Farnsworth, Norman R. (1988). Screening Plants for New Medicines. In: *Biodiversity* (E.O. Wilson and Francis M. Peters, eds.), pp. 83–97. Washington, D.C.: National Academy Press.

Farnsworth, Norman R., O. Akerele, A. S. Bingel, D. D. Soejarto and Z. Guo. (1985). Medicinal Plants in Therapy. *Bulletin of the World Health Organization* 63:965–981.

Ford, Richard I. (1978). The Nature and Status of Ethnobotany. Anthropological Papers, No. 67. Museum of Anthropology, University of Michigan, Ann Arbor.

Greaves, Tom. (1994). *Intellectual Property Rights for Indigenous Peoples.* Oklahoma City: Society for Applied Anthropology.

Grifo, Francesca T. (1996). The Role of Chemical Prospecting in Sustainable Development. In: *Emerging Connections Among Biodiversity, Biotechnology, and Sustainable Development in Health and Agriculture,* (Julie Feinsilver, ed.). Washington, D.C.: Pan American Health Organization.

Hogan, D. (1979). *The Regulation of Psychotherapists* Vol. 1. Cambridge, MA: Ballinger.

IUCN - The World Conservation Union. (1994). Report of the First Global Forum on Environmental Funds. Washington, D.C.: IUCN.

Iwu, Maurice. (1995). *Linking Biodiversity and Socio-economic Development.* Washington, D.C.: Bioresources Development and Conservation Programme.

King, Steven R. (1996). Conservation and Tropical Medicinal Plant Research. In: *Medicinal Resources of the Tropical Forest* (Michael J. Balick, Elaine Elisabetsky, and Sarah Laird, eds.), pp. 63–74. New York: Columbia University Press.

King, Steven R. and Thomas J. Carlson. (1995). Biocultural Diversity, Biomedicine and Ethnobotany: The Experience of Shaman Pharmaceuticals. *Interciencia.* Vol. 20, No. 3:134–139.

Linden, E. (1991). Lost Tribes, Lost Knowledge. *Time,* Sept. 23.

Moran, Katy. (1991). Debt-for-Nature Swaps: U.S. Policy Issues and Options. *Renewable Resources Journal.* Spring:19–24.

Moran, Katy. (1992). Ethnobiology and U.S. Policy. In: *Sustainable Harvest and Marketing of Non-Timber Forest Products* (Mark Plotkin and Lisa Famolare, eds.), pp. 289–301. Washington, D.C.: Island Press.

Moran, Katy. (1994). Biocultural Diversity Conservation Through the Healing Forest Conservancy. In: *Intellectual Property Rights for Indigenous Peoples* (Tom Greaves, ed.), pp. 101–109. Oklahoma City: Society for Applied Anthropology.

Morris, Karen. (1992). *International Directory and Resource Guide.* Oakland, CA: South and Meso American Indian Information Center.

Oldfield, Margery L. and Janis Alcorn. (1991). *Biodiversity: Culture, Conservation and Ecodevelopment.* Boulder: Westview Press.

Poole, Peter. (1995). *Indigenous Peoples, Mapping and Biodiversity Conservation: An Analysis of Current Activities and Opportunities for Applying Geomatics Technologies.* Washington, D.C.: Biodiversity Support Program Peoples and Forests Program.

Prance, Ghillean T. (1991). What Is Ethnobotany Today? *Journal of Ethnopharmacology* 32:209–216.

Reichel-Dolmatoff, G. (1976). Cosmology as Ecological Analysis: A View from the Rainforest. *Man: A Journal of the Royal Anthropological Institute.* Vol. 11, No. 3: 307–318.

Schultes, Richard Evans. (1988). Primitive Plant Lore and Modern Conservation. *Orion Nature Quarterly.* Vol. 7, No. 3:8–15.

Schultes, Richard Evans and Robert F. Raffauf. (1990). *The Healing Forest: Medicinal and Toxic Plants of the Northwest Amazonia.* Portland: Dioscorides Press.

Soejarto, D. and C. Gyllenhaal (eds.). (1992). The Declaration of Belem and the Kunming Action Plan. *International Traditional Medicine Newsletter* No. 4:1.

Spruce, R. (1908). *Notes of a Botanist on the Amazon and Andes.* London: MacMillan.

Tempesta, Michael and Steven R. King. (1994a). Ethnobotany As a Source for New Drugs. In: *Annual Reports in Medicinal Chemistry* (E. Venuti, ed.), pp. 325–330. San Diego: Academic Press.

————. (1994b). Tropical Plants As a Source of New Pharmaceuticals. In: *Pharmaceutical Manufacturing International: The International Review of Pharmaceutical Technology Research and Development* (Pamela A. Barnacal, ed.), pp. 47–50. London: Sterling Publishers.

Ten Kate, Kerry. (1995). *Biopiracy or Green Petroleum? Expectations and Best Practices in Bioprospecting.* London: Overseas Development Administration.

United Nations Environment Program. (1992). Convention on Biological Diversity. Rio de Janeiro, Brazil, June 5.

United Nations Working Group on Indigenous Peoples. (1992). Report on the Intellectual Property Rights of Indigenous Peoples. July 6, 1992. United Nations, New York.

World Resources Institute. (1994). *People and the Environment: Resource Consumption, Population Growth and Women.* New York: Oxford University Press.

PART IV

An Agenda for the Future: Conserving Biodiversity and Human Health

In this section several contributors explore the relationships among biodiversity conservation, development, and human health, and they discuss specific solutions to the challenge of preserving our health and biodiversity. Nongovernmental organizations, multilateral and bilateral donors, and national governments are exploring the value of linking conservation to development. While parks and protected areas are an important component of any conservation plan, the vast majority of nations have set aside no more than 5% of their land for preservation. Integrated conservation and development projects offer a potential tool for conservation of all lands, while meeting basic human needs through economic development.

Several chapters explore opportunities and mechanisms through which "bioprospecting" might offer compensation to source countries and serve as an incentive for biodiversity conservation. Unless benefits begin to accrue to the stewards of biodiversity in the developing world for access to their genetic resources, the raw materials on which therapeutic agents depend will disappear, first politically, and then biologically. Tom Mays and colleagues (Chapter 12) describe the approach that the National Cancer Institute (NCI) is following to compensate source countries for their potential contributions of genetic resources and traditional knowledge to

NCI's drug discovery program. Joshua Rosenthal (Chapter 13) discusses an experimental program that uses multidisciplinary research and equitable benefit-sharing arrangements to generate opportunities and incentives for conservation and development. Dan Janzen (Chapter 14), in his inimitable style, offers a vision that has inspired efforts underway in Costa Rica to explore and utilize every possible element of that country's biota, and therein provide for its conservation. Chuck Peters (Chapter 15) cautions against the uncritical acceptance of the premise that harvesting medicinal plant and other nontimber forest products is necessarily sustainable and urges us to avoid the temptation to accept a quick fix to a complex problem. The section concludes with two treatments that look to the future. Ecologist Walter Reid (Chapter 16) and medical experts Byron Bailey and John Grupenhoff (Afterword) reflect from their different perspectives on the possibilities for continued collaboration among the disciplines of biodiversity conservation and human health.

CHAPTER 12

A Paradigm for the Equitable Sharing of Benefits Resulting from Biodiversity Research and Development

THOMAS D. MAYS, KATE DUFFY-MAZAN, GORDON CRAGG, AND MICHAEL BOYD

Introduction

Scientists frequently learn from observation of nature and through research on naturally occurring compounds. Genetic resources comprising all plants and microorganisms are diverse building blocks that enable biomedical researchers to continue to strive to develop new therapeutic agents. Access to these natural products or source biomaterials is important in fostering and promoting new knowledge. The United Nations Convention on Biological Diversity provides for the equitable sharing of benefits arising from the commercialization of research results based upon collection of naturally occurring biomaterials.

The National Cancer Institute, the largest of the National Institutes of Health (NIH), is an agency of the United States Government. NCI's primary statutory mission is to promote and conduct biomedical research relating to the etiology, diagnosis, treatment and prevention of cancer. It is also one of several federal agencies supporting research relating to diseases associated with infection of the human immunodeficiency virus or HIV, the causative agent of acquired immune deficiency syndrome (AIDS).

In 1955, the NCI launched an extensive program for the procurement and screening of natural products for chemotherapeutic activity. NCI recognizes that well over 50% of the estimated 250,000 plant species found on earth come from tropical forests and, thus, concentrates on these regions. Plant materials have been collected by contractors on behalf of NCI from tropical and subtropical regions of Africa and Madagascar, Central and South America, and Southeast Asia. Marine organisms, fungi, cyanobacteria, and marine anaerobic bacteria and other protists have been collected mainly from the Indo-Pacific region. To date 35,000 plant samples, representing 9,000–10,000 different species, and over 6,000 marine samples have been collected.

Plant-derived anticancer drugs identified through the screening program include the *Vinca* alkaloids, vinblastine and vincristine; etoposide and teniposide, semisynthetic derivatives of epipodophyllotoxin; and Taxol®, isolated from *Taxus brevifolia* and other *Taxus* species (O'Dwyer et al. 1984, 1985; Rowinsky et al. 1992). Other promising agents in clinical development are topotecan and CPT-11, synthetic derivatives of camptothecin from *Camptotheca acuminata* (Kingsbury et al. 1991; Ohno et al. 1990). Two agents, didemnin B from *Trididemnum solidum* and bryostatin 1 from *Bugula neritina,* derived from marine samples, have advanced to clinical trials (Rinehart et al. 1990; Pettit 1991). To date, NCI has identified three specific agents isolated from natural products that demonstrate unique and highly active anti-HIV properties. These compounds are michellamine B, isolated from the plant *Ancistrocladus korupensis* found in Cameroon; conocurvone, isolated from the Australian smokeweed bush *Conosperum incurvum;* and compounds of the families calanolides and costatolides, isolated from the rainforest trees of the genus *Calophyllum* found in Sarawak, Malaysia.

The primary interest of the National Cancer Institute is to relieve pain and suffering and provide a healthier lifestyle for those diagnosed with cancer or acquired immune deficiency syndrome (AIDS). Within the context of its biological prospecting activities, it is also important to the NCI that the biological materials obtained throughout the world be collected under a shared sense of cooperation with the source country and in due consideration of problems associated with commercialization of the source country's biodiversity resources. Without the cooperation of the source countries, there can be little exploration of new biological materials and such limited access could be permanently precluded if certain species become extinct. However, under current patent law, source countries and indigenous populations are not recognized as inventors and thus do not benefit through operation of intellectual property rights. As the developing world is quickly becoming aware of the economic value of its native biological diversity, there is growing concern over the inability to recognize significant contributions of source countries and indigenous populations in the drug discovery process. If mechanisms to recognize their contribution are not identified, there may be little incentive for source countries to preserve their biodiversity for drug discovery, or any other use.

The NCI and other federal laboratories are responsible for transferring federal laboratory research results to the public. Generally, NCI transfers new biomedical research results through scientific publications and conferences. Under the Federal Technology Transfer Act of 1986 (FTTA), the federal laboratories (consistent with their mission) are charged to license their patentable inventions to the private sector for commercial development. The primary goal of the FTTA was to stimulate domestic economic development and increase U.S. competitiveness in international markets.

The challenge arises for NCI to balance at least three somewhat competing interests: to conduct premier biomedical research; to transfer technology to the private sector; and to comply with the provisions of the United Nations Convention on Biological Diversity, including sharing of financial and scientific benefits with other nations. This is somewhat further complicated by current patent statutes that provide for the creation of intellectual property rights on behalf of an inventor. NCI inventors are required to assign their rights to the NCI. NCI is not able to provide intellectual property rights to noninventors except under a patent license or by requiring NCI's patent licensee to enter into a benefits sharing agreement with a source country organization.

In natural products discovery, many creative persons play essential roles in taking a pharmaceutical agent from the plant in the rainforest to the patient in the hospital bed. Botanists, biologists, and others with technical taxonomic training seek out specific plants, based on information provided by indigenous peoples or traditional healers, whose knowledge of the uses of certain plants or other biological materials is based upon information handed down generation to generation since time unrecorded. Next in the chain of discovery are the scientists who take the collected plants and employ elaborate and sophisticated chemical fractionation and isolation procedures, as well as screening, testing, and retesting methodologies, to identify new therapeutic agents for further study. Finally, clinicians and other medical personnel plan, implement, and evaluate new drugs based upon *in vitro* and *in vivo* animal and clinical studies. Each person in this chain of events plays a critical role, without which, some agents would clearly never be discovered.

The contribution of indigenous knowledge may play an essential role in the selection of specific biological materials for testing. However, there is little certainty that inventorship will vest in the shaman or traditional healer who provided the information that led to the selection of the plant for testing, as that information was already known, at least to those who passed it by word of mouth down through the lineage of time. Historically, indigenous knowledge has played a direct role in only a very few of NCI's collections. The recognition of the contributions of indigenous knowledge challenges the fundamental basis of the patent system which rewards contributions of new information (Greaves 1994).

NCI's Letter of Collection Agreement

Several legal instruments are being used, or proposed, to achieve what current patent law cannot. The alternatives range from the use of contracts, trusts, and other fiduciary mechanisms to the creative use of licensing agreements. The National Cancer Institute's letter of collection agreement (or LOC, formerly the "Letter of Intent") is an example of an alternative through which developed and developing countries may "fill-in the gaps" created by patent law.

The LOC is a contractual agreement that permits the recognition and financial reward of indigenous peoples and source countries for their contribution to the identification and collection of natural products with potential therapeutic value in the treatment of cancer and AIDS. The NCI LOC provides for compensation and the transfer of technology to the source country in the form of royalties and scientific exchange. Further, the LOC sets forth a commitment on the part of NCI to utilize its best efforts to ensure that any commercialized products arising from the biological materials of the source country will be manufactured from materials acquired from the source country. It is important to note, however, that the LOC cannot be used as a mechanism to promise future intellectual property rights. As an agency of the U.S. Government, the NCI is prohibited under statute from agreeing to promise or encumber any future intellectual property rights to the source country, including specific royalties or inventorship rights, except as provided under a Cooperative Research and Development Agreement (CRADA) (35 U.S.C. §207, 208 and 209 (1984 and Supp. 1993); 15 U.S.C. §3701a(2)(b) (1986 and Supp. 1993).

A model LOC agreement evolved in the early 1990s to respond to the specific concerns of individual source countries or source country institutions. Structurally, it includes one general introductory clause and 18 composite clauses that define the rights and obligations of the National Cancer Institute, the source country collaborators, and their respective agents. The introductory clause identifies the relevant parties and their agents and describes the NCI's interest in investigating the potential of natural products from the designated source country. In addition, this clause confirms the NCI's commitment to ensuring the transfer of knowledge, expertise, and technology related to drug discovery and development to the source country, the conservation of biological diversity, and the need to compensate source country organizations and peoples in the event of commercialization of a drug developed from an organism collected within their borders.

Thirteen composite clauses define the role and responsibilities of NCI under the collection agreement. These clauses address a number of issues including: conservation and benefits sharing; royalties; technology transfer; patent protection; confidentiality and publication guidelines; licensing; and access to and use of materials. The five remaining composite clauses define the role and responsibilities of the source country's government and/or appropriate source country organizations under the agreement. These clauses address a number of issues including: facilitation of collection and export of materials; use of indigenous knowledge to guide collections; provision of raw materials; provisions for mass production with appropriate measures for conservation; and use of samples by the source country, or source country institution, for other purposes. Source country collaborators are asked to review the model letter of collection, and negotiations follow based on the specific needs and concerns of the country. Additions and modifications may be made to the composite clauses to address specific concerns of the source country or source country institution.

The provisions of the LOC are reviewed below. The provisions outlining NCI's obligations are addressed first and are followed by the provisions outlining the obligations of the source country.

NCI Obligations

Conservation/Benefits (¶1; §9; §10) Through the operation of these provisions the NCI and the source country will cooperate in efforts to promote conservation and ensure that the source country is able to utilize its biodiversity resources to the fullest. Of great concern to many representatives of a source country is the loss of a proprietary claim to the natural product materials after a synthetic means of manufacturing the active agent is developed. While genetic engineering has not yet played a significant role in the manufacture of agents isolated from natural products, the LOC pledges compensation should the agent later be manufactured by a totally synthetic process.

Benefits Sharing: Royalties (¶2; §7 §8;§9;§10;§12;§13) NCI (as an agency of the U.S. Government) must comport with all applicable statutes and regulations relating to the licensing of patent rights vested in the U.S. Government. Any entity may apply for a license to a patent held by the U.S. Government. In drug discovery, applicants often include biotechnology firms and pharmaceutical companies. The Institute considers the requirement upon the licensee to negotiate an agreement with the source country as the best means to permit the country and licensee to reach accord on mutually acceptable terms that address issues of common concern. This ensures that the unique concerns of the source country will be addressed to its own satisfaction, rather than attempting to impose NCI's views on the source country–licensee relationship. NCI recognizes that a licensee should comport with the spirit of the LOC, even though it is not a party to the agreement. By apprising the licensee of the nature and existence of the LOC and requiring the licensee to enter into an agreement with the source country, NCI uses its best efforts to ensure that the concerns of the source country are addressed.

Technology Transfer: Guest Researcher/Scientific Support (¶2; §3;§4) NCI has attempted to assist in the transfer of technology to the source country by providing support for an exchange program, where scientists of the source country are able to develop skills and the knowledge base to establish related drug discovery programs in their own laboratories. As of early 1993, 12 scientists from 9 countries had carried out, or were undertaking, collaborative research projects with scientists in NCI facilities. Over 406 scientists from 18 countries had visited the NCI's drug discovery laboratories as the NCI Frederick Cancer Research and Development Center for periods of one to two weeks. From these exchanges, both the NCI and the source countries benefited by increased scientific cooperation and mutual understanding. These efforts ensure that source countries with drug discovery research programs can continue and expand their scope, as well as provide advanced training in useful skills for source country scientists.

Technology Transfer: Collaboration (§5) In addition to providing general guest researcher training, NCI and source countries may participate in expanded collaboration relating to active agents found during screening of natural products materials. Such collaboration may proceed under a cooperative research and development agreement (CRADA), wherein NCI is authorized under the FTTA to agree to license future patent rights. This mechanism permits a more rapid transfer of technology using streamlined administrative procedures and promotes close collaborative research. Currently, NCI and one source country organization are negotiating a

CRADA under which the two parties will collaborate to further develop a potential therapeutic agent identified under the NCI drug screening program from materials submitted by the source country organization.

Patent Protection (§6;§2) NCI is required to utilize the patent system to promote the transfer of its technology. To those ends, each party agrees that patent applications should be considered, evaluated, and filed where appropriate to successfully protect any active agents identified as active during screening of the source country's biological materials. Such filing could be made in the source country by either NCI or NCI's licensee as provided under the license agreement.

Confidentiality/Publications (§2) NCI is very concerned that the scientific results arising from the drug discovery program be quickly shared with the medical community to promote the rapid dissemination of new therapeutic knowledge. However, this interest is balanced against the preliminary steps needed to protect intellectual property rights prior to publication. A routine review of all manuscripts prior to publication affords the opportunity to determine which should be considered for patent filings.

Access to Data (§1) NCI acknowledges the source country's interest in learning the results of the drug screening assays of its natural products materials. In an effort to further transfer the technology as well as collaborate with the source country, NCI makes every effort to provide this data via the collecting contractor.

Access to Materials (§13) The natural product materials obtained from source countries are valuable scientific resources that enable other (non-NCI) researchers to pursue drug discovery for active agents. NCI provides these materials under a material transfer agreement (MTA) that references and incorporates the terms of the LOC.

Source Country Obligations

Collection Permits (§1) The safeguarding of biodiversity can often be best achieved by source country regulations and the use of collection permits. NCI and source country work together to ensure that regulations, local laws and customs are followed through the use of such permits.

Indigenous Knowledge (§2) The source country and NCI exchange useful information in expediting the drug discovery effort. NCI respects the contribution and acknowledges the value of indigenous knowledge in this research effort and has therefore taken great care to ensure that collaboration between the two parties is not jeopardized by disregarding the insights and traditions of the indigenous peoples, nor are the contributions of the indigenous peoples used without attribution.

Provision of Raw Materials and Mass Propagation (§3; §4) The source country provides additional raw material for studies as needed. In the event that large quantities of the materials are needed efforts are made to investigate the mass propagation of the material within the source country, while conserving the biological diversity of the region and involving the local population in the planning and implementation stages.

Additional Uses of Materials (§5) Both NCI and the source country are able to pursue noncancer and non-AIDS drug discovery research independently of one another. This ensures the efficient use of natural product materials collected for additional scientific research.

In order to gain a clearer perspective of the implementation of the LOC agreement, a brief review of the present NCI Natural Products Program and the current operation of the LOC agreement may be helpful. As of 1996, the LOC and its predecessor agreement, the Letter of Intent (LOI), has served as very useful tools to enable the NCI to tackle its statutory balancing act.

At that time, 24 LOCs/LOI executed with 22 countries and another 8 LOC agreements were under active negotiation. Three specific case studies are illustrative of the operation of the LOC agreement: (1) the michellamine B compounds from the liana, *Ancistrocladus korupensis,* of Cameroon; (2) the calanolide and costatolide compounds from a tree, *Calophyllum lanigerum,* of the Malaysian State of Sarawak; and (3) the conocurvone compounds from the smokeweed bush of the state of Western Australia. Samples of each family of compounds have demonstrated high anti-HIV activity *in vitro* and are undergoing further evaluation for clinical trials.

Implementation of the Letter of Collection

As of 1996, NCI had entered into LOCs with the following countries or organizations within the countries of The Andean/Amazonian region,

Bangladesh, Brazil, Cameroon, China, Costa Rica, Ecuador, Gabon, Ghana, India, Korea, Madagascar, Mexico, Pakistan, Palau, Panama, Philippines, Russia, Sarawak, South Africa, Tanzania, and Zimbabwe. Active negotiations are underway with several other countries. Source country collaborators are asked to review the model LOC, and negotiations follow based on the specific needs and concerns of the country. Additions and modifications may be made to the composite clauses to address specific concerns of the source country or source country institution. The following scenarios highlight specific collaborations with source countries utilizing the LOC.

Michellamine B of Cameroon

In March 1987, Duncan Thomas collected a sample of the stems and leaves of a liana in the Korup rainforest region of Cameroon's southwest province as part of an NCI plant collection contract with Missouri Botanical Gardens (MBG) for collections in Africa and Madagascar. Solvent extracts of this liana showed significant *in vitro* activity in the NCI's anti-HIV screen, and bioassay-guided fractionation of the extract in 1989-1990 by chemists of the NCI Laboratory of Drug Discovery Research and Development yielded the active agent, michellamine B (Boyd et al. 1994).

The liana was initially tentatively identified as *Ancistrocladus abbreviatus,* but was later classified as a new species, *Ancistrocladus korupensis.* This liana mainly grows in the rainforest canopy and is sparsely distributed in the Korup region. In 1992, NCI provided supplemental funding to MBG to perform surveys of the occurrence and abundance of the liana; in late 1993, a contract was awarded to Purdue University to expand this survey and explore the feasibility of cultivation of the plant in the Korup region. In performing these tasks, a productive collaboration has been established between scientists from the University of Yaounde, personnel of the Cameroon government, the World Wide Fund for Nature (a nongovernment organization which coordinates the management of the Korup National Park for the Cameroon Government), Missouri Botanical Gardens, Oregon State University (represented by Dr. Thomas), the NCI Frederick Cancer Research and Development Center (NCI-FCRDC, represented by the NCI contractor, Program Resources, Inc.), and Purdue University. Over 1,000 samples of the leaves of the liana have been collected and analyzed for michellamine B content, and significant progress has been made in cultivation of high-yielding specimens. Extensive use is being made of local labor and expertise in the Korup region in the collection and cultivation projects, and, should michellamine B or an active derivative advance to clinical trials and commercial production, *Ancistrocladus korupensis* will be developed as a cash crop in the Korup region. In such an event, the preliminary extraction of the plant material in Cameroon will also be explored. A significant discovery has been that the fallen leaves of the liana collected from the forest floor also contain acceptable quantities of michellamine B, and the collection of these leaves has provided a nondestructive renewable source of the drug.

An agreement based on an NCI LOC was signed with the University of Yaounde in 1992, and negotiations for a new agreement with the Cameroon government are in progress. NCI is also collaborating with Professor Gerhard Bringman of the University of Wurzburg in Germany who is the world's leading authority on the chemical constituents of the *Ancistrocladus* genus. Professor Bringman has found that compounds related to michellamine B, isolated from *A. korupensis* and related species, exhibit antimalarial activity; this discovery could provide further valuable uses for these plants, and resultant benefits for Cameroon. Professor Bringman has agreed to abide by the NCI policies of collaboration and compensation contained in the LOC.

Michellamine B is currently in advanced preclinical toxicology, and the results of these studies will determine whether or not it advances to clinical trials. Even if michellamine B itself does not qualify as a clinical candidate, it provides a very valuable lead for the synthesis of related active compounds for evaluation as development candidates. In all these developments NCI is committed to abiding by the terms of collaboration and compensation spelled out in the LOC.

Calanolides/Costatolides of Sarawak

In April 1987, John Burley of Harvard University Arnold Arboretum and botanists of the Sarawak State Forestry Department collected a sample of the stems and leaves of the tree *Calophyllum lanigerum* as part of an NCI plant collection contract with the University of Illinois at Chicago (UIC) for collections in Southeast Asia. Solvent extracts of the sample exhibited significant *in vitro* activity in the NCI anti-HIV screen, and chemists of the NCI Laboratory of Drug Discovery Research and Development isolated calanolide A as the major active constituent (Kashman et al. 1992).

Recollections of *C. lanigerum* in the same region failed to yield significant quantities of calanolide A, and since 1993 NCI has funded an extensive survey by botanists of UIC and the Sarawak State Forestry Department in an effort to locate productive specimens of the tree. Thus far only low-yielding specimens have been found, but, during the course of the survey, another species, *Calophyllum teysmanii,* has been found to yield a slightly less active compound, costatolide, closely related to calanolide A. Costatolide is isolated in very high yields (exceeding 20%) from the latex of *C. teysmanii;* isolation from the latex thus provides a nondestructive renewable source of this potential anti-HIV agent. Surveys of the occurrence and abundance of *C. teysmanii* were funded by NCI in collaboration with the Sarawak State Forestry Department. Calanolide A has been synthesized independently by an industrial group, but the synthesis yields material which is mixed with inactive forms of the compound and is therefore less active; while it is possible to separate the active and inactive forms, the scale-up of the procedure to give large quantities is difficult and will require considerable research and development.

In May 1994, NCI signed an agreement, based on the NCI LOC, with the Sarawak State Government for the study of Sarawak natural resources and the development of calanolide A and costatolide. A senior scientist from the University of Malaysia Sarawak spent a year at the Frederick Cancer Research and Development Center collaborating in the isolation of larger quantities of costatolide, and gaining experience and training in the NCI anti-cancer and anti-HIV screening techniques. This scientist returned to Sarawak in April 1995 to establish a screening program at the University of Malaysia and to initiate a project for the isolation of costatolide in Sarawak. Such training and technology transfer is in line with the terms of the NCI–Sarawak State Government agreement and the U.N. Convention on Biological Diversity.

Both calanolide A and costatolide are in early preclinical development at NCI, and larger quantities are being produced to permit performance of advanced toxicology prior to consideration for clinical trials.

Conocurvone of Western Australia

The conocurvone case study was somewhat more controversial following several press reports in Australia. These reports demonstrated that without clear communication among all pertinent parties, a collaboration, which could be very productive, can get embroiled in local and national policies concerning development of products from biodiversity.

In the early 1980s, the NCI through a contract collector, identified an active fraction from the smokeweed bush found in Western Australia. Following a series of correspondence between the NCI and the Conservation and Land Management (CALM) agency of the state of Western Australia, in August of 1993, the National Cancer Institute published a notice in the Federal Register seeking licensees for the development of conocurvone as a drug for the treatment of HIV infection. This notice included the following language concerning NCI's commitment to ensuring that the people of Australia would share in the benefits of any product developed as a result of that license:

> Since this agent, Conocurvone, is isolated from flora indigenous to Western Australia, NCI is concerned that the collection and utilization of the natural plant material comport with all applicable Federal and Australian policies related to biodiversity. In order to comport with such policies, the successful applicant will be required to negotiate and enter into agreements with the appropriate Australian and Western Australian Government agencies. (58 Federal Register 45901, 45902 (August 31, 1993))

This commitment to the source country is an integral component of NCI's Natural Products Program and is provided under a negotiated letter of collection agreement. Specifically, the LOC requires any future licensee of an agent developed from the source country's natural products to negotiate an agreement with the source country organization named in the LOC to ensure that benefits flow back to appropriate individuals and organizations in that country.

On November 29, 1993, the Australia-based company, AMRAD Corporation Ltd., was notified of their selection as the exclusive licensee for this technology. On December 10, 1993, in compliance with United States statute [35 U.S.C. 209(c)(1) and 37 CFR 404.7 (a)(1) (i)] a second Federal Register Notice providing notification of NCI's intent to grant the exclusive license to AMRAD was published. On February 11, 1994, AMRAD was notified that we had received no opposition to the notice of the intent to grant, and license negotiations began. These initial negotiations culminated in a meeting in late April 1994, during which the final issues related to NCI's license with AMRAD were substantially resolved. The license was subsequently signed after AMRAD reached agreement with CALM.

Summary

As NCI moves closer to licensing its first compound identified under the LOC, it will be easier to determine how best to modify the LOC agreement as well as the implementing program and underlying policies. Perhaps through separate agreements between NCI's licensee and source country, the benefits sharing program envisioned under the U.N. convention will result in financial return to source countries. NCI is already working diligently and with good faith to ensure that the source country shares in the transfer of technology and scientific information.

Through the LOC, NCI is striving to achieve an equitable arrangement between a country that is a source of biological materials and a research or commercial party whose research and development efforts may result in the subsequent commercialization of products derived from these biological materials. The collection agreement at a minimum provides the source country an appropriate return of benefits for the use of the material. In addition to a financial return, such as future royalties from sales of a biological product, it is also important to provide the source country a return on intellectual capital by providing scientific training, thus enabling the source country to more efficiently manage its own natural resources. Concomitantly, the research or commercial party must receive a commitment from the source country to permit the uninterrupted access to biological materials from which the biological product is isolated. Finally, the LOC reduces the degree of confusion, disillusionment, or uncertainty among the parties, by clearly stating the rights and obligations of each party so that in the event a pharmaceutical agent or other biological product is identified that demonstrates significant commercial potential, each party clearly understands its role and its obligations as well as what benefits it might receive.

In conclusion, the efficient transfer of federal biomedical research results to the private sector is essential to provide the greatest benefit to the public health. Access to such biomaterials as a source of natural products is an important factor in expediting drug development. Through the use of benefit sharing agreements, such as the NCI LOC and agreements between NCI's licensees and source countries, financial and technical resources can be more effectively leveraged to provide support for bioconservation programs and the creation of accurate biomaterial inventories.

References

Boyd, M.R., Y.F. Hallock, J. H. Cardellina II, K.P. Marfredi, J.W. Blunt, J.B. McMahon, R.W. Beached Jr., G. Bringmann, M. Schaffer, G.M. Cragg, D.W. Thoma, and J.G. Jato. (1994). Anti-HIV Michellamines from *Ancistrocladus korupensis. Journal of Medicinal Chemistry* 37:1740–1745

Greaves, T. (1994). Introduction. In: *Intellectual Property Rights for Indigenous Peoples: A Source Book* (T. Greaves, ed.), pp. ix–xii. Oklahoma City: Society for Applied Anthropology.

Kashman, Y., K.R. Gustafson, R.W. Fuller, J.H. Cardellins II, J.B. McMahon, M.J. Currens, R.W. Beached Jr., S.H. Hughes, G.M. Cragg, and M.R. Boyd. (1992). The Calanolides, a Novel HIV-inhibitory Class of Coumarin Derivatives from the Tropical Rainforest Tree *Calophyllum lanigerum*. *Journal of Medicinal Chemistry* 35:2735–2743

Kingsbury, W.D., J.C. Boehm, D.R. Jakas, K.H. Holden, S.M. Hect, G. Gallager, M.J. Caragna, F.L. McCabe, L.F. Faucette, R.K. Johnson, and R.P. Hertzberg. (1991). Synthesis of Water-Soluble (Aminoalkyl) Camptothecin Analogues: Inhibition of Topoisomerase I and Antitumor Activity. *Journal of Medicinal Chemistry* 34:98–107.

O'Dwyer, P., M. Alonso, B. Leyland-Jones, and S. Marsoni. (1984). Teniposide: A Reveiw of 12 Years of Experience. *Cancer Treatment Reports* 68:1455–1466.

O'Dwyer, P., M. Alonso, B. Leyland-Jones, S. Marsoni, and R. Wittes. (1985). Etoposide (VP-16-213) Current Status of Anticancer Drug. *New England Journal of Medicine* 312:692–700.

Ohno, R., K. Okada, T. Masaoka, A. Kuramoto, T. Arima, Y. Yoshida, H. Ariyashi, M. Ichimaru, Y. Ito, Y. Morishima, S. Yoomaku, and K. Ota. (1990). An Early Phase Study of CPT-11: A New Derivative of Camptothecin, for the Treatment of Leukemia and Lymphoma. *J. Clin. Oncol.* 8:1907–1912.

Pettit, G.R. (1991). The Bryostatins. In: *Progress in the Chemistry of Organic Natural Products* (W. Herz, G.W. Kirby, W. Steglich and C. Tamm, eds.), pp. 153–195. New York: Springer-Verlag.

Rinehart, K.L., T.G. Holt, N.L. Fregeau, P.A. Keifer, G.R. Wilson, T.J. Perun Jr., R. Sakai, A.G. Thompson, J.G. Stroh, L.S. Shield, D.S. Seigler, L.H. Li, D.G. Martin, C.J.P. Grimmelikuhijzen, and G. Gade. (1990). Bioactive Compounds from Aquatic and Terrestrial Sources. *J. Nat. Prod.* 53:771–792.

Rowinsky, E.K., N. Onetto, R.M. Canetta, and S.G. Arbuck. (1992). Taxol: The First of the Taxanes, an Important New Class of Antitumor Agents. *Seminar in Oncology* 19:646–662.

CHAPTER 13

Integrating Drug Discovery, Biodiversity Conservation, and Economic Development: Early Lessons from the International Cooperative Biodiversity Groups

JOSHUA ROSENTHAL

Biodiversity prospecting offers a valuable opportunity to directly address one of the most important relationships between biodiversity and human health—our dependence on nature for medicine. Drug discovery from natural products, if appropriately designed and carried out, can provide economic incentives for the conservation of the diversity of plants, animals, and microorganisms on earth. While natural products drug discovery has a long history (see Laughlin and Fairfield, Chapter 7; Grifo et al., Chapter 6), the idea that this process could yield conservation and development benefits is relatively new. This integrated approach, commonly known as "bioprospecting," has received a great deal of attention in recent years as a potential tool for conservation (for examples see Reid et al. 1993; Eisner and Beiring 1994; Endangered Species Coalition 1995; Goering 1995). Statements regarding the importance and potential of bioprospecting have become common in the conservation literature, as well as in the strategic plans of most major conservation organizations and many governments. Bioprospecting is also a major concern of the United Nations Convention on Biological Diversity. However, despite widespread interest in the concept, significant attempts to carry out fully integrated bioprospecting projects have been very few.

The International Cooperative Biodiversity Groups represents a novel experimental program that is one of the first large-scale attempts to design and execute such a multidisciplinary approach to drug discovery. The program was designed to stimulate the field of bioprospecting, to gather evidence on the feasibility of bioprospecting as a tool for conservation and economic development, and to provide models for its future development. In addition to these ambitious goals, it is hoped that the program will provide insight into important scientific questions regarding the relative efficiency of different scientific modes of drug discovery from natural products. Finally, in the spirit of the United Nations Convention on Biodiversity, the program aims to promote equitable sharing of benefits that flow from both the research process and its potential commercial products. This final goal also represents an important experimental effort for which there are few models in existence.

In the first two and a half years of work the International Cooperative Biodiversity Groups (ICBG) have made significant progress toward several of these long-term goals, and some valuable lessons have been learned regarding benefit-sharing and the requirements of international partnerships. Here I provide a brief overview of the program and the groups that are carrying out this pioneering work, followed by a summary of their progress to date in several areas. I will also describe some of the lessons that have been learned during the establishment and early work of the groups.

International Cooperative Biodiversity Groups

History, Goals, and Program Structure

In 1992 three agencies of the U.S. Government—the National Institutes of Health (NIH), the National Science Foundation (NSF), and the U.S. Agency for International Development (USAID)—launched the International Cooperative Biodiversity Groups program. The goals and design of the program were based on recommendations that emerged from a jointly sponsored and widely attended conference held the year before in Washington, D.C. (Schweitzer et al. 1991).

The ICBG program is based on the premise that appropriately designed natural products research and development can bring both short- and long-term benefits to the countries and communities that are the stewards of genetic resources [Schweitzer et al. 1991; Grifo 1996]. Sharing benefits from both the research process and from any drug discoveries that are made down the road creates incentives for conservation and provides alternatives to destructive use.

The program has three principal goals that reflect the mandates of its three government agency sponsors. The first is to improve human health through the discovery of new therapeutic agents to treat diseases of importance to both developed and developing countries. This includes the preparation of crude materials, bioassay testing, chemical isolation, and pre-clinical evaluation of agents from natural sources to treat or prevent cancer, infectious diseases including AIDS, cardiovascular diseases, malaria, mental disorders, parasitic infections, and other diseases.

The second goal is to conserve biodiversity through valuation of diverse biological organisms and the development of local capacity to manage these natural resources. This goal encompasses creating incentives at all levels for the preservation of intact habitat; increasing the knowledge base upon which conservation activities are based; and developing long-term ecological and economic strategies to ensure more sustainable harvesting of targeted organisms.

The third goal of the program is to promote sustainable economic activity in less developed countries by sharing the benefits of the drug discovery and conservation research processes. This is accomplished in part through benefit-sharing agreements that use novel contractual arrangements to ensure an equitable financial return to the host country, group, or organization that facilitates the discovery drug process. In addition, support for research and capacity-building targeted toward the needs of the source country or countries represented within the Group fulfill this goal.

Each of the International Cooperative Biodiversity Groups is a consortium of academic institutions, local and international private voluntary organizations, and in most cases a private pharmaceutical company. One or more of the partner organizations of each ICBG is based in the source country. The Groups are run by an academic principal investigator, who directs his or her own research program in natural products chemistry, drug development, or ethnobiology and coordinates the activities of several associate programs. Each associate program is charged with one or more of the basic missions of the ICBG—biodiversity inventory, collection and conservation, screening and chemistry, drug development, and economic development. Training and other local capacity-building efforts are usually part of each of the associate programs. The awards that fund the groups are in the form of cooperative agreements, rather than grants. This means that the U.S. Government has continued involvement in the projects through scientific advisory committees that include representatives from each funding agency. The Fogarty International Center of the NIH manages the program and provides policy advice to the groups.

Equitable research and benefit-sharing agreements are key to the conservation and development goals of bioprospecting. Because patent law is unable to reward stewardship of biodiversity and traditional knowledge (Greaves 1994; Mays and Mazan 1996), contractual agreements that guarantee that benefits will flow from the research process and its potential products provide a crucial incentive mechanism to stimulate conservation of these endangered resources [Grifo and Downes 1996; Iwu 1996; Rosenthal (in press)].

Applicants for the ICBG awards were given a description of program goals and intellectual property principles to use in the design of their research proposals and contractual agreements. Formal written agreements that govern treatment of intellectual property and benefit-sharing were required of all applicants prior to making an award. The funding agencies are generally not party to the ICBG research and benefit-sharing agreements. As a result, the U.S. Government representatives are prohibited by federal statute from stipulating terms or structures of those agreements. Rather, the funded parties were asked to develop workable agreements to fit the nature of the organizations, countries, communities, and resources involved, within the general framework of the program's principles.

The ICBG Request for Applications (RFA TW-92-01) and other background papers (Schweitzer et al. 1991; Grifo and Downes 1996; Grifo 1996) describe these principles in detail. In general, they require that full disclosure and informed consent are carried out, that both near- and long-term benefits are shared with appropriate source country communities and organizations, that pertinent international and local laws are followed, that local customs are respected, and that credit is given to local indigenous or other intellectual contributors whenever possible. This approach to intellectual property and contractual agreements is also an experiment from which valuable lessons may be learned.

ICBG Awards

In September of 1993 and 1994, following a multidisciplinary peer review of 34 competitive proposals, five groups were selected for funding at an annual level of $400,000 to $500,000 per group, with an expected duration of five years.

Dr. David Kingston of Virginia Polytechnic Institute and State University (VPISU) is studying rainforest plants in Suriname, in collaboration with the Forest People of Suriname, Conservation International—Suriname, the National Herbarium of Suriname, the Missouri Botanical Garden, Bedrijf Geneesmiddelen Voorziening Suriname, and Bristol-Myers Squibb Pharmaceutical Research Institute.

Dr. Jerrold Meinwald of Cornell University is the group leader for the study of insects and related organisms from the dry tropical forests of the Guanacaste Conservation Area in Costa Rica, in conjunction with the Instituto Nacional de Biodiversidad (INBio) of Costa Rica, the Universidad de Costa Rica, and Bristol-Myers Squibb Pharmaceutical Research Institute.

Dr. Barbara Timmermann and colleagues of the University of Arizona are studying arid land plants in Latin America (Argentina, Chile, and Mexico), in collaboration with the Instituto de Recursos Biológicos de Argentina, the Universidad Nacional de la Patagonia, Pontificia Universidad Católica de Chile, the Universidad Nacional Autónoma de México, Purdue University, G. W. L. Hansen's Disease Center, and the Medical and Agricultural Divisions of Wyeth-Ayerst/American Cyanamid Co.

Dr. Walter Lewis of Washington University is group leader for ICBG research on plants that have been used medicinally for generations in Andean tropical rainforests of Perú. He is collaborating with several organizations of Aguaruna people under the leadership of the Confederacion de Nacionalidades Amazonas del Peru, the Universidad San Marcos, the Universidad Peruana Cayetano-Heredia, and Monsanto-Searle Co.

Dr. Brian G. Schuster leads a group from Walter Reed Army Institute of Research that is focusing on cures for parasitic diseases from rainforest plants of Africa (Cameroon and Nigeria). Their collaborators are the Smithsonian Institution, the Bioresources Development and Conservation Programme, the University of Yaounde in Cameroon, the Biodiversity Support Program, and Shaman Pharmaceuticals.

For more complete information on the initiation of the program and the review process that led to these awards, as well as details of the group members and their objectives, see Grifo (1996).

ICBG Summary Progress and Development

The first three ICBGs (Suriname, Latin America, Costa Rica) have now completed three years of research and development work. The last two ICBGs (Africa and Perú) are completing their second year of funding, but because of various delays in getting started both have completed little more than a year of research and development activity. All together they are working in 8 countries in Latin America and Africa: Costa Rica, Perú, Suriname, Chile, Argentina, Mexico, Cameroon, and Nigeria. I will summarize here some preliminary scientific and development activities of the groups, including biodiversity collections and screening, research methods, and local capacity-building efforts.

Biodiversity Collections and Screening Activities

The initial steps in natural products drug discovery research involve the description, bulk collection, and biological screening of samples for activity (see Grifo et al., Chapter 6). In some projects plant collections are made in close coordination with shamans or other local informants. Following drying and extraction, the samples are run through a battery of bioassays, or screens, for potential activity in the therapeutic areas of interest.

Following initial screens, "positive samples" or "hits" undergo a very lengthy process of retesting, re-collection, and biochemical analysis to ensure that a find is novel, sufficiently active, and biochemically manageable before it is pursued further. At each step most of the samples are eliminated, leaving fewer and fewer samples to study (see Artuso, Chapter 8). The decision to continue with a given sample involves numerous variables beyond those just mentioned. The strategic interests of the partners and the instincts of the investigators are two such variables that can be very important to the process. As a result, many samples that are eliminated continue to be valuable resources for future use by the source country partners as new partnerships form and advances in screening technology take place.

In their first two years, the five ICBGs collected over 3,000 bulk samples representing approximately 2,500 species of plants, insects, and mollusks. These samples produced approximately 7,500 extract samples, most of which have undergone initial testing. Collected samples have been examined in over 120 different bioassays reflecting a dozen disease areas, including cancer, AIDS, fungal, bacterial and viral infections, tuberculosis, malaria, leishmaniasis, heart disease, central nervous system and reproductive disorders, as well as several agricultural uses. All together, this represents over 100,000 separate screening events. These numbers represent primarily the activities of the first three ICBGs—Suriname, Costa Rica, and Latin America—which began more than a year before the Africa and Peru ICBGs.

When biological activity is detected and confirmed for a sample in at least one screen it is considered a positive hit. The hit rates among screening institutions vary widely depending upon the number and types of screens the group is running and their methods. To date, hit rates among the ICBGs range from 5 to 25% of the samples analyzed. Currently, approximately 350 samples are of continuing interest and approximately 35 are considered to be high priority leads for treatment of malaria, leishmaniasis, tuberculosis, drug-resistant bacterial infections, and central nervous system disorders. While a number of these leads currently look very promising, it is quite possible that none will lead to a new drug (see Artuso, Chapter 8; Grifo et al., Chapter 6).

Random, Biorational, and Ethnomedical Approaches to Drug Discovery

The ICBGs use various combinations of three general methods of sample collection and drug discovery research—random, biorational, and ethnomedical. Traditional approaches to natural products drug discovery by chemists and commercial researchers generally begin with collection of samples of plants, fungi, insects, and other biological specimens. Collection is guided principally by the desire to sample the greatest diversity possible. This sampling method is frequently referred to as a "random" approach. Collectors tend to take most identifiable specimens that they encounter, in some cases with a modest bias toward taxonomic groups that are known to be chemically interesting. Biorational approaches to discovery utilize more detailed knowledge of the biology of the specimens. For example, a relatively undamaged plant in a forest in which most plants show considerable damage from insects may be more likely to contain potent chemicals. This approach may increase our efficiency in finding useful chemicals compared with methods that depend on random techniques alone (Eisner 1989). Similarly, many researchers have championed the use of ethnomedical knowledge from traditional societies as a more direct means of finding specimens of utility in developing modern therapies for important human diseases (Cox and Balick 1994; Lewis and Elvin-Lewis 1995; Conte 1996). Many traditional societies have highly evolved systems of herbal medicine that represent valuable, yet rapidly disappearing intellectual resources (see Cox, Chapter 9).

One eventual product of the ICBG program will be information regarding the comparative efficiencies of these approaches. While data to assess these issues are, at present, insufficient, several developments are worth noting.

The Costa Rica ICBG uses a biorational approach in focusing part of their collection and screening efforts on insects that display notable chemical ecology in their interactions with plants and predators. Recently, this group further extended this approach by recognizing the biological differences in the different parts of a given species and physically subdividing specimens of a given species into finer units prior to extraction and screening. This move was based on the knowledge that biologically distinct structures on one individual insect or plant frequently differ chemically as well. One insect species can be divided developmentally into egg, larva, pupa, adult, and it can be divided structurally into head, body, fecal product, etc. Products of the interaction of plants and insects, such as galls, may conceivably also differ chemically from either of their parent organisms.

Similarly, one plant can logically be divided into units such as leaves, flowers, fruits, twigs, stem bark, stem wood, root bark, and root wood. Separation of some of these parts is a component of the research approach of several of the ICBGs. The relative age of the structure as well as the degree and type of microorganism infection it exhibits may also be potentially important variables for the initial screenings.

Sustainable use of these organisms requires that sampling practices make judicious use of structures and developmental stages that might injure the viability of the organism or its population (see Peters, Chapter 15).

Ethnomedical knowledge is utilized in different ways in four of the ICBGs. For at least some of the collections in the Africa, Suriname, and Perú ICBGs this knowledge is used explicitly to help guide the sample collection process, and in more limited ways during screening and subsequent chemical analyses as well. In the Latin America ICBG, ethnomedical knowledge is gathered primarily to help ensure preservation of that knowledge and maximize the potential of rewarding it with financial benefits, even if the link to the product is made after a discovery.

The Suriname and Perú ICBGs have explicit research designs that will attempt to provide data on the relative efficiency of ethnomedically based searches versus more random searches, although data are not yet available to assess this. The answers may depend, in part, on the measures of efficiency used. Number of species examined, collection time, and financial investment per hit, for example, are all reasonable measures depending upon one's interest.

While there have been a number of 'meta-studies' comparing ethnomedical to random and other approaches (Lewis and Elvin-Lewis 1995), it has been difficult to produce truly comparable samples. A precise study should control for geographical, seasonal, and other potential effects on the chemistry of the organisms being studied. One also needs to utilize the same or comparable bioassays. One organizational challenge to the ICBGs attempting this comparison is posed by the fact that large commercial partners continually modify and replace their assays. Because sample collection using ethnomedical techniques tends to be much more time consuming than random collections and does not necessarily involve reproductive specimens, coordinating the collection, identification, and testing of the two sample types is difficult.

Use of biological and ethnomedical information on the species being studied also poses some complex issues for control of the potential intellectual property it represents. Early disclosure of the identities and uses of the species being studied could weaken the control that source country partners have over that information (Laird 1994). Each of the ICBGs has chosen to deal with these issues in slightly different ways. Some of the ICBGs provide some of this information to their drug development partners with the samples, under strict confidentiality and limited use agreements. Others provide only numbered samples initially. In these arrangements, when a sample appears to be of interest, its species name is generally released to research partners, with consent of all source country parties, to help ascertain if the active sample represents a novel compound.

The Numbers Game

Irrespective of the collection approach, drug discovery from natural products is, in part, a numbers game (see Artuso, Chapter 8). Once a lead sample has been identified, skill and insight in natural products chemistry become important. But until that point, screening larger numbers of samples for a greater number of therapeutic possibilities and other uses generally increases the likelihood of finding a valuable lead.

Each of the ICBGs involves more than one institution in the screening process. The collaborating institutions within each group frequently arrange to use different bioassays and generally work in different therapeutic areas to maximize efficiency and avoid potential conflicts of interest. For example, the Latin America Arid Lands ICBG has four different institutions doing bioassays. The industrial partner has two divisions: Wyeth-Ayerst screens for a variety of medical uses and American Cyanamid is searching for agricultural and veterinary end uses. Simultaneously with that search, Purdue University is looking for generally bioactive compounds, the G. W. L. Hansens Center is looking for antituberculosis compounds, and the Universidad Autonoma de México is running cytotoxicity assays.

All of the ICBGs are facilitating the development of additional screening programs in the source countries with which they are collaborating. Where possible, the source country programs and those of other academic partners further enable work on diseases such as tuberculosis, malaria, and leishmaniasis that are rarely within the standard early screening portfolio of major pharmaceutical companies.

Biodiversity Information Resources

A prerequisite for conservation and sustainable use of biodiversity is having information on the identity and distribution of the organisms within a habitat. The plants and insects collected for drug discovery research form only a part of that information gathering process for the ICBGs. Almost all fieldwork activities include collection and documentation of plants and animals throughout the region. Each ICBG is compiling preserved specimens in at least one institution in the source country and one in the U.S. for the development of resources for biodiversity research and management. This information is being recorded in geographic information systems and other computer databases in each country. In several of the ICBGs these databases complement other existing resources in the source countries (Costa Rica, Perú, Suriname ICBGs), and in others (Latin America and Africa ICBGs) entirely new systems have been created to manage ICBG data and make them available to managers. It is hoped that these databases will become more and more integral components of land use decisions within the country in the coming years.

Capacity-Building in the Source Country

Long-term development of the scientific, commercial, and management capacity of source countries may be the single most valuable benefit of bioprospecting research and development work (Juma 1993; Baker et al. 1995; Ten Kate 1995). Capacity-building is a central component of all three goals of the program—drug discovery, conservation, and economic development. For the ICBGs, capacity-building means training, equipment transfers, and infrastructure development in collaborating institutions and communities.

ICBG trainees include source country technicians, graduate, postgraduate and postdoctoral students, and faculty. Training includes long-term and degree program work as well as short technical courses and workshops in biodiversity description and management and biomedical science. To date, over 135 students and technicians from at least 15 developing country institutions have received or are receiving training in association with the

ICBGs. These include numerous exchanges between the United States and host-country universities.

A few examples will illustrate the types of training involved. Two graduate students from Argentina are being trained at GWL Hansen's Disease Center in antituberculosis screening. A postdoctoral researcher from Cameroon is receiving long-term training at Walter Reed Army Institute of Research in natural products chemistry and anti-leishmania screening, as are several students and faculty in Perú. Several technicians in Suriname are being trained in the use of geographic information systems (GIS) to record biodiversity inventories and ethnobotanical information from the Amazon rainforest and its peoples. Parataxonomists are being trained in Costa Rica to identify and raise numerous species of insects in field conditions, interested community members in the Peruvian rainforest are being trained in ethnobotanical collection methods, and more than a dozen African biologists have completed a short course in ethnobotanical methods in Cameroon.

Equipment transfers to source country collaborators come both through government funding and directly from commercial partners. Laboratory equipment related to the preparation, extraction, storage, and microbiological screening of specimens is commonly transferred to the source country. Other equipment purchases include herbarium storage cases, computers, software, and field equipment to aid with biodiversity description and management. Infrastructure development efforts include vehicle purchases, renovation of laboratories, herbaria and a medical clinic, and improvements to a community-managed ecotourism lodge.

Early Lessons from the ICBGs

While the program is still too young to permit an evaluation of the effectiveness of bioprospecting for integrated drug discovery, conservation, and economic development, some valuable lessons can be learned from the experience that the ICBGs have gained in establishing their partnerships.

Benefits-Sharing

Sharing the benefits of the process and product of drug discovery research with source country partners and communities is one of the central elements of the ICBG program. The types of benefits, the identities of the beneficiaries, and the development of benefit-sharing agreements are complex issues that the ICBGs are attempting to address. I will provide a short summary of some early lessons in this important area. More in-depth analyses can be found in Rubin and Fish (1994), Iwu (1996), Grifo and Downes (1996), and Rosenthal (in press).

The types of benefits that flow to source countries and communities from ICBG research and development include monetary and nonmonetary benefits. It is important both to provide benefits in the near-term to help address acute needs and provide for long-term benefits that will help source countries develop options for resource use, while improving quality of life in a lasting way. Specifically, the types of benefits include advance payments, royalty earnings in the event of a commercialized product, capacity-building efforts, research on priority diseases or regions, and the establishment of collaborative relationships with long-term potential.

Advance payments are increasingly considered critical by source countries and communities. These offer a very attractive balance to the risk and delays associated with royalty earnings from bioprospecting ventures, and in some cases may result in a trade-off with royalty rates. In several of the ICBGs, source country partners have received advance payments. To date, these payments have been used primarily to establish trust funds for the disbursement of small community grants for development projects such as medicinal plant cultivation and marketing, for tool purchases, written educational materials, shaman apprenticeship programs, and travel and workshops to build alliances among local community leaders.

Royalty earnings, usually a percentage of income from a commercialized product, are often the initial focus in a discussion of benefits. The four ICBGs that have a commercial licensing option with a pharmaceutical partner all have specified terms or ranges for royalty earnings in their research and benefit-sharing agreements. The division of that royalty among the stakeholders is stipulated either in the same agreements or in associated agreements.

Royalty negotiations should always take place in the context of other elements of the agreements. Examples of such other elements are advance payments and periods of sample use exclusivity. The timing of the negotiations may also be important. Final negotiation of a specific rate after a strong product candidate has been identified may net the source country the best earnings. Whenever possible all parties should have expert technical, commercial, and legal counsel to draw upon during the discussions.

In addition to monetary rewards, partnerships can provide opportunities for numerous nonmonetary benefits that in some cases may have a greater overall impact. As discussed above, capacity-building through training, equipment transfers, and the development of infrastructure to carry on biomedical research and to manage natural resources are fundamentally important benefits to be gained from bioprospecting partnerships. The ICBGs are also focusing research on diseases or geographical regions of importance to the source country, and in so doing are providing a benefit that addresses local priorities. Another important benefit is the development of long-term collaborations among workers and organizations both within and between countries (Iwu 1996).

The identification of appropriate beneficiaries is one of the most complex and important issues that the ICBGs have encountered. Individuals and communities, nongovernmental organizations (NGOs), and governmental organizations are identified variously in the ICBGs as potential beneficiaries for their efforts. The choices are guided by conservation and development objectives, fairness, local laws, concepts of intellectual property, and degree of participation.

Identification of appropriate and representative indigenous organizations has been a major challenge for the Peru ICBG. Amazonian societies are in a continuing process of political reorganization and realignment (Brown 1993). Overlay of this fluid political process on the profoundly difficult task of defining ownership of the intangible cultural resources represented in medicinal plant knowledge (Brush 1994) results in an extraordinarily complex job in identifying appropriate community beneficiaries. One useful tool is the establishment of a trust fund to disburse near- and long-term financial proceeds to worthy projects initiated by community members (see Moran, Chapter 11). The benefits-sharing plan currently being developed by the Peru ICBG involves management of and access to its trust fund initially by the participating organizations and includes a mechanism by which all linguistically and geographically related communities can achieve access over the long-term.

Informed Consent, Negotiation of Agreements, and Consensus-Building

While the principle of prior disclosure and informed consent is widely considered to be of fundamental importance today (Convention on Biological Diversity; Cunningham 1993; Missouri Botanical Garden 1996; Grifo and Downes 1996), the application of this principle to use of the tangible and intangible resources involved in bioprospecting is a relatively recent development, and a consensus on its meaning is as yet unclear. The concept of informed consent was originally conceived with regard to protection of the personal safety of human subjects participating in medical research (NCPHS 1979). Until recent years, informed consent in ethno-biological research was generally interpreted to mean verbal disclosure to the individual regarding the potential uses of his or her knowledge. Today, the possibility that financial benefits may result from this research necessitates complex arrangements with source communities and a very thorough information sharing process. Beyond in-depth discussions with local resource providers, this process can include sharing related contracts and facilitating legal advice during negotiations. It frequently also requires sharing project descriptions, lists of collections, and progress reports on research for review by the individuals, participating organizations, and national government authorities.

The ICBG program principles require that disclosure to source country participants be as complete as possible (Grifo and Downes 1996). The specifics of this disclosure process vary according the cultural and economic context as well as the type of arrangements and resources involved. The process frequently involves all the information described above. Complete disclosure is frequently complicated by the fact that indigenous partners, researchers, and industrial partners all may have relevant information that they consider proprietary. This information usually consists of plant species names, localities, and uses, specific bioassay techniques and chemical compositions, and financial terms in contracts. Such information is usually treated as confidential and is shared regularly only with negotiating and signing partners to written agreements. When outside parties, such as authorizing government officials, require such proprietary information, they are frequently asked to sign confidentiality agreements to protect the interests of the partners.

A thorough disclosure and education process leading up to negotiation of agreements not only empowers local communities in the current partnership, but also trains them to directly evaluate the future themselves and to negotiate on their own terms. To date, NGOs and source country universities have played a major role in mediating the negotiation process in most of the ICBGs. In addition, efforts have been made to facilitate representation of the interests of indigenous partners by qualified legal counsel. The Perú ICBG is currently renegotiating its agreements in a precedent-setting fashion, with more direct discussions between indigenous organizations and commercial partners than, to our knowledge, has been the case in other bioprospecting partnerships.

The principle of disclosure and informed consent, broadly interpreted, can also be a valuable tool for building consensus among stakeholders. Public workshops involving representatives from potential indigenous collaborators, government agencies, environmental NGOs, and researchers can simultaneously provide information, solicit input, and build consensus for the primary objectives of a project. Different versions of such workshops have now been held in all of the ICBGs. Ideally, workshops should take place prior to contract negotiation, allowing time for representatives to consult with their communities. Finally, consent from each and every individual that shares ethnomedical knowledge with ICBG researchers is obtained during the research phase of the projects. This last step also facilitates individual understanding and support for the project goals. Individuals realize that while much of their knowledge is communally derived, their participation and the biodiversity they depend on for their health (and in many cases their livelihood) have worldwide importance and potential financial benefits.

Established and Ongoing Presence in Source Communities

An important prerequisite for effective design and implementation of integrated conservation and development projects is an established and ongoing presence of project-affiliated individuals and organizations in local communities (Brown and Wyckoff-Baird 1992). Such organizations and individuals can provide a direct link between the local and international components of the project. Conservation International–Suriname (Suriname ICBG) and the Bioresources Development and Conservation Programme (Africa ICBG) offer useful models in this regard. These are organizations with community-based approaches to conservation. They are staffed and managed by source country nationals, but have strong ties to international partners.

A previously established organization run by source country nationals is most likely to understand national and local health, biodiversity, and development priorities. Furthermore, it may have credibility among the local community members and can help them to formulate their economic, cultural, and other concerns into specific terms during the design, research, and development phases of the project. Such an organization can also communicate the objectives and concerns of the other partners to the local participants. Bioprospecting researchers, usually taxonomists, ecologists, chemists, and physicians, may not adequately fulfill the community role, even when they are source country nationals. The professional interests and obligations of researchers, and often their cultural backgrounds, tend to restrict them to relatively brief, infrequent, and focused stays in the communities.

Lastly, a focus on defined areas and communities is critical to progress toward conservation and development objectives. The Latin America ICBG, working simultaneously in Chile, Argentina, and Mexico, has an outstanding drug discovery program. However, because of the broad geographical focus and the distance of the U.S.-based conservation and development team from the region, the design and implementation of adequate conservation and development activities have proceeded much more slowly. The group recently held workshops in Buenos Aires and Santiago to help focus these activities and has reorganized its conservation and development efforts under the direction of local institutions.

Source Country Infrastructure and Legal Environment

To take advantage of the possibilities that bioprospecting offers it will be extremely important for governments to facilitate development of institutional capacity and define access legislation. Minimally, governments should define which institution has the authority to issue collection permits and what, if any, its role is to be in the scientific and commercial aspects of research and development. This is taking place in many countries as they work to implement the Convention on Biological Diversity and define their policies with regard to international trade in genetic resources.

The Instituto Nacional de Biodiversidad (INBio) of Costa Rica has established a leading role for source country institutions in the field of bioprospecting due in large part to the outstanding scientific support and well-defined policies of the institution and the national government. These advantages make it a desirable partner to international researchers and pharmaceutical companies, and a promising associate in competitive grant applications. INBio's competitive position is reflected in its association with one of the five ICBGs and two of the recent BOA small grants (discussed in the next section). INBio may not be an appropriate model for all source countries, however, because it works only in government-controlled national parks that do not have large indigenous populations. However, its early successes highlight the importance of a relatively streamlined and stable bureaucracy, as well as technical and commercial sophistication. When international partners seek to carry on bioprospecting work in Costa Rica they most often work directly with INBio, which coordinates the technical details of sample collection and identification where necessary, as well as the contracts and collection permits.

Optimum legislation reduces the uncertainty of scientific and commercial collaborations and stimulates innovative approaches to integrated research and development projects. Legislation that guarantees that adequate benefits accrue to source country partners will facilitate capture of the opportunities that biodiversity prospecting offers.

The ICBG program is currently affecting policy and legislative events in many countries. The governments of Suriname, Perú, Chile, Argentina, and Cameroon are directly utilizing their ICBG experience during the formation of their own resource management and genetic resource policy development. The program also provides a key example of efforts by the U.S. Government to comply with the principles of the United Nations Convention on Biological Diversity even before the treaty has been ratified by the U.S. Senate. In addition, the program has been a case-study for the Organization for Economic Cooperation and Development (OECD) Working Group on Economic Incentives for Conservation [Rosenthal (in press)].

A Daughter Program—the Bioprospecting Opportunity Awards (BOA)

Because of great demand for funding to carry out integrated bioprospecting and conservation research, the Fogarty International Center, in collaboration with the National Science Foundation (NSF), initiated a new small grants program modeled in part on the ICBGs. Following peer review of competitive applications by NSF, four Bioprospecting Opportunity Awards (BOAs) were made in September 1995 to NSF grantees whose biodiversity studies could logically and efficiently be extended to include a chemical screening component. Funded projects include a study of tropical tree chemistry in Costa Rica as a source of environmentally friendly insecticides, a study of macro-fungi in Costa Rica for pharmaceutical potential; screening of tropical trees for pharmaceutical and agricultural potential in Panama; and one focused on deep-sea thermal vent organisms as sources of novel pharmaceutical and industrial agents. It is not known at this time if funding for a second round of BOAs will be available.

ICBG Funding

In fiscal year 1995, total U.S. Government funding for the ICBG program from the three federal agencies (NIH, NSF, and USAID) was $2.3 million. The NIH contribution was $1.75 million, including contributions from the Fogarty International Center, the National Cancer Institute, the National Institute of Allergy and Infectious Diseases, and the National Institute of Mental Health. The NSF contributes one-half million dollars per year, as has USAID until this past year. Due to shifting priorities and diminishing resources USAID reduced its contribution to the ICBG program this past year, and future funding from that agency is in doubt. Groups are likely to receive smaller awards in the remaining two years of this cycle. Contributions to Groups from their private pharmaceutical partners to date exceeds $400,000 of in kind investment in screening and chemistry, and at least $200,000 in advance payments and equipment donations.

Conclusion

The International Cooperative Biodiversity Groups program was designed to catalyze and explore the widely cited potential of bioprospecting as a tool for drug discovery, conservation of biodiversity, and economic development. In its first two years of development, significant advances have been made toward scientific goals of drug discovery and basic chemistry and biodiversity inventory. Source country capacity-building has proceeded rapidly, and near-term financial benefits to communities have begun to flow.

Valuable lessons can be learned from its earliest efforts. The ICBGs have provided models for appropriate structuring of source-country benefits. The experiences have demonstrated the principled and pragmatic value of broad application of information disclosure and consent procedures. Furthermore, experiences have highlighted the importance of an established presence in source communities, as well as the competitive advantage of source countries with established research infrastructure and a clearly identified body with legal authority.

Clear guidelines by national governments and international bodies as well as additional public funds are important to the future of bioprospecting. Each of the ICBGs is challenged to develop its program in an uncertain, changing legal and political context. It is in large part the security of five years of ICBG funding that has offered the opportunity to explore scientific, programmatic, and commercial alternatives of biodiversity prospecting in its infancy. Governmental and other nonprofit sources of funding may continue to be important to obtain the full range of conservation and development benefits that bioprospecting offers. As the Convention on Biological Diversity is implemented around the world, and as working models are elaborated in this new field, the relative success of various approaches to biodiversity prospecting will emerge. In this context the ICBGs play a critical role in providing ambitious working models to help guide policy and program directions in the coming years.

References

Baker, J., R. Borris, B. Carte, G. Cordell, D. Soejarto, G. Cragg, M. Gupta, M. Iwu, D. Madulid, and V. Tyler. (1995). Natural Product Drug Discovery and Development: New Perspectives on International Collaboration. *Journal of Natural Products* 58: 1325–1357.

Brown, M., and B. Wyckoff-Baird. (1992). *Designing Integrated Conservation and Development Projects*. The Biodiversity Support Program—World Wildlife Fund, The Nature Conservancy, World Resources Institute. Landover, Maryland: Corporate Press, Inc.

Brown, M. F. (1993). Facing the State, Facing the World: Amazonia's Native Leaders and the New Politics of Identity. *L'Homme 126-128* XXXIII:307–326.

Brush, S. B. (1994). A Nonmarket Approach to Protecting Biological Resources. In: *Intellectual Property Rights for Indigenous Peoples: A Source Book* (T. Greaves, ed.), pp. 131–143. Oklahoma City: Society for Applied Anthropology.

Conte, L. A. (1996). Shaman Pharmaceuticals' Approach to Drug Development. In: *Medicinal Resources of the Tropical Forest: Biodiversity and Its Importance to Human Health* (M. J. Balick, E. Elisabetsky, and S. A. Laird, eds.), pp. 94–100. New York: Columbia University Press.

Cox, P. A., and M. J. Balick. (1994). The Ethnobotanical Approach to Drug Discovery. *Scientific American* June:82–87.

Cunningham, A. (1993). *Ethics, Ethnobiologal Research and Biodiversity*. Gland, Switzerland: World Wildlife Fund for Nature.

Eisner, T. (1989). Prospecting for Nature's Chemical Riches. *Issues in Science & Technology*, Winter:31–34.

Eisner, T., and E. A. Beiring. (1994). Biotic Exploration Fund—Protecting Biodiversity Through Chemical Prospecting. *BioScience* 44:95–98.

Endangered Species Coalition. (1995). *The Endangered Species Act: A Commitment Worth Keeping*. Washington, D.C.: The Wilderness Society.

Goering, L. (1995). Rainforests May Offer a New Miracle. *Chicago Tribune*, September 12.

Grifo, F.T. (1996). The Role of Chemical Prospecting in Sustainable Development. In: *Emerging Connections Among Biodiversity, Biotechnology, and Sustainable Development in Health and Agriculture* (Julie Feinsilver, ed.). Washington, D.C.: Pan American Health Organization.

Grifo, F., and D. Downes. (1996). Agreements to Collect Biodiversity for Pharmaceutical Research: Major Issues and Proposed Principles. In: *Valuing Local Knowledge: Indigenous People and Intellectual Property Rights* (S. Brush and D. Stabinsky, eds.), pp. 281–303. Washington, D.C.: Island Press.

Greaves, T. (1994). IPR, A Current Survey. In: *Intellectual Property Rights for Indigenous Peoples* (T. Greaves, ed.), pp. 1–17. Oklahoma City: Society for Applied Anthropology.

Iwu, M. (1996). Implementing the Biodiversity Treaty: How to Make International Cooperative Agreements Work. *Trends in Biotechnology* 14:78–83

Juma, C. (1993). Policy Options for Scientific and Technological Capacity Building. In: *Biodiversity Prospecting* (W.V. Reid, S. A. Laird, C. A. Meyer, R. Gamez, A. Sittenfeld, D. H. Janzen, M. A. Gollin, and C. Juma, eds.), pp. 199–222. Washington, D.C.: World Resources Institute.

Laird, S. (1994). Natural Products and the Commercialization of Traditional Knowledge. In: *Intellectual Property Rights for Indigenous Peoples: A Source Book* (T. Greaves, ed.), pp. 145–162. Oklahoma City: Society for Applied Anthropology.

Lewis, W. H., and M.P. Elvin-Lewis. (1995). Medicinal Plants as Sources of New Therapeutics. *Annals of the Missouri Botanical Garden* 82:16–24.

Mays, T. D., and K.D. Mazan. (1996). Legal Issues in Sharing the Benefits of Biodiversity Prospecting. *Journal of Ethnopharmacology* 51:93–109.

Missouri Botanical Garden. (1996). Natural Products Research Policy. Unpublished policy available upon request from MBG, St. Louis, MO.

NCPHS. (1979). The Belmont Report: Ethical Principles and Guidelines for the Protection of Human Subjects of Research. Research, National Commission for the Protection of Human Subjects in Biomedical and Behavioral Research, Washington D.C.

Reid, W.V., S.A. Laird, C.A. Meyer, R. Gamez, A. Sittenfeld, D. Janzen, M.A. Gollin, and C. Juma. (1993). *Biodiversity Prospecting.* Washington, D.C.: World Resources Institute.

RFA—Request for Applications (1992). International Cooperative Biodiversity Groups. NIH, NSF, USAID TW-92-01. Washington, D.C.: Government Printing Office.

Rosenthal, J. P. (in press). Equitable Sharing of Biodiversity Benefits: Agreements on Genetic Resources. *OECD International Conference on Biodiversity Incentive Measures.* Cairns, Australia.

Rubin, S., and S. Fish. (1994). Biodiversity Prospecting: Using Innovative Contractual Provisions to Foster Ethnobotanical Knowledge, Technology, and Conservation. *Colorado Journal of International Evironmental Law & Policy* 5:23–58.

Schweitzer, J., G. Handley, J. Edwards, F. Harris, M. Grever, S. Schepartz, G. Cragg, K. Snader, and A. Bhat. (1991). Summary of the Workshop on Drug Development, Biological Diversity and Economic Growth. *Journal of the National Cancer Institute* 83:1294–1298.

Ten Kate, K. (1995). Biopiracy or Green Petroleum? Expectations & Best Practice in Bioprospecting. Overseas Development Admistration, London.

CHAPTER 14

Causes and Consequences of Biodiversity Loss: Liquidation of Natural Capital and Biodiversity Resource Development in Costa Rica

DANIEL H. JANZEN

Editor's Note: National development of biodiversity to prevent its loss is the subject of this chapter. Costa Rica, the location of some of the best known efforts to conserve biodiversity, is the case study used to illustrate this approach. While the following essay is not specifically directed toward human health, the discussion is relevant to all relationships between humans and the biosphere. Destruction of biodiversity represents tacit acceptance of the notion that any and all long-term consequences of liquidating our natural capital, including loss of potential drugs, unforeseen epidemiological effects, diminished oxygen production, etc., are necessary to meet our near-term needs. Janzen provides us with an alternative model that treats biodiversity as an investment whose interest earnings can simultaneously be maximized for our near- and long-term benefit, and he calls upon us to examine our relationship with biodiversity in this light.

Introduction

The cause of tropical biodiversity loss, by and large, is the very human behavior of liquidating natural capital to generate cash to finance human desires. Some of this cash is invested in the resulting agroscape. Thirty percent of Costa Rica is currently in marginal pasture generated by liquidating its forest capital. A Costa Rican beef cow is effectively a 32-

month "certificate of deposit" with a high penalty for early redemption, earning an unpredictable and declining interest rate in a market suffering 10–30% national inflation and ever rising wages. It isn't surprising, then, that Costa Rica is eagerly seeking the financial and technical means to convert this portion of the agroscape to a multitude of different kinds of woody crops—ranging from fruit trees for fruit juice to biodiversity-rich wildlands for ecotourism, and from designer drugs to sequestered carbon where Costa Rica acts as a green scrubber for carbon emitters.

The consequence of tropical biodiversity loss is that today we ask how wildland biodiversity can be sustainably developed, and thus become a major yet nondamageable resource for national development. If we fail, wildland's destiny is to become yet more conventional agroscape, no matter how low the interest rate on this or that hectare as agroscape.

We need to move away from the situation where biodiversity survival is the outcome of the current global war between the liquidators of biodiversity capital, and those who care about the survival of wildland biodiversity (for a wide variety of reasons). This essay is based on the philosophy that good fences make good neighbors and that sustainable human societies are rife with negotiated settlements.

My example comes from Costa Rica, a small Mediterranean country lying somewhere between Spain, Italy, and Germany. The philosophy results from a lifetime of watching animals and plants, and their interactions, on the tropical countryside. The ideas displayed here have evolved through a succession of contemplations of the perplexing interdigitation of tropical development with the sustainable development of tropical wildlands (e.g., Cotterill 1995; Janzen 1973, 1984, 1991, 1992, 1993, 1994a, 1994b, 1995, 1996; Janzen and Hallwachs 1994a, 1994b; Langreth 1994; Reid et al. 1993; Stone 1994).

Land Use Economics

Costa Rica is the size of West Virginia. Every Costa Rican generally wants the same things that members of developed countries want—good schools and universities, good hospitals and doctors nearby, nice highways, safe streets, enough healthy food, quality jobs, control of one's destiny, etc. But the per capita GNP of the 3.3 million people in Costa Rica is only about 10% of that of developed countries. That is to say, even with very equitable distribution of resources, Costa Rica's national capital is currently generating only enough "interest" to support about 300,000 people as if they were in a developed country. Either the number of people declines, the national capital and/or its interest rates increase, or both.

Costa Rica has two basic categories of capital: land and people. Land capital, for most practical purposes, is of two kinds. One kind is the agroscape. Humans have more than 10,000 years of experience and experiments in how to manage it, sustainably or otherwise. The agroscape is one of the two major publications in *Homo sapiens'* Curriculum Vitae. Quite frankly, this green factory is scraping its productivity ceiling in Costa Rica, if its full management costs are internalized. The other kind of land capital is wildlands that are being conserved for their ability to generate interest income as ecosystem and biodiversity goods and services. This major publication is still an extremely rough draft manuscript. Costa Rica has only just begun to examine how to sustainably shepherd her wild natural capital into sustaining a significant interest rate. This may be described as "use it without damaging it and without converting it to marginal agroscape." The name of the game is "exceed the opportunity costs" created by leaving it as wildlands conserved for nondamaging use.

Both deliberately and fortuitously, Costa Rica has come to a spoken and unspoken contract with itself as to the dynamic of the interaction of these two kinds of land use. About three-quarters of the country is destined to be agroscape (including urban zones), deliberately managed for quality in the production of conventional domestic goods and services, and the quality of life for its occupants. Given Costa Rica's generally poor soils and small area, Costa Rica's real future in the agroscape lies in development of human capital and applying what is known to the goal of national sustainability.

About one-quarter of the country is now conserved (or is to be conserved) for and through wildland biodiversity use. This new kind of "crop" is grouped into eight rather large and somewhat fragmented sections of state-owned lands termed *Conservation Areas*. The actual and anticipated land use here is the sustainable and nondamaging (non–capital-depleting) production of interest income from the production of the wildland "crop." Clearly, the real future also lies in development of human capital, and developing wild biodiversity and the ecosystems that contain it.

Biodiversity Development

Conserved biodiverse wildlands have enormous potential as a resource for sustainable development. However, to achieve this potential (as with its agroscape, health, education, businesses, and other traditional sectors) Costa Rica must undertake a three-step process including:

(1) Designate the raw materials to be used for this end. This has been nearly completed with the past ten years of effort to consolidate the Conservation Areas (National System of Conservation Areas under the Ministry of Natural Resources and Energy). They are about 25% of the country and contain at least 90% of its biodiversity and sustainable fragments of nearly all of its ecosystems.

(2) Determine what is "in the warehouse and what the factory looks like." This means an inventory of ecosystems and their biodiversity in the Conservation Areas—what the species are, where they are, how to get them to hand or eye, and what they do, and nestle this information into computerized GIS-based knowledge bases that can be formatted and reformatted into a wide variety of biocatalogs and roadmaps for users spread across society.

(3) Manufacture and market the products. Calculated and focused action is required to generate biodiversity goods and biodiversity information products, advertise these products, develop markets for these products, and channel the income into more human capital and more natural capital in both the wildlands and the agroscape.

Steps two and three are highly interactive and need to operate in parallel on many fronts. In other words, the wildland biodiversity crop, like the agroscape, requires all those processes that are found in any developed sector in a developed country—planning, investment, administration, zoning, legislation, taxes; inventors, managers, entrepreneurs; research, market development, pricing; local, national and regional coordination, etc.

The rudiments of these processes are already minimally present as a largely serendipitous and evolved mosaic of familiar wildland actions—ecotourism, biodiversity prospecting for drugs, education programs, environmental monitoring, novel crops, sustainable forestry, carbon storage, biological control of pests, green certification, water production and management, management of human diseases, etc. However, each set of practitioners is largely pursuing its own agendas, hunting and gathering in a manner not really much different than our ancestors. Each process has its own long history of environmental train wrecks clashing with this or that subsector of society as ecosystem and biodiversity capital is snatched up for liquidation, under the "finders keepers" rule of thumb for agroscape development.

These are only the preliminary steps in discovering what biodiverse wildlands can offer. But by recognizing wildland biodiversity as another kind of crop (i.e., a legitimate and society-supporting land use), we can take it off the battlefield and put it in the stock market. A large sector of wildland biodiversity and its associated ecosystems can potentially become a coordinated generic land use, with all the economies of scale, commonalities of interest, and coordination of impacts that are expected in a well-developed agroscape or urban center.

Biodiversity Inventory

As mentioned earlier, Costa Rica has determined that 25% of its national territory will be dedicated wildlands that are conserved through and for nondamaging sustainable use. In other words, manage the capital, live off the interest is the only real definition of "sustainable." Here, the word "inventory" also takes on a new meaning. It is a necessary part of the creation of the protocols and information for sustained wildland use. As a result, Costa Rica's Instituto Nacional de Biodiversidad (INBio) is conducting an All Taxa Biodiversity Inventory (ATBI) of the 120,000-hectare Guanacaste Conservation Area (GCA) in northwestern Costa Rica.

This $90 million seven-year investment in Costa Rica's future will formally begin in late 1996 with support from sources ranging from multilateral development banks to local parataxonomists. INBio's ATBI of the GCA will determine, for an estimated 235,000 species, ranging from viruses to beetles to huge trees,

- what species are there and how to tell them apart,
- where they are,
- how to get them to hand or eye when desired, and
- what is their natural history.

The ever-growing knowledge bases will be computerized in the public domain on the Internet. This information will be manipulatable into formats for all sectors of society. Natural history, taxonomy, and even butterfly nets can become essential infrastructure in a major national sustainable industry. The myriad wildland biodiversity information that has been gathered through three decades (and for some groups of organisms, much longer) of national and international field biology in Costa Rica and much of the remainder of the Neotropics, supported by a huge number of sources for many different agendas, suddenly becomes applied as well as basic research of great importance to biodiversity development. Knowledge about biodiversity, and the ecosystems that contain it, becomes the tools and philosophy for its survival through nondamaging use. One has to know what a tool is for, and one has to know at least some of the properties of the raw materials, if one is to construct without damaging or wasting them.

Costa Rica is attempting to integrate its human and natural resources with international science and other parts of society to generate this biodiversity moonshot. The massive flow of specimens and information from the ATBI is being reorganized at INBio through computerization, and through the taxonomic logic long ago established by taxonomy and taxonomists—the international taxasphere. The team of national parataxonomists, university graduates in biodiversity resources, administrators, and advanced degrees work with all sectors of society—local, national, and global. It has not proven to be easy economically, institutionally, or sociologically to move from biodiversity as a "pretty picture on the wall" to biodiversity as national natural capital. However, the process is in motion and available for inspection—on-site on the World Wide Web at http://www.inbio.ac.cr.

This kind of biodiversity inventory, reverse-engineered from the user and the products back to the information management and discovery protocols, tries to offer something to everyone. This all-inclusiveness is one of the ATBI's many goals—to maximize involvement by many social sectors, many interested parties, and many parties who as yet do not even know they are interested. It is certainly not a stand-alone exercise. Portions of the taxasphere, a far larger industry, focus for a few years on this point of the globe, and move on (leaving in place continued taxonomic development of this biodiversity). The pharmaceutical industry pools its ATBI information and samples with the multitude of samples gathered more haphazardly from around the globe. The administrator from another tropical country visits, learns, inspires, and exports what of the process suits her. The ATBI is a mine canary. But its song also ripples far out over the tropical agroscape, bettering the life of an agriculturalist thousands of kilometers away. Indeed, a major reward of the ATBI to Costa Rica is in its being a "proof of concept" pilot project—in a tiny country whose specialty is to be a cutting edge and stimulus for tropical sustainable development.

Biodiversity Products

Many biodiversity products are already well-known, even if not yet developed past the "cottage industry" stage. Other members of this forum are

dwelling on the technology of some of them. I will focus briefly on overlap between the biodiversity industry and the climate change industry.
These two major industries have been developing in parallel for years, to
such a degree that we even have two quite separate global conventions for
them. They often overlap in the area of the (quite legitimate) concern that
climate change will wreck havoc with conserved wildlands because wild
ecosystem islands cannot follow moving climates across the landscape.
However, there are at least two other relatively unexplored and much more
positive areas of overlap between biodiversity and climate change.

Risk in Carbon Sequestration

A major discussion today centers on carbon sequestration through forest regeneration. The investing industry—the greenhouse-gas-generator—invests
in a biological carbon scrubber (a.k.a. regenerating forest). The question
then becomes "What is the nature of my fire department?" That is to say,
where is the social process that guarantees that the carbon being sequestered
in today's regenerating forest will still be there as stored carbon a half a century hence, and not liberated (again) by wildfires or forest clearing?

Biodiversity development can be a highly effective security for that carbon. In short, the array of wildland biodiversity industries and their em-
ployees that depend on that forest—ecotourism, parataxonomists, environmental monitoring, education programs, water factories, gene bankers,
medical researchers, etc.—have a vested interest in seeing that the forest
persists into perpetuity (quite irrespective of its stored carbon). For them,
developed biodiversity products and ecosystem services are "quality carbon," some bits of which may retail at a substantially higher price per gram
than the carbon-generating industry was required to pay per ton of scrub-
bing services. One industry's waste may become another industry's raw
materials. From the national viewpoint, biodiversity and ecosystem development are value-added to the carbon sequestration process, but to the
carbon sequestor, the biodiversity managers are acting as the police department, fire department, and Ministry of Public Works.

Fine-Tuning the Carbon Crop

The value of the carbon crop—tons of sequestered carbon per year per
hectare—dances to a complex market of international trends, legislation,
conventions, supply–demand, ideologies, national needs, etc. For any given

hectare, there will be an optimal carbon sequestration rate and pattern in the context of these externalities. Moving toward this optimum, however, means matching the biodiversity properties of the carbon sequesters—the trees, herbs, fungi, herbivores—with that hectare's soil, climate, human resources, and culture. The information for this match is biodiversity information—a host of small to large natural history details about how organisms store carbon and interact to generate forests of various carbon sequestration characteristics. For example, a forest can be bioengineered to store a smaller total amount of carbon (standing equilibrium biomass) very fast, or to store a much larger total amount of carbon quite slowly. And the recipe varies with factors such as soil type, climate, biological threats, social threats, and secondary users. But these fine-tunings cannot be realized in the absence of varied species and natural history knowledge about those species.

In other words, quite aside from the use of sequestered carbon for biodiversity development in conserved wildlands, there should be a seeking of biodiversity and ecosystem information by those interested in carbon sequestration per se, as part of their efforts to bring the carbon crop to market on time in the right packages for the right consumer. Corn and rice farmers have done the same for millennia.

Conclusion

Product development, integrated industries, market penetration, green certification, trade, energy efficiency, sustainable consumption, decentralization, and many more processes commonplace in a "developed" society are all part of wildland biodiversity development. Putting on these glasses leads to pointed comments.

Perhaps there should be a little cotton tax, and its income should be used to protect the wild strains of cotton, the habitats of wild cotton, the wild strains of boll weevils, the wild strains of the things that eat boll weevils, and the wild biodiversity with genes in it that might be used to make a better cotton factory—otherwise known as the cotton plant. A one cent per cup tax on each cup of coffee sold worldwide would endow the basic management of all tropical wildlands, forever. And the industry would never have noticed this tax if it had been, from the beginning, part of the production cost of converting tropical biodiversity to breakfast coffee.

The "endangered species concept" effectively disappears in a world that is organized around two accepted land uses—that of the agroscape, and wildlands conserved for their ecosystem and biodiversity goods and services. The species that survive in the (necessarily large) conserved wildlands, which should be carefully positioned so as to contain the great majority of a nation's (or major region's) biodiversity, are by definition not endangered since habitat is conserved by designation for that land use. Wild species that survive in the agroscape are among the many tools of wise agroscape management, but the survival of each species per se is not a priority of the agroscape, and many species will likely disappear from it. This heretical view entails paying 10% of a nation's biodiversity to insure that 90% survives into perpetuity. If someone decides that some hapless species living only in the agroscape is to be guaranteed survival, then society purchases its habitat and adds it to the conserved wildlands, rather than decreeing it to be endangered and therefore silently levying a form of omnipresent biodiversity zoning on the agroscape wherever it happens to occur. Here, conserved wildlands are biodiversity that is conserved for nondamaging use, while biodiversity in the agroscape is more a tool—biological control agents, new varieties, pesticide monitors, new genes, soil managers, etc. This is the route that Costa Rica started down somewhat serendipitously and is now following explicitly.

National development of wildland biodiversity is an international industry. The great majority of tropical species have distributions extending across many nations, or in the case of very large nations, across many nation-sized provinces. This means that information gathered about a species at one place is significant to biodiversity managers in far distant places. And through the impressive inferential power of taxonomy and classification, information gathered about one species is information gathered about related species living as far away as the other side of the world. That is to say, what appears at first glance to be a very national act of wild biodiversity development, for example INBio's ATBI of the GCA in Costa Rica, is in fact a megaexample of south–south international collaboration.

Society is made up of peoples organized formally or informally into negotiated agreements of ownership. The globe is rapidly moving from frontier to nearly complete civilization. Biodiversity needs to be admitted into this society. Let's stop arguing over whether to eat our natural capital and get on with figuring out how to invest it in such a manner that no one would even dream of cashing the bond.

References

Cotterill, F.P.D. (1995). Systematics, Biological Knowledge and Environmental Conservation. *Biodiversity and Conservation* 4:183–205.

Janzen, D.H. (1973). Tropical Agroecosystems. These Habitats Are Misunderstood by the Temperate Zones, Mismanaged by the Tropics. *Science* 182:1212–1219.

———. (1984). The Most Coevolutionary Animal of Them All. Crafoord Lectures, pp. 2–20. Royal Swedish Academy of Sciences, Stockholm, Sweden.

———. (1991). How to Save Tropical Biodiversity. *American Entomologist* 37:159–171.

———. (1992). A South–North Perspective on Science in the Management, Use, and Economic Development of Biodiversity. In: *Conservation of Biodiversity for Sustainable Development* (O.T. Sandlund, K. Hindar, and A.H. D. Brown, eds.), pp. 27–52. Oslo: Scandinavian University Press, Oslo.

———. (1993). Taxonomy: Universal and Essential Infrastructure for Development and Management of Tropical Wildland Biodiversity. In: *Proceedings of the Norway/UNEP Expert Conference on Biodiversity, Trondheim, Norway* (O.T. Sandlund and P.J. Schei, eds.), pp. 100–113. Trondheim, Norway: NINA.

———. (1994a). Priorities in Tropical Biology. *Trends in Ecology and Evolution* 9:365–367.

———. (1994b). Wildland Biodiversity Management in the Tropics: Where Are We Now and Where Are We Going? *Vida Silvestre Neotropical* 3(1):3–15.

———. (1995). Neotropical Restoration Biology. *Vida Silvestre Neotropical* 4(1):3–9.

———. (1996). On the Importance of Systematic Biology in Biodiversity Development. *ASC Newsletter* 24:17, 23–28.

Janzen, D.H., and W. Hallwachs. (1994a). Ethical Aspects of the Impact of Humans on Biodiversity. In: *Man and His Environment. Tropical Forests and the Conservation of Species* (G. B. Marini-Bettolo, ed.). Pontificiae Academiae Scientiarum Scripta Varia 84:227–255.

———. (1994b). All Taxa Biodiversity Inventory (ATBI) of Terrestrial Systems. A Generic Protocol for Preparing Wildland Biodiversity for Nondamaging Use. Report of a NSF Workshop, 16–18 April 1993, Philadelphia, Pennsylvania, 132 pp. http://www.inbio.ac.cr/ATBI

Langreth, R. (1994). The World According to Dan Janzen. *Popular Science* 245(6):78–82,112–115.

Reid, W.V., S.A. Laird, R. Gámez, A. Sittenfeld, D.H. Janzen, M.A. Gollin, and G. Juma, eds. (1993). *Biodiversity Prospecting.* Washington, D.C.: World Resources Institute.

Stone, R. (1994). Counting Creatures Great and Small. *Science* 264:191.

CHAPTER 15

Sustainable Use of Biodiversity: Myths, Realities, and Potential

CHARLES M. PETERS

Introduction

A large percentage of the biodiversity used by human societies are non-timber resources such as fruits, nuts, oil seeds, latexes, resins, fibers, and medicinal plants. Most of these resources grow in tropical forests and, to date, most of them are exploited on a subsistence basis by indigenous, forest-dwelling communities. As recently as twenty years ago, the wild species comprising this enormous pool of useful biodiversity were called "minor" forest products and nobody really paid much attention to them. This situation, however, has changed dramatically in the last two decades.

Rainforest biodiversity is big news now and nontimber tropical forest products, or NTFPs as they are currently known, are big business. Little-known Amazonian fruits and nuts are made into ice creams, lotions, and shampoos, condoms are made from natural rubber, and large pharmaceutical companies are painstakingly analyzing native pharmacopeia in the search for new drugs. Indonesia exports about US$200 million of rattan products each year (Manokaran 1990), and over 50,000 metric tons of Brazil nuts are harvested annually from the forests of Brazil, Bolivia, and Peru for sale in international markets (LaFleur 1992).

Numerous projects are currently underway to promote the increased exploitation of nontimber tropical forest products. Much of this work focuses on finding new products (bioprospecting) or developing markets for existing products, implementing local processing and value-added strategies, and insuring an equitable distribution of the revenues generated by the sale of NTFPs. Securing land tenure and usufruct rights for local collector groups has also been an important component of the development of these resources. Clearly, there are good reasons for emphasizing these socioeconomic factors. If you want to collect forest products, you need access to the forest. If you want to sell these products, you need markets. If you want to stay in business, you need to capture as large a percentage of the final selling price as possible.

It is surprising, however, that the ecological factors associated with the exploitation of nontimber forest products have so rarely been addressed. Maintaining a reliable flow of benefits from the exploitation of biodiversity requires that the species on which this flow is based be maintained as well. If these species are depleted through overexploitation, destructive harvesting, or poor management, no new market, cottage industry, or land-tenure system will make very much difference. In the long run, ecology is probably the real bottom line here.

This chapter was written with three main objectives in mind. The first of these is to challenge the common assumption that the commercial harvesting of fruits, nuts, latexes, and medicinal plants has only minimal impact on a tropical forest. The second is to demonstrate how intensive harvesting of NTFPs can permanently affect the structure and regeneration of tropical plant populations. The final objective is to propose a series of steps that can be used to enhance the long-term sustainability of biodiversity exploitation. The discussion focuses primarily on nontimber plant resources,[1] with particular emphasis on tropical tree species in Southeast Asia and Amazonia.

Ecological Impacts of Forest Use: The Myth

Human beings have developed a variety of different ways to use tropical forests and each type of land-use carries with it a particular suite of ecological costs. Perhaps the most intensive and environmentally costly way to use a tropical forest is to cut it down, burn it, and plant something else on the site (e.g., timber trees, agricultural crops, or pasture grasses). The ecological impacts of forest conversion are immediate, highly visible, and, in most cases, very severe. Current research in tropical forests suggests that the most important of these impacts include:

- loss of biomass and species diversity
- release of CO_2 and other greenhouse gases

- disruption of nutrient and hydrological cycles
- soil loss through erosion
- increased local temperatures and decreased local rainfall

To put some of these consequences in perspective, a one-hectare tract of primary forest in the Brazilian Amazon may contain more than 200 tree species ($\geq$ 10 cm in diameter) and comprise an aboveground living biomass of about 300 tons/ha (Brown et al. 1995). Cutting and burning this forest would eliminate most of the biodiversity and would release approximately 150 tons of carbon per hectare in the form of carbon dioxide and other heat trapping gases (Keller et al. 1991). The removal of vegetative cover would increase water movement, soil erosion, and nutrient loss, decrease evapotranspiration and total ecosystem productivity (Jordan 1987), and potentially modify local climatic regimes because of the increased reflectance of solar radiation (Skukla et al. 1990). The site would be characterized by stumps, blackened tree trunks, and, depending on the topography, a growing network of eroding gullies. It would be obvious to the most casual observer that a major ecological disturbance had occurred.

Another common use of forests is to selectively cut and remove the boles of desirable timber trees. Although certainly less damaging than total forest conversion, selective logging is also known to produce a number of ecological repercussions. The most conspicuous of these are:

- loss of some plant and animal species
- damage to residual trees
- soil loss through erosion
- loss of nutrients through stem removal
- change in forest structure and increase in light levels

A major problem with selective logging in tropical forests is that the crowns of many large canopy trees are lashed to those of their neighbors by a profusion of vines, lianas, or climbers. When the timber trees are felled, other canopy species are also pulled down and the whole woody mass crashes through the lower canopy snapping tree boles, breaking branches, and flattening a considerable proportion of the forest understory. Harvesting a small number of stems can destroy up to 55% of the residual forest and seriously damage an additional 3.0 to 6.0% of the standing trees (Burgess 1971; Johns 1988). Associated impacts include soil compaction, decreased infiltration of water, increased rate of soil loss from erosion, disruption of local animal populations, increased susceptibility to fire (Uhl et al. 1988), and nutrient loss from the removal of sawlogs or pulpwood. Commercial tree felling produces a notable impact on a forest ecosystem, and the physical evidence of this disturbance is immediately apparent and persists in the form of logging roads, skid trails, and scattered tree stumps for many years.

A final form of forest use that has attracted a lot of attention in recent years involves the selective harvest of fruits, nuts, latexes, medicinal plants, and other nontimber resources. Although relatively benign when compared with forest clearing and selective logging, this activity also produces a number of ecological impacts including:

• gradual reduction in vigor of harvest plants
• decrease in rate of seedling establishment of harvest species
• potential disruption of local animal populations
• nutrient loss from harvested material

At first glance, these impacts seem relatively insignificant. The harvest of nontimber forest resources does not necessarily kill the plant,[2] compact the soil, increase erosion, or cause a notable change in the structure and function of the forest. A forest exploited for fruits and latex, unlike a logged-over forest, maintains the appearance of being undisturbed. It is easy to overlook the subtle impacts of NTFP harvest and to assume *a priori* that this activity is something that can be done repeatedly, year after year, on a sustainable basis. This ubiquitous idea, or some variant of it, has appeared in books, scientific papers, conference proceedings, grant proposals, magazines articles, newspaper stories, on television and radio shows, in the annual reports of private companies, and even on the back of cereal boxes and ice cream cartons. It is the basis for much of the current excitement over nontimber tropical forest resources. Unfortunately, in the great majority of cases, this assumption is patently incorrect.

Some Facts about Tropical Trees and Forests

Tropical forests exhibit several ecological characteristics which make the sustainable exploitation of nontimber resources a more difficult undertaking than it might first appear. One of the most fundamental and well-known features of these forests is their great species richness, or large number of plant species per unit area. To illustrate this point specifically for trees, floristic data collected from small tracts of tropical forest around the world are shown in Table 15.1. Although there is much variability from site to site, the results from these surveys show that tropical forests are extremely diverse and may contain from one hundred to over three hundred species of trees per hectare. In contrast, a mature northern hardwood forest in the eastern United States contains about ten to fifteen tree species per hectare (Braun 1950).

From a commercial standpoint, the high diversity of tropical forests is a mixed blessing. On the positive side, forests containing a large number of different species usually contain an equally diverse assortment of useful plant resources (i.e., species richness and resource richness are usually correlated). The great interest in tropical forests as a source of undiscovered pharmaceuticals, for example, is largely in response to the magnitude of the species pool in these ecosystems. Unfortunately, an additional correlate to high species diversity is that the individuals of a given species usually occur at very low densities. There is a limit to the total number of trees that can be packed into a hectare of tropical forest. If you have a large number of species, each species can only be represented by a few individuals.

This tendency of high species diversity coupled with low species density is illustrated in Figure 15.1 using inventory data collected from small tracts of tropical forest in Brazil and Sarawak. As shown in the histogram,

TABLE 15.1

Number of Tree Species (≥ 10 cm in diameter) Recorded in Small Tracts of Tropical Forest

Location	Sample Area (hectares)	Number of Species (≥10 cm DBH)	Source
Cuyabeno, Ecuador	1.0	307	Valencia et al., 1994
Mishana, Peru	1.0	289	Gentry, 1988
Lambir, Sarawak	1.6	283	Ashton, 1984
Bajo Calima, Colombia	1.0	252	Faber-Lagendoen and Gentry, 1991
Sungei Menyala, Malaysia	2.0	240	Manokaran and Kochummen, 1987
Wanariset, East Kalimantan	1.6	239	Kartawinata et al., 1981
Gunung Mulu, Sarawak	1.0	225	Procter et al., 1983
Rio Xingu, Brazil	1.0	162	Campbell et al., 1986
Barro Colorado, Panama	1.5	142	Lang and Knight, 1983
Oveng, Gabon	1.0	123	Reitsma, 1988

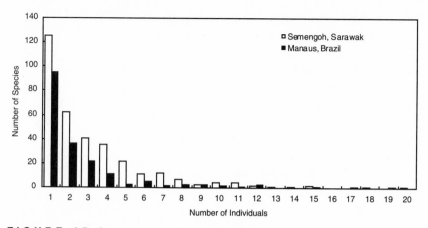

FIGURE 15.1

Densities of different tree species within small tracts of tropical forest. Inventory data from Semengoh, Sarawak, based on a 4.0 hectare sample plot (Ashton, 1984); Manaus, Brazil, data based on a 1.0 hectare sample plot (Prance et al., 1976).

the great majority of the species at each site are represented by only one or two trees; less than ten percent of the species exhibited densities greater than four trees/hectare. Although there may be an abundance of resources in tropical forests, most of them are scattered throughout the forests at extremely low densities. Low density resources are difficult for collectors to locate, they require lengthy travel times, produce a low-yield per unit area, and they are extremely susceptible to overexploitation. None of these are desirable traits in a forest resource.

A second characteristic of tropical trees that represents a stumbling block to sustainability concerns the way that they move their pollen and disperse their seeds. The low density and scattered distribution of individuals in many tropical tree populations greatly complicates the process of pollination. Given that the distance between conspecific individuals may be greater than 100 meters in some cases, moving pollen from the flowers of one tree to another can be a tricky proposition. Many tropical trees have overcome this problem by coevolving relationships with a variety of animals, ranging from tiny thrips and midges to bees and large bats, that act as long-distance pollen vectors. These relationships can be quite specific, with one type of insect being solely responsible for pollinating the flowers of a particular species, or even genus, of forest trees (e.g., Wiebes 1979). The use of biotic vectors to transfer pollen is apparently the norm in tropical forests, and recent studies in Costa Rica (Bawa et al. 1985) suggest that over 96% of the local tree species are pollinated exclusively by animals.

Animals also play a very important role in dispersing the seeds produced by tropical trees. Studies conducted in Rio Palenque, Ecuador (Gentry 1982), for example, have shown that 93% of the canopy trees produce fruit adapted for consumption by birds and mammals, while Croat (1978) estimates that 78% of the canopy trees and 87% of the subcanopy trees at Barro Colorado Island in Panama have animal-dispersed fruits. These animals may either remove fruit and seeds directly from the tree (primary dispersers), or they may forage on fruits that have already fallen to the ground and split open (secondary dispersers).

The important lesson here is that the production of fruits, seeds, and seedlings in tropical forests unavoidably involves the collaboration of animals. Although it is very easy to overlook this fact, or to view forest animals solely as pests that damage large quantities of fruit and compete with commercial collectors, sustainable resource use in tropical forests depends on the continual availability of pollinators and seed dispersers. Stated in simple terms, no pollination means no fruits, no fruits and no dispersers means no established seedlings, and no established seedlings means no next generation, no products, no profits—and no sustainability.

A final characteristic of many tropical tree species is that they have a very difficult time recruiting new seedlings into their populations. Even given abundant pollination, fruit set, and dispersal, there is still a very small probability that a seedling will become successfully established in the forest. The seed must avoid being eaten, it must encounter the appropriate light, soil moisture, and nutrient conditions for germination, and it must be able to germinate and grow faster than the seeds of all other species that are competing to establish themselves on the same site. The young seedling must then stay free of pathogens, be able to recuperate from the damage caused by herbivores, avoid falling branches and other hazards, and continue to photosynthesize and push its way upward into the forest canopy. Not surprisingly, mortality during the early stages of the life cycle of a tropical plant is extremely high.

A graphic example of the intense seedling mortality experienced by tropical trees is provided by the four survivorship curves shown in Figure 15.2. *Brosimum alicastrum* is a widely distributed canopy tree from the neotropics (Peters 1990a); *Shorea curtisii* and *Shorea multiflora* are dominant tree species in Southeast Asia (Turner 1990); and *Grias peruviana* is an abundant lower canopy tree in western Amazonia (Peters 1990b). As is illustrated in these histograms, seedling mortality in these four species during

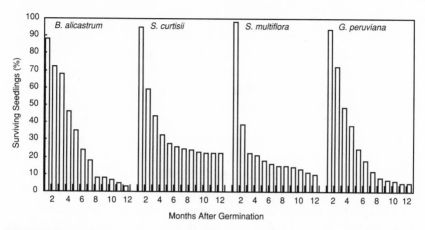

FIGURE 15.2

Seedling survivorship curves for *Brosimum alicastrum, Shorea curtisii, Shorea multi-flora,* and *Grias peruviana.* Histograms show the percentage of seedlings surviving during the first year following germination. Data for *B. alicastrum* were collected in Veracruz, Mexico (Peters 1990a), *S. curtisii* and *S. multiflora* were studied in Peninsula Malaysia (Turner 1990), and *G. peruviana* data were collected in Peruvian Amazonia (Peters 1990b).

the first twelve months following seedfall ranges from a high of 22% for *S. curtisii* to a low of 3% for *B. alicastrum.* Half-lives, or the time required to kill off 50% of the initial cohort, vary from two to five months. Taking into account seed predation and germination failure, less than 0.1% of the seeds produced by *B. alicastrum* become established seedlings. Only a very small fraction of these (approximately 1 in 1.5 million)[3] will ever make it to the canopy and start producing fruit. Data such as these, which are by no means atypical for tropical trees, provide perhaps the most convincing demonstration of how difficult it is for a species to maintain itself in the forest—even in the absence of any type of resource harvest.

The Reality of NTFP Harvest

Given the low density of tropical forest species, their reliance on animals for reproduction, and the difficulty experienced in establishing their seedlings, the harvest of any type of plant tissue will invariably have an effect on the target species. The delicate ecological balance maintained in a tropical forest is easily disrupted by human intervention, and extractive activities that at first glance appear benign can later have a severe impact on the structure and dynamics of forest tree populations. This impact may not be immediately visible to the untrained eye—but it is definitely occurring.

In general, the ecological impact of NTFP utilization depends on the nature and intensity of harvesting and on the particular species and type of resource being exploited. Sporadic collection of a few fruits or the periodic harvesting of leaves for cordage or chemical extraction may have little impact on the long-term stability of a tree population. Intensive, annual harvesting of a valuable market fruit or oil seed, on the other hand, can gradually eliminate a species from the forest. The felling of large adult trees can produce a similar ecological result in a much shorter period of time.

Although the fact is seldom mentioned, a large number of nontimber forest resources are actually harvested destructively. Uncontrolled felling for fruit collection has virtually eliminated the valuable aguaje palm (*Mauritia flexuosa*) from many parts of Peruvian lowlands (Vazquez and Gentry 1989). Destructive harvesting has also seriously reduced the local abundance of the ungurahui palm (*Jessenia bataua*), the babassu palm (*Orbignya phalerata*), and a wide variety of other important Amazonian fruit trees such as *Parahancornia peruviana, Couma macrocarpa,* and *Genipa americana* (Peters et al. 1989). Gharu trees (*Aquilaria malaccensis*) in Southeast Asia are routinely cut to harvest the resinous heartwood (Jessup and Peluso 1986), and the collection of "damar" from *Dipterocarpus* trees in Peninsula Malaysia involves hacking a large box in the trunk of the harvest tree and then building a fire inside this cavity to stimulate the flow of oleo-resin (Gianno 1990). There are numerous examples of species that are killed or fatally wounded by the harvest of nontimber, vegetative tissues such as rattan, palm heart, *Lonchocarpus* roots,[4] thatch, and an assortment of barks, stems, and leaves that are used medicinally (e.g., Cunningham and Mbenkum 1993).

Even in the absence of destructive harvesting, the collection of commercial quantities of fruit and seeds can still have a significant ecological impact. In terms of simple demographics, if a tree population produces 1,000 seeds and 95% of the new seedlings produced from these seeds die during the first year, the population has still recruited 50 new individuals. If, on the other hand, commercial harvesting removes all but 100 of these seeds from the site prior to germination, the maximum number of seedlings that can be recruited into the population is reduced to only 5. This tenfold shortfall in recruitment can cause a notable change in the structure of the population.

In reality, this example is probably overly optimistic. First, it is assumed that all of the seeds left in the forest are positioned in precisely the right spot for germination and early growth. Second, there is always the possibility that the fruits and seeds left in the forest will experience a rate of mortality that is higher than 95%. Commercial collectors, in effect, are competitors with forest frugivores, and their activities reduce the total supply of food resources available. In response to the reduced abundance of fruits and seeds, frugivores might be forced to increase their foraging to obtain sufficient food. The net result would be an increase in the total percentage of seeds destroyed.

All of these factors interact in a synergistic fashion to reduce the number of new seedlings that become established in a plant population. Over time, this lack of seedling establishment will alter the size–class distribution of the population being harvested. If commercial collection continues uncontrolled, the harvest species can be gradually eliminated from the forest.

This process of gradual population disintegration is illustrated in Figure 15.3 using demographic data for *Grias peruviana* and the results from computer simulations using a transition matrix model (Peters 1990b). Size classes 1 to 4 are based on height measurements, while classes 5 through 12 reflect a 5.0 cm DBH (diameter at breast height) interval. For the purpose of the simulation, the intensity of harvest was set at 85% of the total annual fruit production. Each time interval shown represents 20 years. Note the change of scale in the latter three time periods to compensate for the decrease in population size.

As is shown at time 0, the *G. peruviana* population initially displays the inverse J-shaped, or negative exponential,[5] size–class distribution of a shade tolerant canopy tree with abundant reproduction. After two decades of fruit collection, however, the structure of the population has changed notably. The infrequency of seedling establishment has caused a reduction in the smaller size classes. The greater number of stems in the intermediate size classes reflects the growth of saplings that were established prior to exploitation. By time 2, the population has been even further degraded by the chronic lack of regeneration. There are intermediate size classes that contain no individuals at all, and it appears that the existing level of saplings and poles is insufficient to restock these classes. Finally, the size–class histogram shown for time 3 represents the culmination of a long process of overexploitation. The population consists of only large, old adult trees, none of which are regenerating. In the absence of remedial action, it is only a matter of time before *G. peruviana* becomes locally extinct.

The important, and distressing, point to be gained from this simulation is that at no point during the process of overexploitation is there any dra-

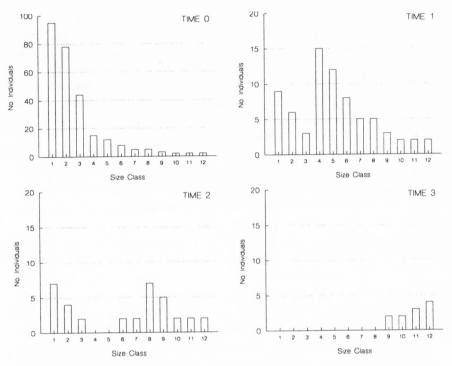

FIGURE 15.3

Simulated change in the population structure of *Grias peruviana* in response to excessive fruit collection. Results based on stepwise analyses using a transition matrix model and demographic data reported in Peters (1990b). Harvest intensity set at 85% of the total annual fruit production. Note change in scale in the latter three time periods to account for progressive decrease in population size.

matic visual evidence (e.g., dead or dying trees) that something is going wrong. Even during the latter stages, the forest still contains a considerable number of *G. peruviana* trees that are producing fruit. Harvesting would undoubtedly continue unabated until these adult trees begin to senesce, at which point collectors would be forced to move into a new area of forest in search of *Grias* fruits.

The example shown in Figure 15.3 represents an extreme case of uncontrolled overexploitation, and does not necessarily imply that every level of NTFP harvest leads directly to species extinction. The simulation is very useful, however, because it shows that even though the ecological impacts of this type of resource use are gradual, very subtle, and essentially invisible, in the long run they can be just as devastating as logging in terms of disrupting local plant populations and causing species extinctions.

Finally, in addition to its impact on seedling establishment and population structure, the collection of nontimber forest products can also affect the genetic composition of a plant population (Peters 1990c). A population of forest fruit trees, for example, will usually contain several individuals that produce large succulent fruits, a great number of individuals that produce fruits of intermediate size or quality, and a few individuals that produce fruits that, from a commercial standpoint, are inferior because of small size, bitter taste, or poor appearance.[6] If this population is subjected to intensive fruit collection, the "inferior" trees invariably will be the ones whose fruits and seeds are left in the forest to regenerate. Over time, the selective removal of only the best fruit types will result in a population dominated by trees of marginal economic value. This process, although more subtle and occurring over a longer period of time, is identical to the "high-grading" or "creaming" of the best tropical timbers that occurs in many logging operations.

Monitoring to Minimize Ecological Impact

The exploitation of nontimber forest products could provide innumerable social, economic, and ecological benefits. If practiced on a sustainable basis, the commercial harvest of these products is a unique way to generate revenues from species-rich forests and still conserve most of the biological diversity and ecosystem functions of the forest (e.g., protect soil fertility, prevent erosion, control run-off, regulate climate). No other form of land-use practiced in the tropics has the potential to do this.

Given the "boom and bust" cycles that have historically characterized the exploitation of nontimber forest products (Padoch 1988; Homma 1992), it seems unlikely that the interaction of markets, commercial collectors, and tropical forest species will automatically produce a sustainable form of resource use. Achieving this objective will require more than blind faith in the productive capacity of tropical trees and an unwavering trust in a free market system. Sustainable use of tropical forests will require a concerted management effort. It will require careful selection of species, resources, and sites. It will require controlled harvesting and periodic monitoring of the regeneration and growth of the species being exploited. More than anything, however, it will require a greater appreciation of the fact that ecology and forest management are the cornerstones of sustainability.

From an ecological standpoint, one of the most essential ingredients required to achieve a sustainable level of resource use is information—information about the density and distribution of resources within the forest,

information about the population structure and productivity of these resources, and information about the ecological impact of differing harvest
levels. An overall strategy for collecting this information, and for applying
it in such a way as to guarantee that the plant populations being exploited
will maintain themselves in the forest over time, is presented as a flow chart
in Figure 15.4. The overall concept and sequence of operations is adapted
from Peters (1994).

As is shown in Figure 15.4, the complete process comprises six basic
steps: (1) species selection, (2) forest inventory, (3) yield studies, (4) regeneration surveys, (5) harvest assessments, and (6) harvest adjustments. Taken
together, these operations accomplish three fundamental management
tasks. The species or resources to be exploited are first selected. Baseline
data about the current density and productivity of these resources are then
collected. Finally, the impact of harvesting is monitored and harvest levels
are adjusted as necessary to minimize this impact.

The basic concept here is to provide a constant flow of diagnostic information about the ecological response of the species to varying degrees
of exploitation. Sustainability is achieved through a continual process of
reciprocal feedback (i.e., the demographic reaction of the target species
must result in a corresponding adjustment in harvest levels). The exact nature of this "fine-tuning" process will depend on the site, the judgment of
the resource manager, the precision of the diagnostic data collected, the effectiveness of harvest controls, and perhaps most importantly, the ecological behavior of the plant population selected for management.

Species Selection

The decision of which plant resources to harvest will be based largely on
economic concerns. Those resources possessing the highest current market
price or the greatest potential for future market expansion will usually be
chosen first. Social factors can also come into play. Some forest resources
may have a long history of extraction or traditional use in the region, and
local people may have a strong cultural preference toward continuing to
exploit these resources. Other resources (e.g., medicinals or plants of cere-
monial importance) may be subject to certain taboos that prohibit commercial exploitation.

In addition to economic and social factors, a third set of criteria that
should also be considered is the overall potential of the resource to be
managed on a sustained-yield basis. Although the fact is rarely appreciated,
some species are inherently better able to withstand the continual pertur-

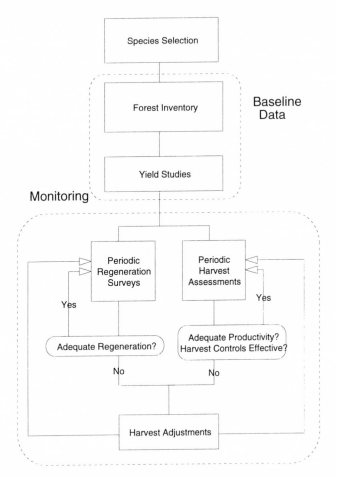

FIGURE 15.4

Flow chart of basic strategy for exploiting nontimber tropical forest plant resources on a sustained-yield basis. Complete process comprises six steps: (1)
species selection, (2) forest inventory, (3) yield studies, (4) regeneration surveys,
(5) harvest assessments, and (6) harvest adjustments. See text for explanation of
each management operation.

bations caused by resource extraction than others. Important ecological
factors to consider include the life cycle characteristics of the species (e.g.,
phenology of flowering and fruiting, pollination, and seed dispersal), the
type of resource produced (e.g., fruits, stems, barks), the abundance of the
species in the forest, and the size–class distribution of natural populations.
The basic idea here is quite simple. Given a group of resources with sim-

ilar economic profiles, why not select those that are the easiest to manage and have the highest potential for sustainable exploitation?

Forest Inventory

Density and size–class structure data are the most fundamental pieces of information required for management. Just as foresters need to know how many cubic meters of timber occur in a particular forest, the management of nontimber resources also relies on estimates of the distribution and abundance of different species. These estimates can only be obtained through a quantitative forest inventory. Inventories also provide the baseline data necessary to monitor the impact of harvesting. Without some knowledge of initial density and size–class structure, the population could slowly go extinct with each successive harvest and never be noticed.

Forest inventories are time-consuming, somewhat costly, and extremely tedious to conduct. It is strongly recommended, therefore, that a professional forester or inventory specialist be involved in the planning and execution of this fieldwork. In general, the inventory should be designed to provide the following types of information:

• A reasonably precise estimate of the total number of harvestable trees per hectare (i.e., the resource density) in different forest types. For fruit and oil seed species, this means the total number of adult trees. For latex-producing species, medicinal plants, and rattans, some juvenile trees may also need to be included.

• Data on the current population structure or size–class distribution of adult trees. Collecting these data requires that the diameter (cm DBH) of all stems be measured. Height measurements can be substituted in the case of herbaceous plants, small understory palms, or woody shrubs.

• A preliminary assessment of the regeneration status of the species. Does the species appear to be maintaining itself in the forest? Are there a sufficient number of small trees to replace the inevitable death of adult trees? To begin answering these questions, smaller, nonproductive individuals must also be counted and measured in the inventory.

Yield Studies

Given an understanding of the density and size–class distribution of a forest species, the next question that needs to be addressed is "How much of the desired resource is produced by natural populations of the species?" Suppose 250 kilograms of fruit is harvested from the forest. Is this level of harvest sustainable? Well, that depends. How many fruits does the population produce? Is this only 10% of the total population seed production, or were 95% of all fruits removed? Clearly, it makes a difference. Just as foresters (theoretically) use growth data to avoid cutting timber faster than it is produced in the forest, the sustained-yield management of nontimber resources also requires information about the productive capacity of the species being exploited. This information is obtained through yield studies.

The basic objective is to obtain a reasonable estimate of the total quantity of resource produced by a species in different habitats or forest types. In view of the fact that larger plants are invariably more productive than smaller plants the relationship between plant size and productivity is of particular interest. Probably the easiest way to obtain these data is to train local collectors to weigh, count, or measure the quantity of resource produced by different sample trees during their normal harvest operations. These studies should be repeated every few years using the same group of sample plants to monitor the variation in yield over time.

Regeneration Surveys

The baseline data collected in the forest inventory and yield studies provide an estimate of the *total* harvestable yield from the forest. Based on the discussion in the preceding sections, however, it is clear that not all of this material can be harvested from the forest for very long. What we really want to know is the *sustainable* harvest from the forest. How much of the resource can we harvest from the forest without damaging the long-term stability of the plant populations being exploited? To answer this question, we require information about the ecological impact of differing harvest levels.

The first signal that a plant population is being subjected to an overly intensive level of harvest is usually manifested in the size–class distribution of that population. For most species, the effects of overharvesting are most clearly visible in the seedlings and small sapling stage. Harvesting may kill a large number of adult plants (e.g., rattan, gharu, or palm hearts), may lower individual tree vigor to the point that flower and fruit production is effected (e.g., leaf or bark harvest, or the tapping of plant exudates), or may remove an excessive number of seeds from the forest. From a population standpoint, the net results of these activities is the same—all reduce the rate at which new seedlings are established in the population. This impact can be detected, and hopefully avoided, by periodically monitoring the density of seedlings and saplings in the populations being exploited. To use a medical analogy, these data are the vital signs by which to assess the health or infirmity of the population.

Harvest Assessments

Harvest assessments are an additional type of monitoring activity used to gauge the ecological impact of resource harvest. These are visual appraisals of the behavior and condition of adult trees that are conducted concurrently with harvest operations. In many cases, these quick assessments can detect a problem with reproduction or growth before it becomes serious enough to actually reduce the rate of seedling establishment. The sample plants selected and marked for the yield studies are perfect subjects for these observations. Examples of the type of information to be recorded during these assessments include: overall vigor of the plant, wounding caused by harvesting, trampling of seedlings by collectors, evidence of insect pests or fungal pathogens, and abundance of fallen flowers and immature fruits under the crown.[7]

Harvest Adjustments

The monitoring operations are used to appraise the sustainability of current harvest levels. The seedling and sapling densities recorded in the original regeneration survey represent the *threshold values* by which sustainability is measured. As long as densities remain above this threshold value—and no major problems are detected in the harvest assessments—there is a high probability that the current level of exploitation can be sustained. If, however, seedling and sapling densities are found to drop below this value, immediate steps should be taken to reduce the intensity of harvest. The effectiveness of this harvest reduction will be verified during the next regeneration survey. Further reductions in harvest levels may be warranted if seedling and sapling densities fail to stabilize, or drop even lower, during the five-year period.

In actual practice, achieving a sustainable yield in this manner will invariably involve a considerable number of harvest adjustments. There is frequently a lag time in a population's response to disturbance, and after several cycles of apparently stable results from the regeneration surveys, the population may exhibit a drastic fluctuation in seedling and sapling densities. The important thing is that these fluctuations do not go unnoticed. By gradually lowering, or even raising in some cases, the intensity of resource extraction, the level of seedling establishment should eventually approximate the threshold value established for the population. From an ecological perspective, this situation represents a verifiable example of sustainable resource exploitation.

In a perfect world, baseline data about the size–class structure and yield characteristics of NTFPs would be collected, regeneration surveys would be conducted as a matter of routine, and harvest levels would be adjusted as necessary to insure the long-term sustainability of resource exploitation. This has rarely, if ever, been the case. In spite of all the recent interest in the conservation and financial value of biodiversity, it is surprising that so little attention has been focused on actually monitoring the sustainability of the resource base from which these benefits accrue. There are basically two ways to exploit biodiversity. We can manage and use it wisely—or we can use it up. In terms of bioprospecting and the development of new pharmaceuticals, the former course of action has the potential to forge a major and unprecedented link between biodiversity and human health.

Notes

1. Crocodiles, butterflies, iguanas, turtles, bird's nests, and a variety of other animal resources and products collected from tropical forests are also usually included in the definition of NTFPs.

2. The harvest of some NTFPs, for example, rattan, certain species of palm heart, gharu wood (*Aquilaria malaccensis*), and most root and barks, does, in fact, kill the plant.

3. This number is based on an estimate of the total reproductive output of a *B. alicastrum* tree during its life time (Peters 1989) and the fact that only a one to one replacement is necessary to maintain the population at its present density.

4. The roots of this leguminous plant contain rotenone, an extremely potent natural insecticide (Acevedo-Rodriguez 1990). In the 1930s a significant export trade for this plant was developed in many parts of Amazonia (Padoch 1987). Within a very short time, however, the species had to be cultivated in the region in response to commercial collectors digging up, and not replacing, a significant percentage of the *Lonchocarpus* plants found in the forest.

5. Several authors (e.g., Leak 1965; Meyer 1952) have reported that diameter distributions conforming to a negative exponential are characteristic of stable self-maintaining populations.

6. Similar examples could be included for almost any other type of nontimber forest product. It is always the best damar producing trees that are tapped the most frequently, the longest and strongest rattan canes that are the first to be harvested, and the plants producing the seeds with the highest oil content that experience the highest intensity of exploitation.

7. In many cases, these fallen flowers will be aborted or unpollinated reproductive structures. A drastic increase in the quantity of fallen flowers beneath a tree could indicate a lack of pollinators or resource limitations.

References

Acevedo-Rodriguez, P. (1990). The Occurrence of Piscicides and Stupefactants in the Plant Kingdom. *Advances in Economic Botany* 8:1–23.

Ashton, P.H. (1984). Biosystematics of Tropical Woody Plants: A Problem of Rare Species. In: *Plant Biosystematics* (W.F. Grant, ed.), pp. 497–518. New York: Academic Press.

Bawa, K.S., S.H. Bullock, D.R. Perry, R.E. Coville, and M.H. Grayum. (1985). Reproductive Biology of Tropical Lowland Rain Forest Trees. II. Pollination Systems. *American Journal of Botany* 72:346–356.

Braun, E.L. (1950). *Deciduous Forests of Eastern North America*. Philadelphia: Blakiston.

Brown, F.I., L.A. Martinelli, W.W. Thomas, M.Z. Moreira, C.A. Cid Ferreira, and R.A. Victoria. (1995). Uncertainty in the Biomass of Amazonian Forests: An Example from Rondonia, Brazil. *Forest Ecology and Management* 75: 175–189.

Burgess, P.F. (1971). Effect of Logging on Hill Dipterocarp Forest. *Malaysian Nature Journal* 24:231–237.

Campbell, D..G., D.C. Daly, G.T. Prance, and U.N. Maciel. (1986). Quantitative Ecological Inventory of Terra Firme and Varzea Tropical Forests on the Rio Xingu, Brazilian Amazon. *Brittonia* 38:369–393.

Croat, T.B. (1978). *Flora of Barro Colorado Island*. Palo Alto: Stanford University Press.

Cunningham, A.B., and F.T. Mbenkum. (1993). Sustainable Harvesting of *Prunus Africana* Bark in Cameroon: A Medicinal Plant in International Trade. *People and Plants* Working Paper 2, UNESCO.

Faber-Langendoen, D., and A.H. Gentry. (1991). The Structure and Diversity of Rainforests at Bajo Calima, Chocó Region, Western Colombia. *Biotropica* 23:2–11.

Gentry, A.H. (1982). Patterns of Neotropical Plant Species Diversity. *Evolutionary Biology* 15:1–84.

———. (1988). Tree Species Richness of Upper Amazonian Forests. *Proc. Natl. Acad. Sci.* 85:156–159.

Gianno, R. (1990). Semelai Culture and Resin Technology. Connecticut Academy of Arts and Sciences.

Homma, A.K.O. (1992). The Dynamics of Extraction in Amazonia: A Historical Perspective. *Advances in Economic Botany* 9:23–31.

Jessup, T.C., and N.L. Peluso. (1986). Minor Forest Products as Common Property Resources in East Kalimantan, Indonesia. In *Common Property Resource Management*. Washington, D.C.: National Academy Press.

Johns, A.D. (1988). Effect of "Selective" Timber Extraction on Rain Forest Structure and Composition and Some Consequences for Frugivores and Folivores. *Biotropica* 20:31–37.

Jordan, C.F. (1987). *Amazonian Rain Forests: Ecosystem Disturbance and Recovery.* New York: Springer-Verlag.

Kartawinata, K., R. Abdulhadi, and T. Partomihardjo. (1981). Composition and Structure of Lowland Dipterocarp Forest at Wanariset, East Kalimantan. *Malayan Forester* 44:397.

Keller, M., D.J. Jacob, S.C. Wofsy, and R.C. Hariss. (1991). Effects of Tropical Deforestation on Global and Regional Atmospheric Chemistry. *Climatic Change* 19:139–158.

LaFleur, J.R. (1992). Marketing of Brazil Nuts. Forest Products Division, Food and Agriculture Organization of the United Nations, Rome.

Lang, G.E., and D.H. Knight. (1983). Tree Growth, Mortality, Recruitment, and Canopy Gap Formation during a 10-year Period in a Tropical Moist Forest. *Ecology* 64:1075–1080.

Leak, W.B. (1965). The J-shaped Probability Distribution. *Forest Science* 11:405–419.

Manokaran, N. (1990). The State of the Rattan and Bamboo Trade. Rattan Information Centre Occasional Paper No. 7. Rattan Information Centre, Forest Research Institute Malaysia, Kepong.

Manokaran, N., and K.M. Kochummen. (1987). Recruitment, Growth, and Mortality of Tree Species in a Lowland Dipterocarp Forest in Peninsula Malaysia. *Journal of Tropical Ecology* 3:315–330.

Meyer, H.A. (1952). Structure, Growth, and Drain in Balanced, Uneven-Aged Forest. *Journal of Forestry* 50:85–92.

Padoch, C. (1988). The Economic Importance and Marketing of Forest and Fallow Products in the Iquitos Region. *Advances in Economic Botany* 5:74–89.

Peters, C.M. (1989). Reproduction, Growth, and the Population Dynamics of *Brosimum alicastrum* Sw. in a Moist Tropical Forest of Central Veracruz, Mexico. Ph.D. dissertation, Yale University, New Haven.

————. (1990a). Plant Demography and the Management of Tropical Forest Resources: A Case Study of *Brosimum alicastrum* in Mexico. In: *Rain Forest Regeneration and Management* (A. Gomez-Pompa, T.C. Whitmore, and M. Hadley, eds.), pp. 268–272. Cambridge: Cambridge University Press.

————. (1990b). Population Ecology and Management of Forest Fruit Trees in Peruvian Amazonia. In: *Alternatives to Deforestation: Steps Toward Sustainable Use of the Amazon Rain Forest* (A.B. Anderson, ed.), pp. 86–98. New York: Columbia University Press.

————. (1990c). Plenty of Fruit but No Free Lunch. *Garden* 14:8–13.

————. (1994). *Sustainable Harvest of Nontimber Plant Resources in Tropical Moist Forest: An Ecological Primer.* Washington: Biodiversity Support Program.

Peters, C.M., M.J. Balick, F. Kahn, and A.B. Anderson. (1989). Oligarchic Forests of Economic Plants in Amazonia: Utilization and Conservation of an Important Tropical Resource. *Conservation Biology* 3:341–349.

Prance, G.T., W.A Rodriques, and M.F. da Silva. (1976). Inventário Florestal de um hectare de mata da terra firme km 30 da estrada Manaus—It a coatiara. *Acta Amazonica* 6(1):9–35.

Procter, J., J.M. Anderson, P. Chai, and H.W. Vallack. (1983). Ecological Studies in Four Contrasting Lowland Rain Forests in Gunung Mulu National Park, Sarawak. *Journal of Ecology* 71:237–260.

Reitsma, J.M. (1988). Vegetation Forestiére du Gabon. *Tropenbos Tech. Series* 1:1–142.

Skukla, J., C.A. Nobre, and P. Sellers. (1990). Amazon Deforestation and Climate Change. *Science* 247:1322–1325.

Turner, I.M. (1990). The Seedling Survivorship and Growth of Three *Shorea* Species in a Malaysian Tropical Rain Forest. *Journal of Tropical Ecology* 6:469–478.

Uhl, C., J.B. Kauffman, and D.L. Cummings. (1988). Fire in the Venezuelan Amazon. 2: Environmental Conditions Necessary for Forest Fires in the Evergreen Rainforest of Venezuela. *Oikos* 53:176–184.

Valencia, R., H. Balslev, and G. Paz y Mino C. (1994). High Tree Alpha-Diversity in Amazonian Ecuador. *Biodiversity and Conservation* 3:21–28.

Vazquez, R., and A.H. Gentry. (1989). Use and Mis-use of Forest-Harvested Fruits in the Iquitos Area. *Conservation Biology* 3:350–361.

Wiebes, J.T. (1979). Coevolution of Figs and Their Insect Pollinators. *Annual Review of Ecology and Systematics* 10:1–12.

CHAPTER 16

Opportunities for Collaboration between the Biomedical and Conservation Communities

WALTER V. REID

Managing Diversity

Collaboration between researchers in the biodiversity and biomedical communities is surprisingly rare. On the surface, close interaction between these communities would seem inevitable—one group studies the distribution and abundance of living organisms and the other spends a substantial amount of time studying how those organisms affect human health. But the relative lack of interaction between these communities is reminiscent of similar barriers that are only now eroding between biodiversity and agricultural sciences, and between biodiversity and economics. Indeed, Western society seems to have spent much of the 20th century trying to convince itself that its new technologies could raise humanity above the messy diversity of life around us, only to now discover that evolution is still a fair match for technology.

When we look back over the past century, we see a series of technological "revolutions," each of which seemed to drive a wedge between the dependence of humanity on the vagaries of living systems and place our destiny more firmly in our own grasp. And though each revolution did partly succeed in this goal, it was far less successful than we thought.

Consider first the industrial revolution, which shifted the primary economic base of society from the fields to the city—from agriculture to industry. It also helped to buffer industrialized societies from the direct environmental and economic costs of the loss of biodiversity and the degradation of biological resources. What need was there to be concerned about environmental management or the management of biodiversity when our economy hardly depended on the environment? This type of thinking undergirded the development of the linear economic models still in use today in which resources enter the economy, are transformed into wealth and waste, and the waste is discharged from the model.

We now know that our economy is anything but linear and that both the economic and social costs of ignoring the impacts of our industrial society on living systems can be profound. People still depend on those living resources, and wealthy societies find tremendous value in the protection of the diversity of life itself, demonstrating a considerable willingness to pay just to know that biological diversity survives. And, we are constantly reminded that we are more dependent on living biological systems than we think. Threats to the productivity of agriculture, forestry, or fisheries represent major threats to regional economies. Periodic floods exacerbated by the loss of wetlands continue to demonstrate the economic folly of the loss of these valuable habitats.

The next revolution was in agriculture. The green revolution seemed to insulate us from the vagaries of agricultural productivity. Through the breeding of "high yielding varieties" that responded to inputs of water, fertilizer, and pesticides, yields shot up around the world, food prices dropped, and farming populations increasingly shifted into the city. Yet the core of this revolution lay in how we managed genetic diversity. The green revolution amounted to a shift from the management of the spatial aspects of agricultural biodiversity to management of diversity through time. Traditional agricultural systems make use of considerable spatial diversity of species and varieties as a means of ensuring high and stable productivity. For example, the Massa of northern Cameroon (who cultivate five varieties of pearl millet), the Ifugao of the island of Luzon in the Philippines (who identify more than 200 varieties of sweet potato by name), and Andean farmers (who cultivate thousands of clones of potatoes, more than 1,000 of which have names) all use highly diversified farming systems.

Green revolution agriculture gains precisely these same benefits of productivity and stability through the management of genetic diversity through time rather than across space. For example, crop yields can be increased by introducing genetic resistance to certain insect pests, but since natural selection often helps insects quickly overcome this resistance, new genetic resistance has to be periodically introduced into the crop just to sustain the higher productivity. In the United States, for example, the average lifetime of a cultivar of cotton, soybean, wheat, maize, oats, or sorghum is between five and nine years (Plucknett and Smith 1986). Pesticides are also overcome by evolution, so another important agricultural use of genetic diversity has been to offset productivity losses from pesticide resistance. Over 400 species of pests now resist one or more pesticides (May 1985), and the proportion of U.S. crops lost to insects has approximately doubled—to 13 percent—since the 1940s, even though pesticide use has increased (Plucknett and Smith 1986).

Thus, the green revolution gave humanity a temporary boost in the evolutionary arms race, but today as much as ever we must be conscious of the need to manage genetic diversity and the associated biodiversity in agricultural systems to ensure the continued productivity of our agricultural systems. Thus far, our management record is not encouraging. Where previously farmers created and maintained diversity as a part of their everyday life, today that diversity must be maintained in costly genebanks subject to financial and management risks. In 1980, for example, experts estimated that between one-half and two-thirds of the seeds collected in past decades had died. In 1991, representatives of 13 national germplasm banks in Latin America reported that between 5 and 100 percent of the maize seed collected between 1940 and 1980 is no longer viable (WRI/IUCN/UNEP 1992). And, because germplasm stored off site is removed from natural selection pressures and thus will not evolve resistance to new pests or new environmental conditions, its potential value to breeders may diminish with time. Thus, the green revolution served, in part, to shift the costs of mismanaging agricultural biodiversity onto future generations.

The third revolution was in health care. Mirroring the pattern of the other two, the health care revolution and miracle drugs seemed once and for all to place health security in our own hands and to insulate us from our unpredictable environment. And so it did, until evolution again caught up. While we may indeed be able to cure some diseases, for many others the battle is likely to be protracted. Even in industrialized countries, the combination of migrating human populations, changing distributions of vectors, changing virulence of disease-causing organisms, emergence of new diseases, and ongoing evolution by these actors is time and again demonstrating that human health is still far more closely linked to our natural world than many people are willing to admit. And the situation in the developing world amplifies this message many times over.

Thus, in the late 20th century, we are in the process of relearning a resource management goal that was almost second nature until the beginning of this century—the need to wisely manage the *diversity* of life as well as its individual components. There is no question that humanity has found ways of buffering its well-being from the vicissitudes of the environment. But we have gone overboard in our reaction to this achievement, often forgetting the nature of our dependence on that diversity of life and forgetting how closely intertwined various components of living systems are.

The way societies manage their resources determines how much diversity survives, and in turn the way that societies manage biological diversity determines the productivity of important resources and ecological services, and ultimately the health and well-being of people. Yet today, most "management" of the diversity and abundance of genes, species, and biological communities is inadvertent. Although we manage specific resources, we rarely design policies, institutions, or management practices to effectively manage the broader base of biodiversity on which those resources and ecological services depend.

Biodiversity and Human Health

More than any other biodiversity-related issue, public health concerns can help restore the need for sound management of the world's biological diversity as an important societal goal. Public health concerns have two important components. First, they embrace the entire scope of biodiversity values and threats, involving (i) the value of the components of biodiversity as resources in the form of important medicinal products; (ii) the value of diversity itself as a source for new pharmaceutical products or as model systems for the study of disease; (iii) the value of biodiversity as an indicator of potential threats to human health; and (iv) the potential public health threats of human-caused changes in biological diversity. Second, all people care deeply about health. Improving public health is a universally shared goal, whereas the notion of conserving and wisely managing biodiversity is a salient issue for relatively few people.

Biodiversity and Traditional Medicinal Products

At one time, nearly all medicines came from plants and animals and today they remain a vital resource. The United Nations estimates that traditional medicines form the basis of primary health care for about 80 percent of people in developing countries. More than 5,100 species are used in Chinese traditional medicine and nearly 2,500 plant species were used for medicinal purposes in the former Soviet Union (WRI/IUCN/UNEP

1992). In countries like India and Indonesia, significant economic sectors are based on traditional health care systems: ayurvedic medicine in India, and Jamu medicine in Indonesia. The Jamu industry in Indonesia for example involves 450 factories and has sales in the hundreds of millions of dollars (Suprana 1994).

As with any consumptive use of biological resources, the use of traditional medicines can threaten biodiversity if the resource is overexploited. In Southern Africa, for example, many villagers can no longer find medicinal plants in part because commercial collectors have overharvested them to meet the demand in cities. In East Asia, the use of rhinoceros in traditional medicine has helped bring several species to the brink of extinction. And the use of bear gall bladders in Southeast Asian markets has helped to endanger the Asiatic black bear. (Similar problems also exist in the case of modern drugs dependent on natural product sources. For example, the Himalayan Yew is being threatened by increased demand as a source of taxol and is being proposed for listing on Appendix II of CITES.)

Biodiversity and Drug Development

In the pharmaceutical industry, after a hiatus in natural products research in the 1970s, interest has intensified over the past decade. Natural products research has been revived by the development of efficient automated receptor-based screening techniques that have increased a hundred-fold the speed with which chemicals can be tested. Although only one in about 10,000 chemicals yields a potentially valuable "lead" (McChesney 1992; Reid et al. 1993), these new techniques have made large natural products screening programs affordable. Researchers are thus returning to such natural sources of biologically active chemicals as plants, insects, marine invertebrates, fungi, and bacteria.

Another and quite different stimulus to natural products research has come from ethnopharmacology—the study of medicines used by traditional communities. Leads based on the use of plants or animals in traditional medicine can greatly increase the probability of finding a commercially valuable drug. For small pharmaceutical companies, drug exploration based on this indigenous knowledge may be more cost-effective than attempting to compete in expensive random screening ventures.

In the United States, some 25 percent of prescriptions are filled with drugs whose active ingredients are extracted or derived from plants. Sales of plant-based drugs in the United States amounted to some $4.5 billion in 1980 and an estimated $15.5 billion in 1990 (Principe, unpublished manuscript). In Europe, Japan, Australia, Canada, and the United States, the market value for both prescription and over-the-counter drugs based on plants in 1985 was estimated to be $43 billion (Principe 1989). In addition, 57 percent of the 150 top prescription drugs sold in the United States in 1993 were in some way linked to natural products. Eighteen percent were used essentially as is, 39 percent were semisynthetic, and 4 percent were developed on the basis of pharmacological properties found in natural products (Grifo et al., Chapter 6).

Technological changes in the pharmaceutical industry suggest that natural products research will become still more important in coming years, primarily as a result of further advances in screening technologies. With the exception of microorganisms which have always been collected in small quantities and cultured in laboratories, natural products drug discovery processes have traditionally required substantial quantities of material. In the case of plants, samples of anywhere from 1 to 10 kg have been required for initial extraction and screening. If an extract shows promising activity, substantially larger samples were then required for activity directed fractionation, isolation, structure elucidation of active compounds, and further biological testing. Because of the complexity of many natural chemicals, their isolation, characterization, and synthesis has been a time-consuming and expensive process.

Today, techniques for extraction, screening, fractionation, and chemical identification are becoming routine and relatively inexpensive (Reid et al. 1995). New chemical and analytical methods help increase the efficiency of natural product screening. Using separation techniques like high performance liquid chromatography (HPLC) and high performance centrifugal counter current chromatography (HPCCC), and analytical methods like high field nuclear magnetic resonance spectrometry, samples can be fractionated and characterized in a small fraction of the time formerly required. Prior to the 1980s, using test-tube and live animal (*in vivo*) assays a lab could screen 100–1000 samples per week. Now, using 96-well microtiter plates and robotics, labs can screen 10,000 samples per week in a broad range of mechanism-based assays in which scientists identify the mechanism by which a disease affects the human physiology and develop an assay to detect compounds that could interfere with that mechanism. For example, hundreds of thousands of microorganism cultures are screened by Merck each year. Where the screening of 10,000 plant extracts would have cost $6 million one decade ago, it can now be accomplished for $150,000. In the next decade, with further microprocessing and high-speed robotics, the throughput could grow by another one or two orders of magnitude.

Tissue and cell culture—the "next frontier" of natural products research—now allows small samples of plants to be collected and screened without any need to return for further supplies. A company can obtain seeds or samples of a variety of plants, establish the material in cell culture, and challenge that culture with various chemicals or environmental stresses to stimulate the expression of new chemicals. The companies Phytera Inc. and Phytopharmaceuticals, for example, base their drug discovery process on plant cell culture methodologies.

Finally, advances in screening methodologies have been accompanied by tremendous improvements in the bioassays themselves, which has significantly decreased the likelihood of false positive "hits." Using cell-based and receptor-based assays, activity can be detected at concentrations several orders of magnitude lower than was previously required for *in-vivo* tests.

Of course, natural products research may be of only short-term importance, soon to be replaced with rational drug design and combinatorial chemistry. But while medical technology may indeed someday make natural products research irrelevant, there is no sign of that occurring soon. In a 1995 survey, leading scientists were asked what they think the future holds in store in their field and Helen Ranney, of Alliance Pharmaceuticals, observed:

> Rational drug design will be possible in the near term only for a few disorders. While modern technologies make possible the tailoring of compounds to block enzymatic actions, or put defective ligands in unoccupied receptors, the intricacies, interdependencies, and redundancies of human physiology will continue to defy simple pharmaceutical solutions to most diseases. . . . The rapid emergence of drug-resistant microbial agents will lead to renewed searches for antibiotics. . . . Rational drug design may play a role in this search, but whether synthetic or natural products will be identified first is far from clear (Ranney 1995).

Biological diversity also plays a significant medical role by providing animal models that help researchers understand human physiology and disease. For example, considerable advances in our understanding of the human nervous system have come from studying other vertebrate and invertebrate nervous systems. Study of the physiology involved in unique behavioral traits, such as hibernation in bears, may help to develop new ways to prevent osteoporosis (Chivian, Chapter 1). And, some species such as sharks rarely develop tumors. Studies aimed at understanding what prevents the onset of tumors in these species may one day lead to means of preventing human cancers (Chivian, Chapter 1).

Biodiversity As an Indicator of Potential Threats to Humans

Biological diversity has often served as an "early warning" system that has foretold threats to human health before sufficient data had been collected to detect direct impacts on humans. Rachel Carson's book, *Silent Spring*, for example, established a strong case against the use of pesticides based primarily on threats to wildlife populations—these same pesticides have since been found to present serious public health risks. Similarly, declines in populations of the common seal in the Wadden Sea and reproductive failure in the Beluga whale in the St. Lawrence River in Canada may stem from the ingestion of PCB-contaminated fish—suggesting that caution be taken to ensure the safety of marine food supplies for human consumption.

More recently, wildlife studies have shown evidence of impacts of various chlorinated organic compounds on the immune system of animals (Repetto and Baliga 1996) and on their reproductive physiology (Colborn et al. 1995). While studies are not conclusive that these compounds have an effect on human physiology, the strong evidence from wildlife studies again helps scientists to identify the need for serious research to ensure that significant public health problems do not arise.

Public Health Threats of Changes in Biodiversity

Ironically enough, it is perhaps the threat that biodiversity holds to human health rather than the vastly greater benefits that it provides to humanity that may now be most likely to stimulate greater public concern about the impacts of human-caused environmental changes on biological diversity.

Human-caused environmental changes have many direct impacts on human health. The neurotoxic effects of pesticides, the developmental impacts of lead poisoning, and respiratory impacts of air pollution are all well-known examples of such direct health effects. But we know much less about indirect impacts of environmental change on human health, wherein biodiversity frequently plays a central role. Human-caused environmental changes can either affect biological systems in such a way as to increase the prevalence of certain diseases or may change human immune defenses in such a way as to increase disease prevalence.

Honduras provides a good illustration of the impacts of human-caused changes in biodiversity on public health (Almendares et al. 1993). In the southern part of the country, expansion of cattle grazing and sugarcane and cotton cultivation increased the ambient temperature, making the area too hot for anopheline mosquitos, and reducing the prevalence of malaria. However, as many of the people who then had little immunity to malaria migrated away from this region into forested areas in the northeast they encountered anopholene mosquitos and malaria incidence increased. The mosquitos, in turn, were largely resistant to pesticides due to indiscriminate use of the pesticides in the agricultural areas. Malaria prevalence increased from 20,000 cases in 1987 to an estimated 90,000 in 1993. Encroachment on forested areas also increased levels of Chagas' disease, as people came into more contact with the insect vectors (*Rhodnius prolixis* and *Triatoma dimitata*) and as the warmer conditions shortened the generation time of the vectors and increased the parasite prevalence.

Similar patterns are found in many other tropical countries. In addition, environmental degradation, population growth, and international travel contribute to the spread to new regions of vectors once confined to the tropics (Rogers and Packer 1993) as well as to the reemergence of diseases once thought to be under control. In the 1980s, incidence of malaria in the Brazilian Amazon rose from 287,000 in 1983 to 500,000 in 1988 (Kingman 1989).Yellow fever has also been increasing dramatically in tropical Africa. And, in Latin America, the mosquitos that carry the disease are bouncing back from efforts to control populations through insecticides. Yellow fever could present a tremendous threat to the rapidly growing urban populations along Latin America's eastern seaboard (Maurice 1993). Indeed, the Asian tiger mosquito that carries yellow fever as well as dengue fever spread from Japan to the United States in 1985 and is now abundant in the United States and Brazil. The resurgence of tuberculosis (TB) is another case in point. Not only is the prevalence of the disease increasing, so too is its resistance to common drugs. Where cases in the United States were declining at a rate of six percent per year until 1985, trends have since reversed in some cities. Cases in New York City doubled between 1985 and 1993. Italy reported a 28 percent increase during 1988–1990, and Switzerland had a 33 percent increase between 1986 and 1990 (WHO 1993). This increase has been due, in part, to the spread of HIV, however prospects are now still more worrisome due to the emergence of strains of TB that are resistant to all known anti-TB drugs.

"Emerging viruses" represent a related threat to public health. Whether due to increased migration of people into regions formerly uninhabited, or the greater likelihood of spread of new diseases due to more rapid transportation, or some combination of these factors, the past decade has seen a remarkable array of new diseases emerging as public health threats, including AIDS, Legionnaires' disease, Lyme disease, hantavirus pulmonary syndrome, and infections from *Eschericia coli* 0157:H7, cryptosporidiosis, multiple drug-resistant pneumococcus, *Helicobacter pylori,* vancomycin-resistant enterococcus (Gellert 1994). Many of these emerging viruses have natural hosts among rodents, birds, and pigs and may have existed relatively harmlessly in remote areas until people migrated into these regions.

Finally, the indirect effects of climate change on human health are likely to enhance these trends still further. For example, strong evidence now links the outbreak of cholera in South America in 1991 to El Niño (Epstein et al. 1993; Stone 1995). The warming of the waters off the coast of South America may have stimulated growth of a plankton harboring the cholera bacterium. While the increased frequency of El Niño in recent years cannot be conclusively tied to human-caused changes in climate, the example demonstrates how ocean current changes that are likely to occur in the event of global warming could have substantial effects on human health. Current models of changes in the distribution of disease vectors under likely future climates suggest that developing countries will see an increase in malaria, schistosomiasis, sleeping sickness, dengue, and yellow fever. Already, some 600 million people are affected by these diseases. Warming temperatures could also lead to migration of mosquitos that can transmit malaria and dengue from Central America into the United States.

Environmental change also indirectly influences the prevalence of disease through its impact on the human immune system. Such impacts are clear in the case of weakened immune systems due to malnutrition. However, evidence is also building that pesticides at very low concentrations—as might be found on treated produce—can weaken immune systems as well. There is considerable laboratory evidence that pesticides damage the mammalian immune system, but very little epidemiological or clinical work has been done to investigate the effects in humans (Repetto and Baliga 1996). If pesticides do impair human immune systems, this (particularly where combined with malnutrition) could increase susceptibility to intestinal parasites, hepatitis, malaria, and respiratory infections.

Opportunities for Enhanced Collaboration

Given this close linkage between biodiversity and human health, where do the opportunities lie for enhanced collaboration between the biomedical and conservation communities? Five particularly strategic areas for collaboration include the following:

1. *Examination of the effects of environmental change on public health, particularly regarding the potential indirect impacts of climate change on health.*

In some respects, the study of linkages between health and biodiversity is about as advanced as was the study of linkages between economics and biodiversity a decade ago. After years of general recognition that a linkage exists, it is now becoming apparent that the consequences for one or the other can be quite profound. Just as the political momentum for biodiversity conservation was dramatically enhanced when the economic costs of the loss of biodiversity became apparent, the same can occur as a result of greater understanding of the public health costs of the mismanagement of biodiversity.

For example, we are well aware of the economic and public health costs of polluted water supplies, and substantial investments are being made to provide clean water for people around the world. Public health concerns thus factor into decisions about the management of water resources and expenditures on those resources. The same cannot be said about policies related to forest resources. People are spreading into tropical forests because the direct economic cost to the individual of transforming those lands is often small and the short-term economic benefits large. We now know that the cost to society is much larger due to the loss of biodiversity and the loss of various ecological services such as the role of the forest in providing clean water or sequestering carbon. If we add to this the additional public health costs incurred by both the individuals and the nation due to greater disease prevalence, the argument for improved management of forest resources grows still stronger.

Agriculture provides another example. We think of the benefits of modern agricultural technology primarily in terms of increased agricultural productivity. But our measurement of productivity is faulty. We count the "goods" produced by modern agriculture—that is, the food—toward productivity without subtracting the "bads" produced by modern agroecosystems from those measures. Several years ago, World Resources Institute (WRI) did a study of the economics of sustainable agriculture in Pennsylvania and found that if the costs of soil erosion were factored in, intensive agricultural systems were economically less productive than "sustainable" agricultural systems (Faeth et. al. 1991). What if we then factor in public health costs? WRI did another study with International Rice Research Institute (IRRI) in the Philippines and found that the public health costs due to pesticide poisonings in rice culture in the Philippines were high enough that farmers were economically better off using natural control methods as opposed to pesticides (Faeth 1993). Now consider the implication that low levels of pesticide residues can weaken the immune system and thereby increase prevalence of other infectious diseases. If this is borne out, the implications are quite profound—if we fully accounted for public health costs of modern agricultural systems, we would often find that the apparent high productivity would vanish, and sustainable systems relying on integrated pest management (IPM) would be far more beneficial economically and environmentally.

Study of the impacts of potential climate-change induced alterations in biodiversity also takes on particular urgency. The potential rapid rate of change, combined with a more mobile human population that could readily transfer diseases or disease vectors substantial distances, holds serious threats to public health. The impact of climate change on health may hold much in common with the devastating public health impacts of the worldwide spread of diseases introduced by Europeans in the 16th and 17th centuries (Crosby 1986). Serious public health consequences are inevitable when diseases are introduced into regions where populations have not built up resistance to the disease. Accurate forecasts of impacts of global change on human society require much more research on these indirect health effects. In turn, adaptation to climate change can be enhanced with better knowledge of where public health problems are likely to occur.

Areas of collaboration of particular importance include epidemiological studies of disease in wild populations; long-term ecological research that monitors human health-related parameters; direct and indirect impacts of pesticides on human health, and exploration of alternative agricultural systems relying less on chemical input and more on associated biodiversity (e.g., IPM); and development of ecological models that could predict the spread of disease-causing organisms or vectors under changed climate regimes.

2. *Enhancement of traditional medicinal care, through linked efforts to promote the conservation of medicinal resources and better integrate traditional medicines into public health care systems.*

Both Western health-care systems and traditional systems have much to offer—medically, economically, and culturally—but the integration of these two systems has had a poor track record. In the Philippines, for instance, medicinal plants have provided a low-cost alternative to Western medicines for many ailments. Yet, in the early 1980s the government actively promoted the abandonment of traditional systems and their replacement with Western medicines (Tan 1986).

Traditional systems can be enhanced both through efforts to improve the efficacy of traditional medicines and to reduce the potential threats to biodiversity. One example of an effort to enhance efficacy is the TRAMIL program in the Caribbean. In this program, physicians assess the effectiveness of traditional remedies through ethnopharmacological surveys and classify traditional herbal remedies as either toxic, indeterminant, or beneficial/innocuous. The program has produced a manual, "Elements for a Caribbean Pharmacopeia," that health-care workers use as a guide to the region's many useful traditional medicinal treatments.

The use of traditional medicines can threaten biodiversity. Accordingly, strengthening such systems requires taking steps to ensure the sustainability of resource use. For medicinal plants, the best insurance against overexploitation is generally to promote their sustained cultivation, looking to agricultural extension, botanic gardens, and arboreta for information and advice (WRI/IUCN/UNEP 1992). For many vertebrates, however, solutions are much harder to find. The Convention on International Trade in Endangered Species (CITES) has helped reduce pressure on some species overexploited for medicinal uses, but this must be buttressed by public education on the problems created by some medicinal uses and by national bans of the sales of medicines derived from endangered or threatened species.

3. *Exploration of opportunities to use ecological knowledge to enhance the search for new pharmaceutical products.*

The search for natural products for use in the pharmaceutical industry has involved two strategies: random collection, and collection based on ethnopharmacological information. A third source of information has barely been tapped—collection based on ecological information. The world's biodiversity has been developing chemical solutions to pest and disease problems far longer than people have and this should create opportunities for more selective searches for new drugs. For example, many plant varieties show immunity to plant viruses that may be serious pests in related varieties. Such resistant varieties would be a logical place to start looking for antiviral chemicals.

More generally, many interactions between organisms reveal evidence of particular toxins or types of disease resistance that may be sufficiently analogous to similar interactions in humans so as to serve as guides to potential therapeutic agents. While ecologists are studying these chemical interactions, their research questions are rarely influenced by pharmacological concerns. In turn, biodiversity prospecting by pharmaceutical companies rarely taps this ecological knowledge.

Other species also can serve as animal models for the study of human disease. Along with chemical compounds, studies of other organisms may reveal genetic traits that confer resistance to particular diseases.

4. *Development of policies that effectively link the pharmaceutical and biotechnology sector's commercial interest in biodiversity with improved management of biodiversity, technology transfer, and improved livelihoods among people living with that biodiversity.*

The value of biodiversity as a raw material for pharmaceutical and biotechnology industries is only a portion of its value to society. It makes good economic sense—and often meets ethical norms—for countries and communities to conserve biodiversity whether or not they become biodiversity prospectors. Indeed, it is entirely possible—and sometimes highly appropriate—for nations to invest in biodiversity conservation without ever seeking to commercialize genetic and biochemical resources. But, there is an urgent need to ensure that the commercialization already under way supports conservation and development (Reid et al. 1993). In particular, three problems must be overcome if biodiversity prospecting is to contribute to national sustainable development and the long-term survival of wildland biodiversity.

First, growing commercial interest in biodiversity will not necessarily fuel increased investment in resource conservation. Genetic and biochemical resources are often described by economists as "nonrival public goods." In other words, their use by one individual does not reduce their value to others who use them. Because any user benefits from investments in their conservation, market forces will lead to less conservation of the resource than its value to society warrants. In fact, unregulated biodiversity prospecting and drug development could speed the destruction of the resource.

Second, there is no guarantee that the institutions created to capture the benefits of biodiversity will contribute to economic growth in developing countries. Quite the opposite has been the case historically. The chief commercial beneficiaries of genetic and biochemical resources found in developing countries have been the developed countries that are able to explore for valuable resources, develop new technologies based on the resources, and commercialize the products. The Convention on Biological Diversity provides a framework that may boost developing countries' negotiating strength and foster needed investments in conservation, but it will be up to individual nations to pass the laws and establish the regulations needed to achieve these benefits. From a conservation standpoint, unless developing countries do realize benefits from these resources, summoning the political will to conserve them will be difficult.

Finally, biodiversity prospecting is just one of many forms of biodiversity development that could take place in rural areas to help raise living standards. In most countries, the people living side by side with wildland biodiversity—farmers and villagers, indigenous peoples, forest dwellers, medicinal healers, and fisherfolk—hold the key to its survival. If local and national citizens do not get something out of maintaining wildland habitats, the habitats will be converted to timber plantations, farms, or other productive uses harmful to biodiversity. Yet, in many cases sustainably managed wildlands won't yield enough direct economic benefits to support large local populations, so governments will have to ensure that a share of the national benefits from activities such as biodiversity prospecting are used to meet rural development needs. How well biodiversity prospecting institutions contribute to sustainable development thus ultimately depends on how effective local and national government policies for conservation and development are.

Countries are now beginning to fill the vacuum of legislation and national policies governing the use of genetic resources. The policy formulation and experimentation taking place today will establish a new legal framework for biodiversity trade which is likely to remain for decades to come. A window of opportunity now exists to ensure that this framework meets economic, social, and environmental objectives consistent with sustainable development and the human rights of affected sectors. The key point of leverage for influencing the development of this legal and institutional structure is the design and enforcement of "access legislation"—

the legislation governing access by both citizens of a country and foreign commercial interests to the nation's genetic resources.

With little precedent for access legislation governing trade in genetic resources and with significant differences among countries in their biodiversity and technology assets, the development of appropriate legislation will necessarily involve considerable national experimentation and adaptation. The most fruitful approach is likely to involve (i) the exploration of appropriate policy options at a national level; (ii) sharing of information among developing countries about the solutions they have developed for specific policy needs; and (iii) establishment of a dialogue with potential genetic resource users (researchers, industry, botanical gardens, etc.) to ascertain how they will response to different options.

5. *Joint public awareness efforts that more convincingly demonstrate the ties between public and environmental health.*

Finally, for this collaboration to serve both public health and conservation needs, the message must reach concerned citizens. The changes in the management of biodiversity needed to address these public health and conservation needs require sufficient changes in the status quo whereas politically powerful stakeholders and constituencies are threatened and will seek to block those changes. The pharmaceutical industry, for example, has a vested interest in a rapid transition from traditional and essential medicines to modern pharmaceutical consumption. Timber companies would be threatened by changes in land tenure in forested areas that seek to produce a fairer balance between the economic and public health costs and benefits of forest development. And, perhaps the ultimate test, the changes needed in the energy base of all economies to meet the year 2000 CO_2 targets agreed to under the Framework Convention on Climate Change (much less to achieve the 60% reduction in CO_2 emissions needed for early stabilization) are already raising powerful opposition to change.

Conclusion

We are nearing the end of a century of hubris in which humanity sought to lift itself above the natural world only to find that the environmental links to both our economy and public health were far stronger than we thought. And we will soon be entering a century of humility in which humanity must come to terms with the fact that we are just one species among many and our health and welfare depend on the rather challenging need to manage a vast and chaotic jumble of life that we call biodiversity.

More than any other biodiversity-related issue, public health concerns can help restore the need for sound management of the world's biological diversity as an important societal goal.

References

Almendares, J., M. Sierra, P.K. Anderson, and P.R. Epstein. (1993). Critical Regions, a Profile of Honduras. *The Lancet* 342:1400–1402.

Colborn, T., D. Dumanoski, and J.P. Myers. (1995). *Our Stolen Future.* New York: Dutton.

Crosby, A. W. (1986). *Ecological Imperialism: The Biological Expansion of Europe 900–1900.* Cambridge, U.K.: Cambridge University Press.

Epstein, P. R., T.E. Ford, and R. R. Colwell. (1993). Marine Ecosystems. *The Lancet* 342:1216–1219.

Faeth, P. (ed.). (1993). *Agricultural Policy and Sustainability: Case Studies from India, Chile, the Philippines, and the United States.* Washington, D.C.: World Resources Institute.

Faeth, P., R. Repetto, K. Kroll, Q. Dai, and G. Helmers. (1991). *Paying the Farm Bill: U.S. Agricultural Policy and the Transition to Sustainable Agriculture.* Washington, D.C.: World Resources Institute.

Gellert, G.A. (1994). Preparing for Emerging Infections. *Nature* 370:409–410.

Kingman, S. (1989). Malaria Runs Riot on Brazil's Wild Frontier. *New Scientist* 12 August: 24–25.

Maurice, J. (1993). Fever in the Urban Jungle. *New Scientist* 16 October: 25–29.

May, R.M. (1985). Evolution of Pesticide Resistance. *Nature* 315:12–13.

McChesney, James. (1992). Biological Diversity, Chemical Diversity and the Search for New Pharmaceuticals. Paper presented at the Symposium on Tropical Forest Medical Resources and the Conservation of Biodiversity, Rainforest Alliance, New York, January 1992.

Plucknett, D.L., and N.J.H. Smith. (1986). Sustaining Agricultural Yields. *BioScience* 36:40–45.

Principe, P.P. (1989). The Economic Significance of Plants and Their Constituents as Drugs. In: *Economic and Medicinal Plant Research, Volume 3* (H. Wagner, H. Hikino, and N.R. Farnsworth, eds.), pp. 1–17. London: Academic Press.

————. Unpublished ms. Monetizing the Pharmacological Benefits of Plants.

Ranney, H.M. (1995). Through the Glass Lightly. *Science* 267:1611.

Reid, W.V., C.V. Barber, and A. La Viña. (1995). Translating Genetic Resource Rights into Sustainable Development: Gene Cooperatives, the Biotrade, and Lessons from the Philippines. *Plant Genetic Resources Newsletter.* Rome, Italy: Food and Agriculture Organization of the United Nations.

Reid, W.V., S.A. Laird, C.A. Meyer, R. Gamez, A. Sittenfeld, D.H. Janzen, M.A. Gollin, and C. Juma (eds.). (1993). *Biodiversity Prospecting: Using Genetic Resources for Sustainable Development.* Washington, D.C.: World Resources Institute.

Repetto, R., and S. Baliga. (1996). *Pesticides and the Immune System.* Washington, D.C.: World Resources Institute.

Rogers, D.J., and M.J. Packer. (1993). Vector-Borne Diseases, Models, and Global Change. *The Lancet* 342:1282–1284.

Stone, R. (1995). If the Mercury Soars, So May Health Hazards. *Science* 267:957–958.

Suprana, Jaya. (1994). The Problems of the Initiatives by Jamu-Industry in Managing the Sustainable Use of Genetic Resources. Paper presented at the Stockholm Environment Institute conference: "Assessment, Conservation, and the Sustainable Use of Genetic Resources," Oct. 10–13, 1994, Bogor, Indonesia.

Tan, Michael. (1986). The Chronic Crisis of the Health Care System. In: *Caring Enough to Cure* (M. Tan and R. Isberto, eds.), pp. 15–50. Manila, Philippines: Council for Primary Health Care.

WHO [World Health Organization]. (1993). WHO Attacks Global Neglect of Tuberculosis Crisis. Press release. November 15, Geneva, Switzerland.

WRI/IUCN/UNEP [World Resources Institute, World Conservation Union (IUCN), and United Nations Environment Programme]. (1992). *Global Biodiversity Strategy: Guidelines for Actions to Save, Study, and Use Earth's Biotic Wealth Sustainably and Equitably.* Washington, D.C.: World Resources Institute, World Conservation Union (IUCN), and UNEP.

A Proposal for a National Council on Biodiversity and Human Health

BYRON J. BAILEY AND JOHN T. GRUPENHOFF

There is a deepening concern developing among scientists, including healthcare professionals and environmentalists, about the lack of understanding by the public as well as policymakers at all levels of government of the simple concept that human health is inseparable from the health of the natural world.

A problem: physicians and their healthcare colleagues do not have environmental issues as their primary professional concern. At this time no system is available to the healthcare community generally, or through their organizational structures, for education and action about biodiversity. However, there are some opportunities available for their involvement.

The National Institutes of Health report to the U.S. Senate on December 20, 1994, "Medical Products from the Natural World and the Protection of Biological Diversity" stated:

> Perhaps the best way to illustrate the importance of biodiversity is by analogy to the diversity of human knowledge stored in books. When the library in Alexandria was consumed by fire in 391 AD, when Constantinople was sacked in 1453, or when Maya codices were burned in the 16th century auto-da-fe, thousands of works of literature were destroyed. Hundreds of works of genius are now known to us only by their titles, or from quoted fragments. Thousands more will never be known; several millennia of collective human memory have been irretrievably lost.
>
> Like books, living species represent a kind of memory, the cumu-

353

lative record of several *million* millennia of evolution. Every species has encountered and survived countless biological problems in its evolutionary history; molecules, cells and tissues record their solutions. Because we are biological beings ourselves, nature offers a vast library of solutions to many of our current health, environmental and economic problems. Unfortunately, that precious and irretrievable information is now being destroyed at an unprecedented rate.

Physicians, who themselves are well-trained scientists, and many of their healthcare colleagues can readily understand the issue of biodiversity protection when the information is presented in this way. The 500,000 U.S. physicians are the most widely geographically distributed, scientifically trained professionals in the nation. If they and their healthcare colleagues can obtain the benefits of the knowledge, experience, and wisdom developed over many years by the scientific and environmental community, in a systematic, organized way, and can be included as equal partners in a national effort, they could add significantly to the public educational efforts under way.

The prescription drug/natural products matrix discussed elsewhere in this volume will aid in the educational process considerably. Medical specialists can understand that many of the pharmaceuticals they use to treat the diseases or conditions of their patients have a direct connection to the natural world, and thus there is a need for protection of biological diversity in all of its manifestations.

Physician interest in biodiversity protection is increasing nationally and worldwide. The American Medical Association (AMA) House of Delegates passed a resolution about biodiversity at their national meeting in December, 1995, which included the following statement:

> RESOLVED, That the American Medical Association urge physicians and healthcare professionals to become more aware of the importance of the protection of biological diversity and its relationship to human health, especially in terms of the development of drugs and biologicals to treat diseases that are derived from plants and animals and other elements of the natural world, and to work with environmental, educational, healthcare and scientific communities to educate the public about this matter.

This resolution, which has become public national policy of the AMA for its more than 350,000 members, is very similar to a policy statement by the International Society of Doctors for the Environment (ISDE). This

rapidly growing organization, made up of more than 30 national organizations of physicians concerned about environmental issues, during the annual 1994 meeting in Rio de Janeiro, included a statement that the ISDE
must work for the development of the educational effort regarding human
health and biodiversity and

> that the ISDE make this interest known to the international health-
> care, environmental and scientific communities, and especially to in-
> ternational conferences considering environmental and health issues;
> the greatest need in this effort is the development of scientific and
> educational materials on these issues, and their translation for use in
> many countries of the world; therefore ISDE must work for the cre-
> ation of an International Council for Human Health and Biodiver-
> sity, to bring together physicians and other healthcare colleagues, sci-
> entists and environmentalists to further these aims.

Other healthcare professionals can also become involved. Pharmacists,
who dispense drugs and medicinal product prescriptions by the millions,
undoubtedly understand that many pharmaceutical and biological products are either derived from the natural world or are synthetically created
from molecular structure discovered in the natural world, and that source
of undiscovered benefits must be protected. They could educate the public by distributing educational material on the issue along with prescriptions, perhaps on the advisory notes now increasingly used which warn
about side effects.

Veterinarians have the same concerns for the bodies and well-being of
animals as their human healthcare colleagues have for their patients; environmental impacts are similar on the physiological systems of most mammals, including humans. Clients who bring their animals for care are likely
also to have concerns for the natural world and its protection; a systematic
program of education by veterinarians and their organizations could have
a major impact on the public's understanding of the issue.

Registered nurses (over two million in the United States) are frequently
the long-term contacts patients have with the healthcare system; they often
are significant educators of patients; and they have a deep nurturing attitude toward all life. They are a positive force in disease prevention programs, and they will help to inform the public of the need for the protection of the health of the natural world.

A model for cooperation between physicians and their healthcare colleagues with botanical gardens, zoos, arboreta, aquariums, natural museums,
and other centers of biological display and education is now being devel-

oped. The purpose of this model is to provide information sources and products and avenues of dissemination about the issue so that scientifically sound actions can be taken.

The development of a "National Council for the Protection of Biodiversity and Human Health," bringing together the expertise of scientists, environmentalists, and physicians and their healthcare colleagues, would be invaluable to move this regional effort and others forward quickly. Such a "Council" could act as a convener, a transfer agent of information, a forum for debate and consensus-building among healthcare professionals, other scientists, and environmentalists on the issues. It could work to develop physician, healthcare, and public information and education programs on this subject, and to develop a communications system for distribution.

A beginning has been made: a focused collaboration between a major botanical garden (Moody Gardens) and an association of medical societies (National Association of Physicians for the Environment), in Galveston, Texas.

In 1993, Moody Gardens opened the Rainforest Pyramid which is a large, glass-enclosed, replication of a tropical rainforest. More than 2,000 species of tropical plants and animals are exhibited. The rainforest pyramid is immediately adjacent to an IMAX Theatre which feature films with an important environmental message. Moody Gardens attracted more than one millions tourists during 1994, including over 50,000 schoolchildren and their teachers.

The first project includes the development of educational curricula for grades K–12 for use in schools on the subject of medicines that are derived from plants. Sets of color slides and text material are being developed to support presentations by physicians and other health professionals to local schools and civic clubs. Videotapes and posters will be created. A physician and healthcare professional communications network is being established.

Contributors

ANTHONY ARTUSO is assistant professor of public policy at the University of Charleston, where his work focuses on environmental economics and policy analysis.

BYRON J. BAILEY is the Wiess Professor and chairman of the Department of Otolaryngology at the University of Texas Medical Branch at Galveston, Texas. He is also chairman of the Committee on Biodiversity of the National Association of Physicians for the Environment (NAPE).

JENSA BELL is a second-year medical student at Mount Sinai School of Medicine in New York.

BHASWATI BHATTACHARYA is a medical student at Rush Medical College in Chicago.

MICHAEL BOYD is chief of the Laboratory of Drug Discovery Research and Development at the U.S. National Cancer Institute. His research is focused on the elucidation of novel anticancer and antiviral compounds from natural products.

MARY S. CAMPBELL is a second-year medical student at Columbia University in New York.

ERIC CHIVIAN is director of the Center for Health and Global Environment and assistant clinical professor of psychiatry at Harvard Medical School.

PAUL ALAN COX is dean of general education and honors and professor of botany at Brigham Young University, where his work focuses on ethnobotany and rainforest conservation.

GORDON CRAGG is chief of the Natural Products Branch of the National Cancer Institute, where his work focuses on the discovery and development of novel anticancer and anti-HIV agents from natural sources.

ANDREW DOBSON is assistant professor in the Ecology and Evolutionary Biology Department at Princeton University. His research focuses on the long-term dynamics of parasites and pathogens in wild animal populations.

KATE DUFFY-MAZAN is senior technology development and patent specialist with the Office of Technology Development at the National Cancer Institute.

ROBERT ENGELMAN is director of the Population and Environment Program of Population Action International (PAI), a nonprofit organization based in Washington, D.C. His program studies and provides information on the impacts of human population dynamics on natural resources, the environment, and other aspects of human well-being.

PAUL R. EPSTEIN is on the faculty of Harvard Medical School and the Harvard School of Public Health.

ALEXANDRA S. FAIRFIELD is senior scientist in the Opportunistic Infections Research Branch, Division of AIDS, National Institute of Allergy and Infectious Diseases of the National Institutes of Health (NIH).

FRANCESCA GRIFO is director of the Center for Biodiversity and Conservation at the American Museum of Natural History.

JOHN T. GRUPENHOFF is co-founder and executive vice president of the National Association of Physicians for the Environment (NAPE).

DANIEL H. JANZEN is professor of biology at the University of Pennsylvania and an ecologist specializing in tropical animal–plant interactions. He is also technical advisor to Costa Rica's INBio and Guanacaste Conservation Area.

CATHERINE A. LAUGHLIN is chief of the Virology Branch in the Division of Microbiology and Infectious Diseases, National Institute of Allergy and Infectious Diseases of the National Institutes of Health (NIH).

THOMAS D. MAYS is director of the Office of Technology Development at the National Cancer Institute (NCI) of the National Institutes of Health. The Office of Technology Development is responsible for the negotiation of agreements for natural product collection and benefit sharing on behalf of the NCI's Natural Products Program.

ROBERT S. MCCALEB is founder and president of the Herb Research Foundation, a nonprofit research and educational organization established in 1983 that provides scientific information on medicinal herbs and other botanicals to doctors, scientists, and the general public.

KATY MORAN is executive director of the Healing Forest Conservancy, a nonprofit foundation established by Shama Pharmaceuticals Inc. to develop and implement a process to compensate indigenous people and biodiversity-rich countries for their contribution to drug discovery.

DAVID NEWMAN is a chemist in the Natural Products Branch at the National Cancer Institute (NCI), where he is currently responsible for the marine and microbial collection programs. He has over 30 years experience in working with natural products as sources of pharmaceutical entities in both the United States and the United Kingdom.

CHARLES M. PETERS is the Kate E. Tode Curator of Botany at the Institute of Economic Botany of the New York Botanical Garden.

WALTER V. REID is vice president for program at World Resources Institute (WRI), where he conducts policy research in fields of biodiversity conservation, sustainable agriculture, and biotechnology.

JOSHUA ROSENTHAL is director of the Biodiversity Program at the Fogarty International Center of the National Institutes of Health.

JOHN VANDERMEER has been working in tropical ecology and agroecology for the past 20 years. He has been an advisor to the Nicaraguan Ministry of Agriculture and the Center for Research and Documentation of the Atlantic Coast in Nicaragua; a visiting professor at various institutions in Mexico, Nicaragua, and Costa Rica; and a professor at the University of Michigan.

Index

Acid rain, 14. *See also* Pollution, airborne
Acyclovir, 154, 251
Aedes, 66, 98
Agricultural practices. *See also* Crops; Nontimber tropical forest products harvest; Pesticides; Small scale farmers
  arable land use, 48, 50–52
  biodiversity loss in and around agroecosystems, 111–125, 310
  emerging infectious diseases and, 100
  fertilizers, 11, 48, 69, 335
  intensification gradient of, 113–117
  medicinal plant farming, 234, 237
  organic farming, 117, 123–124
  sustainable, 345
Agricultural revolution, 335
Aguaje palm, 320
AIDS, 87. *See also* HIV
  annual U.S. mortality, 129
  fungal infections in immunocompromised individuals, 171–172
  immunosuppression with, 62
  incubation time, 89–90
  root causes of, xv

Algal blooms, 11, 69. *See also* Phytoplankton
  bacteria and, 73
  die-offs from, 69, 70
  effects of upwelling on, 70–71
  effects on freshwater ponds, 74
  global climate change and, 11, 70
  marine mammal mortality rates and, 18
Alkaloids, xvii, 23
Allergies, origins of top pharmaceuticals for, 140, 146–147, 150–153, 158–159
Allium, 224–225, 228
All Taxa Biodiversity Inventory (ATBI), 306–307
Alphitonia zizyphoides, 214–215
American Medical Association, resolution on biodiversity, 354–355
4–Aminoquinoline, 176
Amphibians, xvii, 18–19, 22–23, 45
Amphotericin, 21, 173
Anabaena variabilis, 73
Analgesics, 22, 140, 146–147, 158–159
Ancistrocladus, 268, 275–276
Ancylostoma duodenale, 103
Angiogenesis-inhibiting factor, 24